T0302184

Finite Element Analysis of Solids and Structures

Finite Element Analysis of Solids and Structures

Sudip S. Bhattacharjee

CRC Press
Taylor & Francis Group
Boca Raton London New York

CRC Press is an imprint of the
Taylor & Francis Group, an **informa** business

First edition published 2021
by CRC Press
6000 Broken Sound Parkway NW, Suite 300, Boca Raton, FL 33487-2742

and by CRC Press
2 Park Square, Milton Park, Abingdon, Oxon, OX14 4RN

CRC Press is an imprint of Taylor & Francis Group, LLC

ISBN: 978-0-367-43705-3 (hbk)
ISBN: 978-1-032-04158-2 (pbk)
ISBN: 978-1-003-02784-3 (ebk)

Typeset in Times
by SPi Global, India

Access the Support Material: https://www.routledge.com/Finite-Element-Analysis-of-Solids-and-Structures/Bhattacharjee/p/book/9780367437053

Contents

Preface

Theories and methods for analysis of solids and structures have developed over several centuries. The academic curricula in most engineering programs cover those ideas through multiple courses starting from the basic course on statics to the very advanced courses on nonlinear solid mechanics and dynamics of structures. Integration of all these ideas, in the form of computer-aided matrix method analysis of multi-degree-of-freedom systems, has evolved into the finite element method – which is an indispensable part of current engineering practice for design and analysis of solids and structures. Many books have been produced over the past half century with a great deal of reference information on analytical formulations and relevant numerical implementation details. These reference books definitely serve as good references for the developers of finite element software packages. The theory manuals of actual software implementations tend to mimic similar analytical details of the reference textbooks. Other software-related documents, namely the users' manuals and example manuals, tend to focus on the computer graphics-based techniques for efficient model preparation and post-processing of results. The successful use of finite element simulation technology, in actual engineering problem solutions, requires the mastery of all relevant subjects and tools – starting from the basic understanding of solids mechanics principles to the computer user skills for post-processing the right response results as obtained from finite element simulation models. The large swath of published materials, available in the form of textbooks and software manuals, and the easy availability of computing equipment and software products have made the learning of finite element subjects accessible and difficult at the same time. Aspiring stress analysts in today's world face the difficult challenge of taking a long suit of academic and non-academic courses to build the bridge between theory and software-oriented engineering practice. It is quite normal that graduates in the structural analysis field appear with disconnected knowledge bins, often with good understanding of the mechanics field without the knowledge of how to approach a practical analysis problem; and occasionally with good user skills of specific software features without the knowledge of what goes inside those tools.

From the synthesis of 30+ years of hands-on work in research, programming, teaching, and practical use of the finite element simulation method, the author has developed this book as an introductory reference to the key ideas of the technique. The attempt is by no means a substitute for existing books and reference materials. It is rather a complimentary guidebook for navigating through the complex learning path of finite element subject, for both academic students and self-learning practicing engineers. Comprehensive theory-to-practice coverage of the essential subject matters also makes the book an excellent reference for use in corporate training programs for engineering graduates who lack the formal rigor of academic preparation in stress analysis domain. The analytical formulations have been presented by using matrix and vector notations, unlike the tensor notation as often used in many other reference books and software manuals. This is intended to make an explicit link between theory and actual analysis projects that obviously involve input and output of data in matrix and vector forms. The analytical and numerical implementation details of finite element methods have been discussed side-by-side with user options available in commercial software packages. The example demonstration of software user features has been discussed with specific references to the analysis capabilities of the ABAQUS package. The discussions on software-aided model preparation and quality checks have been presented with example references to the HyperMesh software package. These software-specific feature descriptions will need to be adapted if different software products are used to solve the practice problems.

This book is structured to progressively build the basic expertise in finite element analysis methods – following the same sequence that the author has used in teaching of one-semester graduate courses over the past many years. Chapters 1 and 2 present the theory of elasticity topics that form the foundation of linear elastic finite element methods. Practice problems are presented, with pre-built analysis model files, to test the basic stress–strain analysis theories by using finite element analysis software as a virtual experimentation tool. Chapter 3 introduces the basics of finite element formulations, including the numerical details as implemented in software tools, for analysis of solids that can be represented by 2D stress fields. Practice problems are presented to use software analysis tools with manual preparation of finite element model input files. The objective is to achieve clear knowledge of the input data structure required for an error-free analysis model preparation. This is an essential skill to debug model errors that often appear while preparing more complex analysis models by using computer-graphics-based model preparation tools. Chapter 4 focuses on how to produce good quality finite element models of two-dimensional solids by using general-purpose model pre-processing software. Although specific references are made to HyperMesh software features for model build operations, and to ABAQUS software for actual FEA solutions, the discussions on key aspects of quality model preparation are equally applicable to other software products for model preparation and validation of results. Three-dimensional elasticity problems, having an axis of symmetry, are discussed in Chapter 5, followed by the introduction of 3D finite element formulations that are required for analysis of general 3D solids. Chapter 6 is dedicated to the elastic deflection and stress–strain analysis of beams for bending, transverse shear and torsional load effects. The size of this chapter, with comprehensive coverage of both theoretical and finite element implementation aspects, is understandably large for a single class learning session. A reduced presentation can be formulated by focusing on finite element implementation methods, with selected references to the analytical methods, if desired. Chapter 7 presents the plate and shell element formulations for the analysis of 3D thin-walled structures. Analysis problems, discussed in Chapters 1–7, represent components that can be modeled with single finite element types. Chapter 8 introduces special numerical techniques that are required to simulate the behavior of joints and interface contacts in multi-component model assemblies. The basic purpose of finite element simulation, i.e. the interpretation of stress analysis results for engineering decisions, is discussed in Chapter 9, with special focus on strength, durability, and integrity assessment of solid products. Chapter 10 presents a comprehensive review of the analytical and numerical methods for vibration frequency analysis which is an important topic in design and analysis of structures for cyclic load effects. Chapter 11 is dedicated to the analysis of structural response for noncyclic dynamic load events. The use of finite element simulation models to predict design response spectra, frequency response function, and time-domain response histories of structures is specifically discussed in this chapter. Finally, Chapter 12 presents a review of key analysis techniques relevant for predicting the nonlinear response of structures. At the end of each chapter, practice problems have been presented for solving with finite element analysis models, and for verification of the results by using simplified analytical prediction models. These studies are intended to reinforce the importance of conducting minimum quality checks of the finite element simulation results. For comprehensive learning experience, some practice problems will require the use of electronic data files that can be downloaded from the site: https://www.routledge.com/Finite-Element-Analysis-of-Solids-and-Structures/Bhattacharjee/p/book/ 9780367437053. PowerPoint slides of the materials covered in this book, and the solution manual of practice problems, are available upon request from academic instructors.

Author

Dr. Sudip S. Bhattacharjee is a supervisor for vehicle crashworthiness engineering in Ford Motor Company, USA. He is also an occasional graduate course instructor at the University of Windsor, Canada, and the University of Michigan, Dearborn. Prior to joining Ford in 2000, Dr. Bhattacharjee was a faculty member at the University of Windsor, a consulting engineer with SNC-Lavalin in Montreal, and a postdoctoral research fellow at Ecole Polytechnique of Montreal. Dr. Bhattacharjee received his Ph.D. from McGill University, Montreal in 1993. Prior to that, he was a lecturer of Civil Engineering at Bangladesh University of Engineering and Technology (BUET), where he also received his Bachelor and Masters degrees in Civil Engineering.

1 Introduction to Stress Analysis of Solids and Structures

SUMMARY

Theories of stress analysis for solids and structures have evolved over many centuries – leading to the development of finite element method in the current time of abundant digital computing power. A short list of the key developments in solid mechanics, by no means a comprehensive review of the subject matter, is presented in Section 1.1 as a form of small tribute to the history. Section 1.2 briefly reviews the present-day product development process where finite element model-based virtual simulation plays a key role in optimization and validation of the product design. Before the appearance of digital computing technology, analysis of structures, however, largely relied on hand calculation of external force effects based on the solution of static equilibrium equations. Two primary alternative methods, namely the force and displacement methods, have initially contributed to the development of matrix method of structural frame analysis. It is the displacement method (or alternatively known as the stiffness method) that has eventually evolved into the present form of finite element analysis (FEA) method for solids and structures. Following a brief review of the flexibility method in Section 1.3, a more detailed description of the stiffness-based matrix equilibrium equations is presented in Section 1.4. The important properties of the stiffness-based matrix analysis method are reviewed in this section with an easy-to-understand example of elementary springs. The finite element discretization technique of solids is briefly introduced in this section to highlight the underlying similarity between the direct stiffness-based matrix method and the more advanced finite element method of structural analysis.

Section 1.5 is devoted to the definition of internal body stress components – as direct extension of describing the force effects on 3D solids. The state of stress-equilibrium and the stress-boundary conditions, important analytical foundation blocks of advanced solid mechanics principles, are presented in Sections 1.6 and 1.7. Finally, two practice problems are presented in Section 1.8 primarily to help with the learning of powerful graphical processing tools for visualization of stress analysis results. Pre-prepared finite element simulation models are presented for analysis with ABAQUS software (Dassault Systems 2020a). However, similar practice models can be prepared (or the presented model files can be converted) for analysis with other commercially available software packages. The objective at this initial step is not to introduce the technique of finite element model building process; it is to use the pre-built models as virtual experimental tools with visualization of the stress fields and boundary conditions.

1.1 INTRODUCTION – A BRIEF SUMMARY OF KEY HISTORICAL DEVELOPMENTS

Structural analysis techniques to determine the effects of forces have evolved simultaneously with the progression of human civilization through millennia. Surviving ancient structures and historical records provide evidence of thoughtful human endeavor to build safe structures based on the understanding of effects of natural forces. The following are examples of few early historical milestones:

3000 BC	Use of beams and columns as structural elements in Pyramids	Egyptians
582 BC	Center of gravity of bodies, and rules of statics	Archimedes (Greek)
100 BC	Use of arch concept in structures	Romans

Subsequent advancements in the understanding of effects of forces on materials and bodies can be attributed to the following individuals:

15th century	Behavior of members in tension, compression, and bending	Da Vinci
17th century	Strength of materials	Gelileo Galilei
17th century	Relation between force and deformation	Robert Hooke
17th century	Laws of motion and calculus	Isaac Newton

Industrial revolution in Europe and the advent of rail-road drove the development of generic analysis techniques to design structures for industrial applications. Notable developments of that era can be summarized as follows:

17th century	Deflection of elastically bent bar	James Bernoulli
18th century	Buckling of columns under axial load	Leonhard Euler
18th century	Theory of friction	C.A. Coulomb
18th century	Generalized forces for the solution of complex structures	J.L. Lagrange

19th and 20th centuries experienced a flurry of advanced theoretical developments forming the discipline of "mechanics of materials":

19th century	Elastic modulus of materials	Thomas Young
19th century	Theory of bent plates and shells	L.M. Navier
19th century	Theory of elasticity and plasticity	Poisson, Cauchy, Lame, Kelvin, St. Venant
19th century	Deflection of plates	Kirchoff
19th century	Failure theory – maximum shear stress yield criterion	Tresca
19th century	Theorem of least work	Castigliano
20th century	Plasticity theory and material failure	von Mises
20th century	Elastic solutions of problems in solid mechanics	Timoshenko

Modern era of structural mechanics has been characterized by the development of numerical methods to solve problems in solids and structures. Some of the key contributors in this still-evolving field can be listed as follows:

20th century	Approximate solution of continuum mechanics	Ritz
20th century	Concepts of analysis of frames and continuum	Argyris, Turner, Clough, and others
20th century	Concepts of finite element methods	Clough, Wilson, Bathe, Zienkiewicz, Hughes, and others

Historical developments listed in the above are few from numerous other individuals and groups that have contributed to the development of discipline what we call today mechanics of solids or structural mechanics. A comprehensive review of historical developments is beyond the scope of this chapter. However, key ideas and individuals are listed here as a tribute to the centuries of theoretical developments that form the foundation of today's finite element analysis method. Modern-day finite element analysis method is the embodiment of centuries of theoretical development in mechanics – making the structural analysis techniques readily available to 21st-century engineers through digital computing software products.

1.2 ROLE OF ANALYSIS/SIMULATION IN PRODUCT ENGINEERING

Externally applied forces cause internal stress and deformation (strain) to materials that have limited resistance capacity. Figure 1.1(a) gives an example of ductile material resistance properties defined by yield strength S_y and failure strain ε_f. Resistance capacity of a structural member may also be exhausted before reaching the material

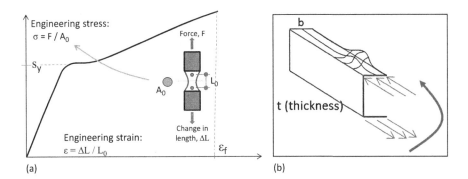

FIGURE 1.1 (a) Resistance capacity of a ductile material (yield strength = S_y and failure strain = ε_f); (b) buckling resistance of a thin wall member subjected to axial compressive stress.

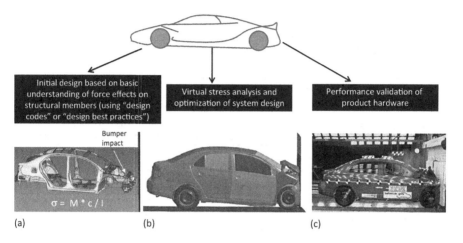

FIGURE 1.2 Example of present-day automotive product development process. Simulation model and crash test pictures are adapted from 2010 Toyota Yaris analysis report (NHTSA 2020).

strength, for example, due to buckling of thin wall members under axial compressive stresses as shown in Figure 1.1(b). Checking internal stress is, thus, an important step in product engineering process to determine the safety of a product.

Product design codes and best practices generally provide guidelines to define initial geometric proportions of members assuming simple stress distributions inside a member when subjected to idealized external forces and boundary conditions. Preliminary member designs, based on design best practice rules, are not necessarily fully optimized design of structures for complex loading and boundary conditions. Advanced techniques of structural analysis, such as finite element method, provide the ability to predict stress–deformation response of structures before manufacturing the hardware products. Virtual assessment and design iterations help to avoid highly inefficient design-test-correct iteration loops. Figure 1.2 shows an example of present-day automotive product development process, starting with: (a) preliminary design of components based on basic principles of engineering mechanics; followed by, (b) detail finite element analysis of structures for upfront performance assessment and design optimization; and (c) final testing of the prototype product for performance validation.

Simulation models are not always intended to replace the actual hardware testing. Results of simulation models often need to be assessed and used within the limitations of assumptions made in the model building process. For example, analysis efficiency may require the use of linear elastic material behavior assumption. So, the results of such analysis model should be assessed with due consideration to that assumption. Figure 1.3 lists some of the limitations of simulation models that engineers need to be specifically aware of.

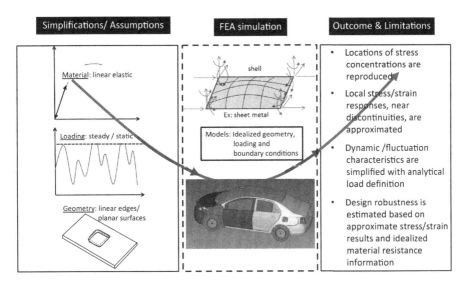

FIGURE 1.3 Potential limitations of simulation models and results.

1.3 STATIC EQUILIBRIUM OF STRUCTURES AND ANALYSIS OF FORCES – STATICALLY DETERMINATE AND INDETERMINATE SYSTEMS

First step in stress analysis of a structure is the determination of external forces and boundary conditions. A statically determinate system is one where external forces and reactions acting on the structure can be determined from the analysis of static equilibrium condition. Figure 1.4 shows the bumper beam of an automotive body

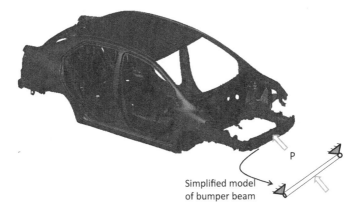

FIGURE 1.4 External force "*P*" acting on an automotive bumper beam. Image of automotive body structure is prepared from Toyota Yaris vehicle model available in NHTSA (2020).

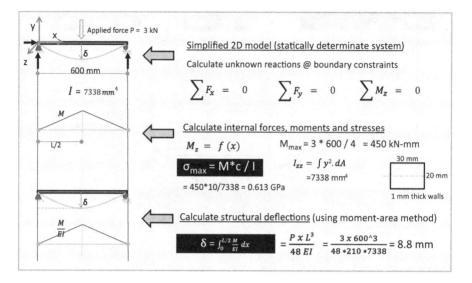

FIGURE 1.5   Response analysis of the simplified model of an automotive bumper beam.

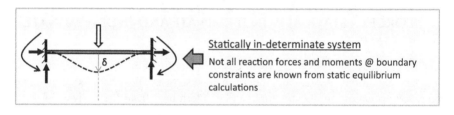

FIGURE 1.6   Example of a statically indeterminate beam model.

structure subjected to an external force, P. End constraints (boundary conditions) can be idealized to make the system statically determinate as shown in the figure. Elementary bending theory leads to complete solution including maximum stress and maximum deflection values for the example analysis problem (Figure 1.5).

However, when the end conditions of bumper beam are assumed to be moment-resistant rigid connections (as shown in the example illustration in Figure 1.6), equations of static equilibrium alone will not be sufficient to determine the unknown reaction forces. Several analysis techniques exist to calculate unknown boundary forces in statically indeterminate beams (Kennedy and Madugula 1990). Two widely used analysis techniques are (i) force method (also known as flexibility method) and (ii) displacement method (also known as stiffness method). Force (or flexibility) method of structural analysis involves the following key steps:

- Make the structure statically determinate by eliminating sufficient number of unknown constraints
- Calculate responses of determinate system
- Calculate the response of "determinate" structure for unit value of "eliminated" force constraint

- Calculate the magnitude of "eliminated" force by using principle of superposition that makes the structural displacement compatible with actual boundary constraint.

Flexibility method of structural analysis is a very systematic step-by-step procedure, but it is generally not used in computer-aided analysis of complex structural systems. Stiffness method (as described in the following section) has become the analysis method of choice in most structural analysis software products.

1.4 STIFFNESS (DISPLACEMENT) METHOD OF STRUCTURAL ANALYSIS

Stiffness of a structural element, defined as the resistance to unit deformation, is used to determine the displacement response from equilibrium condition. An elementary structural example is represented by the spring element in Figure 1.7(a) where one end of the spring is attached to a rigid boundary, while the other end is subjected to an external force P. Displacement response at the load application point is identified by u. Figure 1.7(b) shows an example force–deformation response of the spring element. Assuming linear elastic behavior for the spring, slope of the force–deformation line in Figure 1.7(b) represents the spring stiffness k. Physically, stiffness k represents the resistance of the spring against unit axial deformation ($u = 1$). At the load application point in Figure 1.7(a), total spring resistance is given by $k \times u$, where u is the unknown displacement caused by the externally applied load P. Equilibrium of forces at load application point provides the following simple but very powerful equation for structural response analysis:

$$k.u = P \qquad (1.1)$$

For this example of single-degree-of-freedom (DOF) spring system (Figure 1.7(a)), displacement response u can be predicted by using the equilibrium equation (1.1) for any given magnitude of force P, provided that the stiffness property k is known. Stiffness property described above can now be used to derive the general 2×2 stiffness matrix of a spring element that has axial stiffness value k_1 and two degrees of freedom (u_1, u_2) as shown in Figure 1.8. In this 2-DOF system, stiffness term k_{ij} is

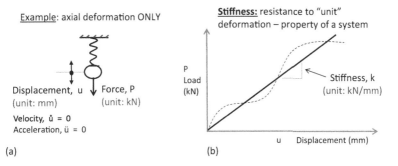

FIGURE 1.7 (a) Single-degree-of-freedom (DOF) spring and (b) linear-elastic force–deformation response.

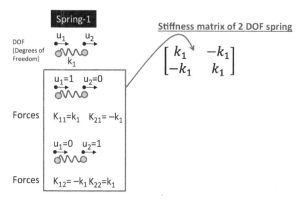

FIGURE 1.8 Stiffness matrix of a 2-DOF spring element (spring-1).

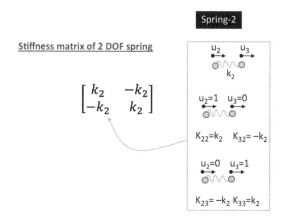

FIGURE 1.9 Stiffness matrix of a 2-DOF spring element (spring-2).

defined as the spring resistance at ith degree of freedom for unit deformation at jth degree of freedom. Similarly, 2×2 stiffness matrix of another spring element (spring-2), having stiffness property equal to k_2 and connecting to two degrees of freedom (u_2, u_3), is derived in Figure 1.9. A structural assembly can be formed by joining the two spring elements, 1 and 2, at common DOF u_2. Stiffness matrix of the example 3-DOF system assembly is shown in Figure 1.10.

Stiffness contributions of element-1 are assigned to k_{ij} terms ($i = 1, 2$ and $j = 1, 2$) in the 3×3 stiffness matrix; and those of element-2 are assigned to k_{ij} terms ($i = 2, 3$ and $j = 2, 3$). Stiffness contributions of both elements are added to define the stiffness parameter k_{22} corresponding to the common degree of freedom u_2. Single-DOF system equilibrium, given by equation (1.1), can be expanded to define the equilibrium of 3-DOF spring assembly as follows:

$$\begin{bmatrix} k_1 & -k_1 & 0 \\ -k_1 & k_1 + k_2 & -k_2 \\ 0 & -k_2 & k_2 \end{bmatrix} * \begin{Bmatrix} u_1 \\ u_2 \\ u_3 \end{Bmatrix} = \begin{Bmatrix} P_1 \\ P_2 \\ P_3 \end{Bmatrix} \qquad (1.2)$$

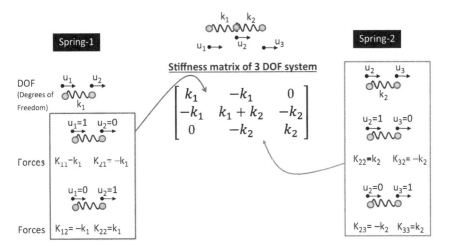

FIGURE 1.10 Stiffness matrix of a 2-spring assembly.

Solution of matrix equilibrium equations (1.2) gives a unique displacement response, {u}, provided that stiffness matrix [k] is positive definite (Crandall 1956). Positive definiteness of the stiffness matrix implies that for any displacement vector {u} we have (Bathe 1996):

$$\{u\}^T \cdot [k] \cdot \{u\} > 0 \tag{1.3}$$

For any displacement vector {u}, equation (1.3) implies that the strain energy, which is one-half of the quantity on the left side term of the equation, is also positive. Note that an unrestrained structural element has rigid body motions implying that the stiffness matrix of such unrestrained system will be semi-definite. The stiffness matrix of a structure is rendered positive definite by eliminating rows and columns that correspond to the restrained degrees of freedom, i.e. by eliminating the possibility of the structure to undergo rigid body motions. An example of modification to equilibrium equations (1.2), with the introduction of boundary conditions $u_1 = u_3 = 0$, is represented by the system of equations (1.4) where the modified stiffness matrix is positive definite:

$$\begin{bmatrix} 0 & 0 & 0 \\ 0 & k_1 + k_2 & 0 \\ 0 & 0 & 0 \end{bmatrix} * \begin{Bmatrix} 0 \\ u_2 \\ 0 \end{Bmatrix} = \begin{Bmatrix} P_1 \\ P_2 \\ P_3 \end{Bmatrix} \tag{1.4}$$

After inserting the boundary constraints in matrix equilibrium equations, as explained in the above, the remaining set of equations can be solved to determine the unknown displacement u_i for an applied set of forces P_i. Key analysis steps for the solution of multi-DOF system example, presented above, can be summarized as follows:

- Calculate the stiffness matrix, $[k]$
- Identify known forces and displacement constraints

- Modify stiffness matrix to take account of known boundary constraints
- Solve equations of equilibrium to determine unknown displacements: $[k] \times \{u\} = \{P\}$
- Calculate member internal forces: $F_i = k_i \times (u_j - u_i)$
- Calculate average internal stress: $\sigma = F_i/A_i$

Solution procedure described above considers member stiffness (and internal material stress) for resistance to axial deformation only. Internal stresses for bending, shear, and torsional loading effects can be calculated following the solid mechanics principles. Figure 1.11 shows beam internal stresses under the actions of axial force P, moment about z-axis M_z, shear force V_y, and torsion T. Additional stresses can also be calculated similarly for moment about y-axis, M_y, and shear force V_z. Beam element stiffness calculations for resistance to bending, shear, and torsional effects will be discussed in Chapter 5.

Relationships among member internal stresses and external forces are straightforward for skeletal structural members such as the beam example shown in Figure 1.11. With simplified stress distribution assumptions, mechanics of solids provides analytical expressions to calculate internal stresses when resultant forces are known from structural analysis. General shaped solid bodies, such as the example shown in Figure 1.12, require special analytical techniques to relate internal stresses with external loads and boundary conditions. Finite element analysis method, to be introduced in Chapters 2–8, builds that bridge between the matrix equilibrium equations of load–displacement structural response, and the theory of elasticity principles describing the internal stress fields at differential elements of general solid bodies. In that process, a general 3D structural body is imagined to be a virtual assembly of standard geometric shape smaller elements that are interconnected through common nodal DOF (Figure 1.13). Special numerical techniques are used to calculate stiffness properties corresponding to various forms of element response mechanisms.

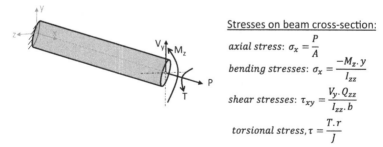

Stresses on beam cross-section:

axial stress: $\sigma_x = \dfrac{P}{A}$

bending stresses: $\sigma_x = \dfrac{-M_z \cdot y}{I_{zz}}$

shear stresses: $\tau_{xy} = \dfrac{V_y \cdot Q_{zz}}{I_{zz} \cdot b}$

torsional stress, $\tau = \dfrac{T \cdot r}{J}$

Here, A=cross-sectional area, I_{zz}=moment of inertia about neutral axis zz, Q_{zz}=1st moment about neutral axis 'zz' for the area beyond the point where shear stress τ_{xy} is calculated, 'b' is the width of beam section at the point of shear stress calculation, 'r' is the distance of torsional stress point from beam axis, and J is the polar moment of inertia of beam section.

FIGURE 1.11 Stresses on beam section under the actions of axial force, P, bending moment about z-axis M_z, shear force V_y, and torsion T.

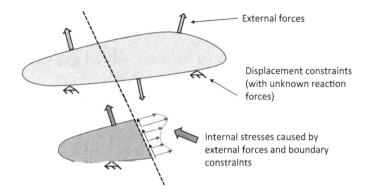

FIGURE 1.12 General 3D body subjected to external forces and boundary constraints.

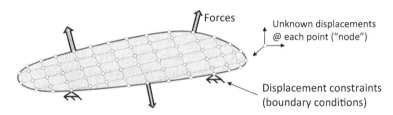

FIGURE 1.13 A solid body imagined to be an assembly of finite size elements of standard geometric shape.

Like the multi-spring system, equilibrium of the "finite element model" is expressed in the following familiar form (equation 1.5):

$$\begin{bmatrix} k_{11} & \cdots & k_{1n} \\ \vdots & \ddots & \vdots \\ k_{n1} & \cdots & k_{nn} \end{bmatrix} \begin{Bmatrix} u_1 \\ .. \\ u_n \end{Bmatrix} = \begin{Bmatrix} P_1 \\ .. \\ P_n \end{Bmatrix} \tag{1.5}$$

where n is the number of DOF, $\{u\}$ is the vector of nodal displacements, and $\{P\}$ is the vector of applied nodal forces. Some of the variables in vectors $\{u\}$ and $\{P\}$ are known, while others are calculated by solving the system equilibrium equations. Once the nodal displacements, $\{u\}$, are known from the solution of equations (1.5), the calculation of internal stresses follows the element-specific internal formulations. Details of the finite element properties will be gradually introduced in next chapters. The remainder of Chapter 1 presents generic descriptions of the internal stress components and their inter-relations. Chapter 2 will describe the internal deformations (strains) in differential solid elements, and the inter-relations among stress and strain variables – leading eventually to the derivation of stiffness matrix and equilibrium equations for general solids.

1.5 COMPONENTS OF STRESSES IN A 3D BODY

Figure 1.14 shows an infinitesimally small area, ΔA, with its normal direction parallel to the x-axis. Normal and tangential forces acting on that area are represented by ΔF_x,

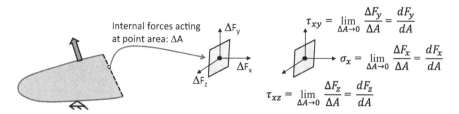

FIGURE 1.14   Components of internal forces (and stresses) acting on an infinitesimal area ΔA.

ΔF_y, and ΔF_z. Intensity of force, at the limit condition of ΔA approaching 0, is defined as the "stress" acting at the measurement point. Normal and shear stress components, corresponding to normal and shear forces acting on the plane ΔA, are defined as σ_x, τ_{xy}, and τ_{xz} in Figure 1.14.

The stress measurement point inside a 3D body, as shown in Figure 1.14, can be considered to reside inside an infinitesimally small volume, ΔV, that is bounded by three mutually perpendicular planes – each plane being perpendicular to an axis of orthogonal x–y–z system (Figure 1.15). Expanding the definition of three stress components acting on 2D x-plane of Figure 1.14, nine stress components acting on three mutually perpendicular planes (x, y, and z) are identified in Figure 1.15.

Normal stresses, acting on opposite surfaces of "infinitesimally" small volume, are assumed equal in magnitude, but opposite in direction (variation of normal stress is ignored) (Figure 1.16). The relationship between shear stresses (τ_{xy} and τ_{yx}), acting on orthogonal faces, can be derived by considering the rotational equilibrium of the entire element about z-axis:

$$\sum M_z = \left(-\tau_{xy}.dy.dz\right).dx + \left(\tau_{yx}.dx.dz\right).dy = 0 \qquad (1.6)$$

Equation (1.6) shows that $\tau_{xy} = \tau_{yx}$. Similarly, considering rotational equilibrium about x- and y-axes, we get, respectively, $\tau_{yz} = \tau_{zy}$ and $\tau_{xz} = \tau_{zx}$. Shear stresses on mutually perpendicular planes are of equal magnitude. This means that stresses at a

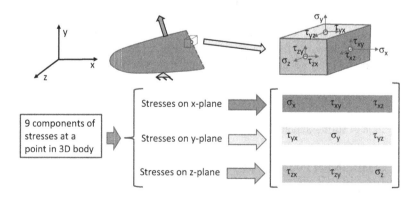

FIGURE 1.15   Components of stresses acting on an infinitesimally small 3D volume.

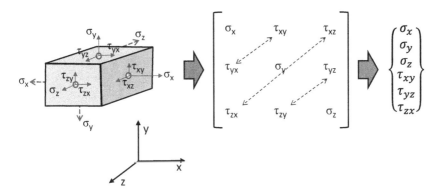

FIGURE 1.16 Vector of 6 independent stress components at a point in 3D body.

point in 3D body can be represented by a vector of six independent components (σ_x, σ_y, σ_z, $\tau_{xy} = \tau_{yx}$, $\tau_{xz} = \tau_{zx}$, $\tau_{yz} = \tau_{zy}$) (Figure 1.16).

1.6 VARIATION OF STRESSES AND DIFFERENTIAL EQUATIONS OF EQUILIBRIUM

Assuming no variation of stresses in thickness direction, variation of stresses in 2D x–y plane of an infinitesimal element ($dx.dy$) is expressed by truncated Taylor's expressions in Figure 1.17. F_x and F_y denote body forces per unit volume acting in directions x and y, respectively. Considering unit thickness of the differential element in normal direction to x–y plane, equilibrium of forces in the x-direction gives:

$$\sum F_x = \left(\sigma_x + \frac{\partial \sigma_x}{\partial x}.dx\right).dy - \left(\sigma_x\right).dy + \left(\tau_{xy} + \frac{\partial \tau_{xy}}{\partial y}.dy\right).dx \qquad (1.7)$$
$$- \left(\tau_{xy}\right).dx + \left(F_x\right).dx.dy = 0$$

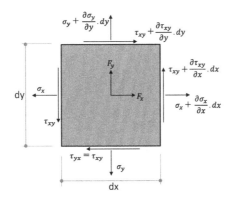

FIGURE 1.17 Variations of stresses in a differential element in x–y plane.

Upon simplification of above equation, and considering $dx.dy \neq 0$, differential equation of stress equilibrium in the x-direction is given by

$$\frac{\partial \sigma_x}{\partial x} + \frac{\partial \tau_{xy}}{\partial y} + F_x = 0 \qquad (1.8)$$

Similarly, by considering the equilibrium of forces in the y-coordinate direction, the following additional differential equilibrium is obtained:

$$\frac{\partial \sigma_y}{\partial y} + \frac{\partial \tau_{xy}}{\partial x} + F_y = 0 \qquad (1.9)$$

Expanding the two-dimensional stress-equilibrium state of differential element (Figure 1.17) to three-dimensional space, and adding additional stress components and body force component F_z, the following general equations of 3D stress equilibrium is obtained (Timoshenko and Goodier 1982, Ugural and Fenster 2012):

$$\begin{aligned}
\frac{\partial \sigma_x}{\partial x} + \frac{\partial \tau_{xy}}{\partial y} + \frac{\partial \tau_{xz}}{\partial z} + F_x &= 0 \\
\frac{\partial \sigma_y}{\partial y} + \frac{\partial \tau_{xy}}{\partial x} + \frac{\partial \tau_{yz}}{\partial z} + F_y &= 0 \\
\frac{\partial \sigma_z}{\partial z} + \frac{\partial \tau_{xz}}{\partial x} + \frac{\partial \tau_{yz}}{\partial y} + F_z &= 0
\end{aligned} \qquad (1.10)$$

Three stress-equilibrium equations (1.10) contain six independent stress variables. So, additional relationships will be required to determine the internal stress field in a statically indeterminate general 3D body such as the one shown in Figure 1.12. Chapter 2 will present those additional relationships to solve the stress analysis problems both analytically and numerically. Stress-equilibrium equations (1.10) can also be written in the following compact form:

$$\left[B \right]^T . \{\sigma\} + F_i = 0 \qquad (1.11)$$

where $\{\sigma\}$ and [B] represent the following:

$$\{\sigma\} = \begin{Bmatrix} \sigma_x \\ \sigma_y \\ \sigma_z \\ \tau_{xy} \\ \tau_{yz} \\ \tau_{zx} \end{Bmatrix} \qquad (1.12)$$

$$[B] = \begin{bmatrix} \dfrac{\delta}{\delta x} & 0 & 0 \\[2mm] 0 & \dfrac{\delta}{\delta y} & 0 \\[2mm] 0 & 0 & \dfrac{\delta}{\delta z} \\[2mm] \dfrac{\delta}{\delta y} & \dfrac{\delta}{\delta x} & 0 \\[2mm] 0 & \dfrac{\delta}{\delta z} & \dfrac{\delta}{\delta y} \\[2mm] \dfrac{\delta}{\delta z} & 0 & \dfrac{\delta}{\delta x} \end{bmatrix} \tag{1.13}$$

Vector and matrix notations of equations (1.12) and (1.13) will appear repeatedly in next chapters on the description of finite element analysis methods. In absence of body forces, or when body forces are negligible, equation (1.11) implies uniform stress state inside the differential element of Figure 1.17, i.e. σ_x is constant in the x-direction, σ_y in the y-direction, and τ_{xy} is constant in the x–y plane. Stresses inside a large solid body or along boundaries do not need to be uniform or constant. Integration of the differential element stresses can equilibrate the resultant effects of bending, shear and torsional loading conditions on member boundaries – as shown by elementary beam example in Figure 1.11. A more general formulation of stress-boundary condition is presented in Section 1.7.

1.7 STRESS BOUNDARY CONDITIONS

Stress components of equation (1.12), measured at any point inside a body, not only will satisfy the differential equations of equilibrium (1.10), but also will satisfy the external boundary conditions. Let us consider N be the normal to a boundary surface point (Figure 1.18), and p the external stress acting at that point with components $p_x, p_y,$

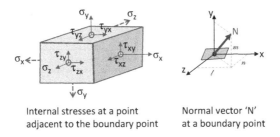

Internal stresses at a point   Normal vector 'N'
adjacent to the boundary point   at a boundary point

FIGURE 1.18 Normal at boundary point and internal stresses at a point adjacent to the boundary.

and p_z in Cartesian coordinate directions x, y, and z, respectively. Direction cosines of the vector N in the x-, y-, z-coordinate systems are given by l, m, and n (projections of unit area in directions x, y, and z). Internal stresses at a point adjacent to the boundary point are (σ_x, σ_y, σ_z, τ_{xy}, τ_{yz}, and τ_{xz}). Considering external stresses (p_x, p_y, p_z) as continuation of internal stresses, we get the stress-boundary conditions of equations (1.14):

$$p_x = \sigma_x.l + \tau_{xy}.m + \tau_{xz}.n$$
$$p_y = \tau_{xy}.l + \sigma_y.m + \tau_{yz}.n \qquad (1.14)$$
$$p_z = \tau_{xz}.l + \tau_{yz}.m + \sigma_z.n$$

Equations (1.14) give the general stress-boundary conditions that must be always validated by the calculated stress response of a given structural analysis problem. Software tools for graphical post-processing of finite element model results can be used effectively to visualize the calculated stresses and verify the boundary conditions. While detailed descriptions of advanced stress analysis methods are introduced in next chapters, Section 1.8 in the following introduces practice examples solely for the learning objective of using software graphics tools for stress field visualization. Interesting insights can be gained from the comparison of elementary beam theory results with those obtained from detail 3D finite element analysis models.

1.8   PRACTICE PROBLEMS ON STRESS FIELD VISUALIZATION WITH CAE TOOLS

Computer-aided-engineering (CAE) tools used in design and analysis of solid products can be generally classified in four groups (Figure 1.19): (1) computer-aided-design (CAD) tools for geometric design, (2) finite element model preparation tools (model pre-processors), (3) finite element analysis tools (solvers), and (4) result visualization tools (post-processors). Many of the listed tools also provide cross-functional capabilities. For example, CAD software tools also come bundled with modelling, analysis and post-processing tools, while general-purpose finite element

(1) CAD tools	(2) Model pre-processor	(3) FEA solvers	(4) Post-processing tools
Catia http://www.3ds.com/products-services/catia/	ANSA http://www.ansa-usa.com/products/ansa	ABAQUS http://www.3ds.com/products-services/simulia/products/abaqus/	HyperView http://www.altairhyperworks.com/Product,11,HyperView.aspx
Unigraphix Nx https://www.plm.automation.siemens.com/global/en/products/nx/	HyperMesh https://www.altair.com/hypermesh/	ANSYS https://www.ansys.com/products/structures	Meta https://www.beta-cae.com/meta.htm
AutoCAD https://www.autodesk.com/products/autocad/overview	PRIMER https://www.oasys-software.com/dyna/software/primer/	NASTRAN https://www.mscsoftware.com/product/msc-nastran	..
Solidworks https://www.solidworks.com/	FEMAP https://www.plm.automation.siemens.com/global/en/products/simcenter/femap.html	LS-DYNA https://www.lstc.com/	..

Meshing and model building   Solutions: [K]{u}={P}   Results visualization

FIGURE 1.19   Partial list of CAE tools for design and analysis of solids.

solvers, listed in group (3), also come bundled with solver-specific pre- and post-processing tools. In the large-scale industrial product engineering setup, execution efficiency is derived from the use of specialized tools for modularized work flow of design, modelling, analysis and post-processing. The following practice problems are presented based on that idea of modular process execution.

Practice Problem-1: A statically indeterminate solid beam, subjected to idealized load and boundary conditions, is shown in Figure 1.20. Skilled students of the subject may attempt to prepare a finite element analysis model of the problem. Alternatively, the ABAQUS finite element model (SOLIDBEAM_COARSE_MESH.inp), provided in the data download site as listed in the preface of this book, can be used to practice the post-processing part of this example. A coarse mesh model, with number of nodes less than 1000, can be run with student version of ABAQUS (Dassault Systems 2020a). Successful execution of the ABAQUS model creates a binary results database file, SOLIDBEAM_COARSE_MESH.odb, that can be post-processed with ABQUS/CAE (Dassault Systems 2020a) or with general-purpose post-processor HyperView (Altair University 2020). Questions to answer:

  a. Identify nonzero stress components on the free-body diagram of a differential element adjacent to the bottom surface of the solid body.
  b. Post-process the simulation model results to plot the shear stress distribution τ_{xz} (identified as S_{13} in ABAQUS results) in the solid body. Verify that the simulation model results are consistent with the zero stress-boundary conditions assumed in (a).
  c. Cut a x-section at $x = 100$ and clip the model on one side; plot the contour of normal stresses σ_x (identified as S_{11} in ABAQUS results)
  d. Hand calculate normal stresses at the top and bottom of mid-span section by using the beam bending theory.
  e. How do hand calculated stresses compare with simulation model results?

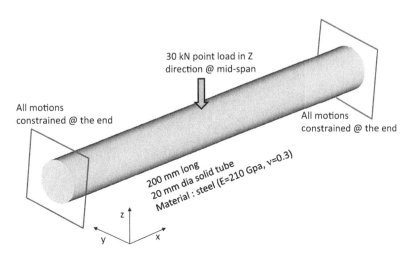

FIGURE 1.20   A statically indeterminate solid beam subjected to idealized load and boundary conditions.

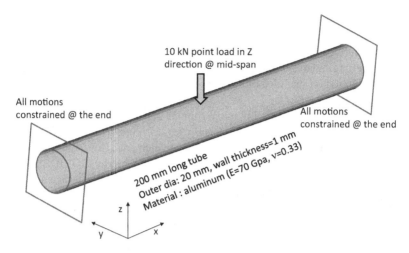

FIGURE 1.21 A statically indeterminate hollow tube beam subjected to idealized load and boundary conditions.

Practice Problem-2: A statically indeterminate hollow tube beam, subjected to idealized load and boundary conditions, is shown in Figure 1.21. Once again, skilled students of the subject may attempt to prepare a finite element analysis model of this problem. Finite element model for ABAQUS analysis is provided in file TUBE_COARSE_MESH.inp. Run this model in student version of ABAQUS (Dassault Systems 2020a), and post-process the results file (TUBE_COARSE_MESH.odb) to answer the following questions:

a. Plot the contour of deformed shape of the tube (z-displacement)
b. Cut a x-section at $x = 100$ and clip the model on one side; plot the contour of normal stresses σ_x (identified as S_{11} in ABAQUS results) on the deformed plot of tube
c. Hand calculate normal stresses at the top and bottom of mid-span section, and the beam deflection at mid-span, by using the beam bending theory.
d. How do hand calculated stress and deflection results compare with simulation model results?

2 Strain–Displacement Relationship and Elasticity of Materials

SUMMARY

Solid mechanics principles that form the foundation of advanced stress analysis methods are reviewed in this chapter. The basic definitions of normal and shear strains, for infinitesimally small deformation problems, are introduced in Section 2.1. Section 2.2 develops very important strain–displacement relationships that are critical for the development of finite element formulations in the next steps. Inter-relationship among strain and displacement components, defining the well-known compatibility condition, is presented in Section 2.3. Any derived solution for strain/displacement field must satisfy the deformation compatibility condition – a fundamental requirement of the solid mechanics principles. Section 2.4 presents yet another important formulation – the generalized stress–strain relationships based on Hookes' law. All these inter-relationships among stress, strain, and displacement components are finally expressed through redefinition of the compatibility condition in terms of stress components. The stress-based compatibility definition, with stress equilibrium and boundary conditions defined in Chapter 1, presents the possibility of solving the unknown stress field problems by using analytical stress functions. Elementary example solution, based on stress function approach, is presented in Section 2.5. The alternative solution method, based on the use of displacement function approach, is presented in Section 2.6. Although the stress function-based formulation technique is not the preferred implementation method in standard finite element software packages, the method however remains potent for specialty problem solutions that will be revisited in Section 12.7. Section 2.7 presents the core formulation of finite element stiffness properties, using the strain–displacement and stress–strain relationships derived in earlier sections of this chapter. This core formulation will be used repeatedly in subsequent chapters to derive the stiffness properties of 2D and 3D solid elements. Detailed steps of the stiffness formulation technique, based on linear displacement variation functions, are demonstrated with simple two-node truss element in Section 2.8. Section 2.9 introduces higher order truss/cable element – capable of reproducing quadratic variation of displacement response over the element domain. Key building blocks of a finite element simulation model are described in Section 2.10 – with ABAQUS specific description of data input syntax. Other solver software products have their own syntax rules that a user should be aware of while preparing a similar analysis model. Two practice problems are presented in Section 2.11 with the learning objective of manually preparing simple finite element model input files. Understanding the syntax rules of finite element data files is an essential skill for diagnosis of syntax errors that often emerge during the software-driven preparation of complex finite element models in real engineering applications. Absence of this understanding often forces the unnecessary reformulation of complex models that could be easily fixed by simply reviewing the text input files.

2.1 MEASUREMENT OF DEFORMATION INTENSITY (STRAIN)

Material deformation behavior is often tested under uniaxial loading condition. Figure 2.1 shows a material specimen subjected to a uniaxial load, F, applied at a very slow rate (termed as "quasi-static" testing). Initial gage length of the specimen, Δx (before loading), changes by incremental quantity Δu under loading action. Rate of change of deformation in direction x, x being normal to the cross-section plane A_0, is defined as "normal" strain in the material:

$$\varepsilon_x = \lim_{\Delta x \to 0} \frac{\Delta u}{\Delta x} = \frac{\partial u}{\partial x} \tag{2.1}$$

The above definition of strain can be expanded to describe bi-directional normal strains inside a differential element in x–y plane (Figure 2.2(a)). Material may also undergo distortional deformation, i.e. initial position of segment OA (Figure 2.2(b)) may rotate by angle $\partial v/\partial x$, and segment OB by $\partial u/\partial y$. Total angular distortion to initial right angle AOB, referred to as shear strain of the material in x–y plane, is given by

$$\gamma_{xy} = \gamma_{yx} = \left(\frac{\partial u}{\partial y} + \frac{\partial v}{\partial x} \right) \tag{2.2}$$

By definition, shear deformation in the material occurs without accompanying length changes, i.e. without affecting normal strains in the material. Shear strains, unlike the normal strains, are not generally measured directly in test specimens. Instead, normal strains are measured by using electric-resistance strain gages – short

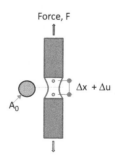

FIGURE 2.1 Measurement of strain in one-dimensional loading condition.

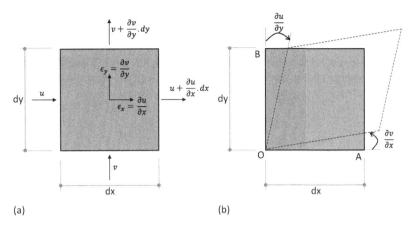

(a)   (b)

FIGURE 2.2 Measurement of strains in bi-directional loading condition: (a) normal strains; and (b) shear strain.

length of insulated wires glued to a target surface. When stretching occurs, electric resistance of the wire increases indicating the deformation intensity in the direction of measurement. Generally, 3 gages, placed 45° or 60° apart, are used in a cluster, called "strain rosette" (Figure 2.3). Relationships among measured normal strains $(\varepsilon_a, \varepsilon_b, \varepsilon_c)$ and the strains ε_x, ε_y, γ_{xy} (in the general reference coordinate system x,y) are given by equations (2.3) that can be solved to determine the normal and shear strain values in $(x–y)$ reference system:

$$\varepsilon_a = \varepsilon_x \cos^2 \theta_a + \varepsilon_y \sin^2 \theta_a + \gamma_{xy} \sin \theta_a \cos \theta_a$$
$$\varepsilon_b = \varepsilon_x \cos^2 \theta_b + \varepsilon_y \sin^2 \theta_b + \gamma_{xy} \sin \theta_b \cos \theta_b \qquad (2.3)$$
$$\varepsilon_c = \varepsilon_x \cos^2 \theta_c + \varepsilon_y \sin^2 \theta_c + \gamma_{xy} \sin \theta_c \cos \theta_c$$

Three strain components $(\varepsilon_x, \varepsilon_y,$ and $\gamma_{xy})$ describe the complete deformation state in the 2D plane of solids.

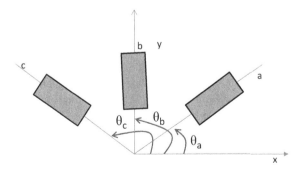

FIGURE 2.3 Normal strain measurement with an array of strain gages ("strain rosette").

22 Finite Element Analysis of Solids and Structures

2.2 GENERAL DESCRIPTION OF DEFORMATION STATE IN 3D SOLIDS

Strain components defined in Section 2.1 for 2D reference plane $(x–y)$ can be expanded to define strains in a 3D differential element as in the following equation (2.4):

$$\varepsilon_x = \frac{\partial u}{\partial x}, \quad \varepsilon_y = \frac{\partial v}{\partial y}, \quad \varepsilon_z = \frac{\partial w}{\partial z}$$

$$\gamma_{xy} = \gamma_{yx} = \frac{\partial u}{\partial y} + \frac{\partial v}{\partial x}$$

$$\gamma_{yz} = \gamma_{zy} = \frac{\partial v}{\partial z} + \frac{\partial w}{\partial y} \tag{2.4}$$

$$\gamma_{xz} = \gamma_{zx} = \frac{\partial u}{\partial z} + \frac{\partial w}{\partial x}$$

Similar to the definition of stress components, symmetry applies to shear strain components as well (i.e. $\gamma_{xy} = \gamma_{yx}$, $\gamma_{yz} = \gamma_{zy}$, and $\gamma_{xz} = \gamma_{zx}$). Equations (2.4) can be rewritten in the following matrix and vector forms:

$$\begin{Bmatrix} \varepsilon_x \\ \varepsilon_y \\ \varepsilon_z \\ \gamma_{xy} \\ \gamma_{yz} \\ \gamma_{zx} \end{Bmatrix} = \begin{bmatrix} \frac{\delta}{\delta x} & 0 & 0 \\ 0 & \frac{\delta}{\delta y} & 0 \\ 0 & 0 & \frac{\delta}{\delta z} \\ \frac{\delta}{\delta y} & \frac{\delta}{\delta x} & 0 \\ 0 & \frac{\delta}{\delta z} & \frac{\delta}{\delta y} \\ \frac{\delta}{\delta z} & 0 & \frac{\delta}{\delta x} \end{bmatrix} \cdot \begin{Bmatrix} u \\ v \\ w \end{Bmatrix} \tag{2.5}$$

Differential operator matrix on the right side of equation (2.5), defining the strain–displacement relationships, is same as the one shown in equation (1.13) for stress equilibrium condition. Re-writing equation (2.5) using matrix and vector notations gives the following:

$$\{\varepsilon\} = [\mathrm{B}].\{u\} \tag{2.6}$$

where $\{\varepsilon\}$ is the vector of six strain components $(\varepsilon_x, \varepsilon_y, \varepsilon_z, \gamma_{xy}, \gamma_{yz}, \gamma_{zx})$ and $\{u\}$ is the vector of three displacement components (u, v, w). Equation (2.6) presents a general form of strain–displacement relationship that will appear repeatedly throughout the subsequent discussions on finite element formulations.

2.3 COMPATIBILITY OF STRAIN (DEFORMATION FIELD) IN A BODY

Kinematic relations (equations 2.5) relate six components of strains to three deformations u, v, and w. All six components of strain are, therefore, not completely independent of one another. Differentiating ε_x in equation (2.4) twice with respect to y, ε_y twice with respect to x, and γ_{xy} with respect to x and y, we get

$$\frac{\partial^2 \varepsilon_x}{\partial y^2} = \frac{\partial^3 u}{\partial x \partial y^2}, \quad \frac{\partial^2 \varepsilon_y}{\partial x^2} = \frac{\partial^3 v}{\partial x^2 \partial y}, \quad \frac{\partial^2 \gamma_{xy}}{\partial x \partial y} = \frac{\partial^3 u}{\partial x \partial y^2} + \frac{\partial^3 v}{\partial x^2 \partial y} \tag{2.7}$$

Combining the three parts in equation (2.7), we get

$$\frac{\partial^2 \varepsilon_x}{\partial y^2} + \frac{\partial^2 \varepsilon_y}{\partial x^2} = \frac{\partial^2 \gamma_{xy}}{\partial x \partial y} \tag{2.8}$$

Similarly, on (y,z) and (x,z) planes, we get

$$\frac{\partial^2 \varepsilon_y}{\partial z^2} + \frac{\partial^2 \varepsilon_z}{\partial y^2} = \frac{\partial^2 \gamma_{yz}}{\partial y \partial z} \tag{2.9}$$

$$\frac{\partial^2 \varepsilon_z}{\partial x^2} + \frac{\partial^2 \varepsilon_x}{\partial z^2} = \frac{\partial^2 \gamma_{xz}}{\partial x \partial z} \tag{2.10}$$

Similar to previous steps, additional relationship can be derived by differentiating ε_x with respect to y and z, ε_y with respect to z and x, and ε_z with respect to x and y:

$$2 \frac{\partial^2 \varepsilon_x}{\partial y \partial z} = \frac{\partial}{\partial x} \left(-\frac{\partial \gamma_{yz}}{\partial x} + \frac{\partial \gamma_{xz}}{\partial y} + \frac{\partial \gamma_{xy}}{\partial z} \right) \tag{2.11}$$

$$2 \frac{\partial^2 \varepsilon_y}{\partial z \partial x} = \frac{\partial}{\partial y} \left(\frac{\partial \gamma_{yz}}{\partial x} - \frac{\partial \gamma_{xz}}{\partial y} + \frac{\partial \gamma_{xy}}{\partial z} \right) \tag{2.12}$$

$$2 \frac{\partial^2 \varepsilon_z}{\partial x \partial y} = \frac{\partial}{\partial z} \left(\frac{\partial \gamma_{yz}}{\partial x} + \frac{\partial \gamma_{xz}}{\partial y} - \frac{\partial \gamma_{xy}}{\partial z} \right) \tag{2.13}$$

Inter-relationships among strain components, shown in equations (2.8)–(2.13), define compatibility conditions among strains inside a body. Strain components, calculated from deformation variables (u, v, w), satisfy all six compatibility equations in a 3D problem – an essential requirement that the finite element formulations, to be presented later, must meet.

2.4 STRESS–STRAIN RELATIONSHIPS (HOOKE'S LAW)

Uniaxial test specimens of ductile materials (Figure 2.4) experience changes in cross-section area and gage length during the loading phase. However, it is general

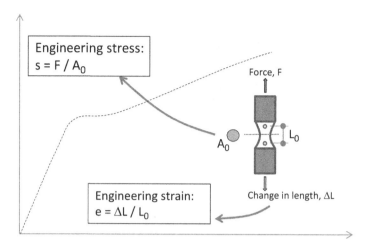

FIGURE 2.4   Engineering stress–strain response of a ductile material.

engineering practice to convert the measured force vs deformation response (F vs ΔL) to engineering stress vs strain relationship by using initial cross-section area (A_0) and initial gage length (L_0) (Figure 2.4). True stress at any given moment of loading can be defined by F/A; and true strain by $\Delta L/L$, where A and L are, respectively, instantaneous area and gage length. Assuming no overall volume changes in the material during deformation, i.e. $A_0 L_0 = AL$, true stresses and strains can be calculated from the engineering measurements as shown in Figure 2.5. The stress level S_y defines a significant change in the material behavior. When applied stress $\sigma < S_y$, normal stress σ is linearly proportional to normal strain ε. This linear elasticity relationship between stress and strain is known as *Hooke*'s law (Robert Hooke 1635–1703). Proportionality constant, E, is known as elastic modulus or *Young*'s modulus (Thomas Young 1773–1829). E is unique for each material – generally same in

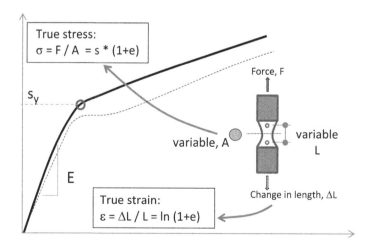

FIGURE 2.5   True stress–strain response of the material.

tension and compression. Engineering unit of E is same as that of stress σ (for example, GPa or kN/mm^2). Denoting the normal direction to cross-sectional area as x (same as loading axis), Hooke's law defining the normal stress–strain relationship is given by equation (2.14):

$$\varepsilon_x = \frac{\sigma_x}{E} \tag{2.14}$$

Experimental data show that elastic axial strain (in tension or compression) is accompanied by lateral strain (contraction or expansion), and the two are related by a constant of proportionality, ν (known as Poisson's ratio after S.D. Poisson 1781–1840):

$$\varepsilon_y = \varepsilon_z = -\nu \frac{\sigma_x}{E} \tag{2.15}$$

For most metals: $\nu = 0.25$–0.35. Stress–strain relationships (2.14) and (2.15), defined for x-directional loading, can also be applied to y- and z-directional loading effects as follows:

$$\varepsilon_y = \frac{\sigma_y}{E}$$
$$\varepsilon_x = \varepsilon_z = -\nu \frac{\sigma_y}{E} \tag{2.16}$$

$$\varepsilon_x = \varepsilon_y = -\nu \frac{\sigma_z}{E}$$
$$\varepsilon_z = \frac{\sigma_z}{E} \tag{2.17}$$

Assuming that the deformations are very small (without any effect on external loads), principle of superposition provides the following generalized Hooke's law for normal stresses and strains in 3D homogenous isotropic material:

$$\varepsilon_x = \frac{1}{E}\left[\sigma_x - \nu\left(\sigma_y + \sigma_z\right)\right]$$
$$\varepsilon_y = \frac{1}{E}\left[\sigma_y - \nu\left(\sigma_x + \sigma_z\right)\right] \tag{2.18}$$
$$\varepsilon_z = \frac{1}{E}\left[\sigma_z - \nu\left(\sigma_x + \sigma_y\right)\right]$$

To derive the shear stress–strain relationship of elastic materials, a special case of biaxial loading on a rectangular element ($\sigma_y = \sigma$, $\sigma_x = -\sigma$, $\sigma_z = 0$) is considered in Figure 2.6. A square element "abcd" is cutout from the center where sides ab, bc, etc., are at 45° angle to reference axes x and y. Considering the equilibrium of triangular element "obc", we get the relationship in equation (2.19):

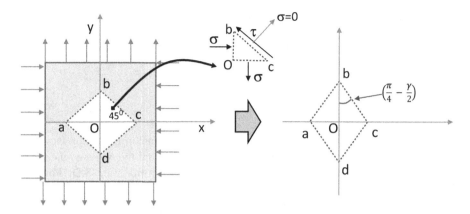

FIGURE 2.6 Internal shear deformation in a differential element under tension–compression biaxial loading.

$$\left(\tau * \sqrt{2}\right) * \sin 45° = 1 * \sigma \qquad \rightarrow \qquad \tau = \sigma \tag{2.19}$$

Under the tension–compression combination of stresses, diagonal "bd" elongates and "ac" shortens – causing a change in angle "abc" by "γ" giving the following equation (2.20):

$$\tan\left(\frac{\pi}{4} - \frac{\gamma}{2}\right) = \frac{oc}{ob} \tag{2.20}$$

Expanding the trigonometric terms on the left-hand side of equation (2.20), and assuming small angular distortion, tan $(\gamma/2) \approx \gamma/2$, equation (2.20) can be re-written as

$$\frac{1 - \dfrac{\gamma}{2}}{1 + \dfrac{\gamma}{2}} = \frac{oc}{ob} = \frac{1 + \epsilon_x}{1 + \epsilon_y} \tag{2.21}$$

Inserting $(\sigma_y = \sigma, \sigma_x = -\sigma, \sigma_z = 0)$ in equations (2.18) and combining that with equation (2.21), we get

$$\frac{1 - \dfrac{\gamma}{2}}{1 + \dfrac{\gamma}{2}} = \frac{1 - \dfrac{1 + v}{E}\sigma}{1 + \dfrac{1 + v}{E}\sigma} \tag{2.22}$$

Substituting ($\sigma = \tau$) from equation (2.19), equation (2.22) leads to the following relationship between shear stress and shear strain:

$$\frac{\gamma}{2} = \frac{1+v}{E}\tau \quad \rightarrow \quad \tau = \frac{E}{2(1+v)}.\gamma = G.\gamma \tag{2.23}$$

G is called shear modulus or modulus of rigidity. Shear stress–strain relationship of equation (2.23) derived for element in 2D plane can be expanded to define the 3D relationships as follows:

$$\gamma_{xy} = \frac{\tau_{xy}}{G}$$

$$\gamma_{yz} = \frac{\tau_{yz}}{G} \tag{2.24}$$

$$\gamma_{zx} = \frac{\tau_{zx}}{G}$$

Normal stress–strain relations of (2.18) and shear stress–strain relationships of (2.24) define the complete stress–strain relationships for 3D problems. Inverting equations (2.18) and (2.14), we get the following generalized Hooke's law for 3D stresses and strains:

$$\begin{Bmatrix} \sigma_x \\ \sigma_y \\ \sigma_z \\ \tau_{xy} \\ \tau_{yz} \\ \tau_{zx} \end{Bmatrix} = \frac{E}{(1+v)(1-2v)} \begin{bmatrix} (1-v) & v & v & 0 & 0 & 0 \\ v & (1-v) & v & 0 & 0 & 0 \\ v & v & (1-v) & 0 & 0 & 0 \\ 0 & 0 & 0 & \frac{(1-2v)}{2} & 0 & 0 \\ 0 & 0 & 0 & 0 & \frac{(1-2v)}{2} & 0 \\ 0 & 0 & 0 & 0 & 0 & \frac{(1-2v)}{2} \end{bmatrix} \begin{Bmatrix} \varepsilon_x \\ \varepsilon_y \\ \varepsilon_z \\ \gamma_{xy} \\ \gamma_{yz} \\ \gamma_{zx} \end{Bmatrix} \tag{2.25}$$

Re-writing equations (2.25) with vector and matrix notations gives the following:

$$\{\sigma\} = [C] * \{\epsilon\} \tag{2.26}$$

where $\{\sigma\}$ is the vector of six stress components in 3D element, $\{\varepsilon\}$ is the vector of corresponding strain components, and [C] is the "elasticity" property matrix defining the stress–strain relationships for 3D solid elements. Like the general strain–displacement relationship of equation (2.6), general stress–strain relationship of equation (2.26) will also appear repeatedly throughout the subsequent discussions on finite element formulations. The component details of [C], however, will vary depending on the response mechanism of different finite element formulations.

Stress–strain relationships of equation (2.25) can also be written in another well-known form as follows (Timoshenko and Goodier 1982):

$$\sigma_x = \lambda e + 2G\varepsilon_x$$
$$\sigma_y = \lambda e + 2G\varepsilon_y \qquad (2.27)$$
$$\sigma_z = \lambda e + 2G\varepsilon_z$$

where e is the volumetric strain given by $e = \varepsilon_x + \varepsilon_y + \varepsilon_z$; and λ and G are called Lame's constants. Shear modulus G has been defined in equation (2.23) and λ is defined as follows:

$$\lambda = \frac{vE}{(1+v)(1-2v)} \qquad (2.28)$$

For special case of uniform hydrostatic pressure, $\sigma_x = \sigma_y = \sigma_z = -p$, relationship between pressure p and volumetric strain e is given by equation (2.29), where K is called the bulk modulus of elasticity:

$$-p = \frac{E}{3(1-2v)}.e = K.e \qquad (2.29)$$

For incompressible materials, e approaches 0 implying that Poisson's ratio approaches a value of 0.5. Stress–strain relations, defined in terms of λ and G, are often used in finite element literature related to the modeling of incompressible material behavior.

2.5 SOLUTION OF ELASTICITY PROBLEMS USING STRESS DISTRIBUTION FUNCTIONS

Three equations of stress equilibrium (equations 1.10), six equations of stress–strain relations (equation 2.25), and six equations of strain compatibility conditions (equations 2.8–2.13) give a total of 15 equations that are sufficient to determine 15 response variables (six stress components, six strain components, and three displacement variables). However, compatibility equations (2.8–2.13), expressed in terms of strain derivatives, can be re-written in terms of stresses by using the Hooke's law (2.25). Combining those with the stress equilibrium equations (1.10), the following set of equations for stress compatibility conditions can be obtained for the special case of no body forces, i.e. $F_x = F_y = F_z = 0$ (Timoshenko and Goodier 1982):

$$\left(1+v\right)\nabla^2\sigma_x+\frac{\partial^2\vartheta}{\partial x^2}=0 \quad \left(1+v\right)\nabla^2\tau_{xy}+\frac{\partial^2\vartheta}{\partial x\partial y}=0$$

$$\left(1+v\right)\nabla^2\sigma_y+\frac{\partial^2\vartheta}{\partial y^2}=0 \quad \left(1+v\right)\nabla^2\tau_{yz}+\frac{\partial^2\vartheta}{\partial y\partial z}=0 \qquad (2.30)$$

$$\left(1+v\right)\nabla^2\sigma_z+\frac{\partial^2\vartheta}{\partial z^2}=0 \quad \left(1+v\right)\nabla^2\tau_{zx}+\frac{\partial^2\vartheta}{\partial z\partial x}=0$$

where $\nabla^2 - \dfrac{\partial^2}{\partial x^2}+\dfrac{\partial^2}{\partial y^2}+\dfrac{\partial^2}{\partial z^2}$ and $\vartheta=\sigma_x \mid \sigma_y \mid \sigma_z$

Stress distributions in an isotropic body must satisfy all three conditions: compatibility equations (2.30), equilibrium equations (1.10), and stress boundary conditions (1.14). These 12 equations are generally sufficient to determine the stress field without ambiguity. Equations of compatibility contain only second derivatives of the stress components. Hence, if the external forces are such that the equations of equilibrium and the stress boundary conditions are satisfied by taking stress components either as constants or as linear functions of the coordinates, the equations of compatibility will be automatically satisfied – thus giving correct solution of the stress distribution.

EXAMPLE: STRETCHING OF A PRISMATIC BAR BY ITS OWN WEIGHT

Figure 2.7 shows a bar stretched by its own weight downward from the upper support at $z = \ell$. Assuming ρ is the density of the material, and g is the acceleration due to gravity, body forces per unit volume are given by

$$F_x = F_y = 0; F_z = -\rho g \qquad (2.31)$$

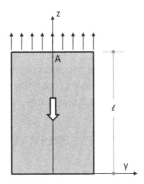

FIGURE 2.7 Stretching of a prismatic bar by its own weight.

From observation, stress distribution functions inside the body can be defined as follows:

$$\sigma_z = \rho g z; \quad \sigma_x = \sigma_y = \tau_{xy} = \tau_{yz} = \tau_{zx} = 0 \tag{2.32}$$

Stress components defined in equation (2.32) readily satisfy the differential equations of stress equilibrium (1.10). Stress boundary conditions of zero value on the surface, zero at the free end ($z = 0$), and a uniformly distributed reaction of $\sigma_z = \rho g \ell$ at the upper end are also satisfied by equations (2.32). Compatibility conditions (2.30) are also satisfied. So, equations (2.32) give correct description of the stress functions in the body under the action of self-weight only. Strains in the body can be found by inserting the stress functions (equations 2.32) in Hooke's laws (2.18 and 2.24). Integration of the strain–displacement relationships (2.4) provides the following equations for displacement responses (Timoshenko and Goodier 1982):

$$
\begin{aligned}
u &= -\frac{\nu \rho g x z}{E} \\
v &= -\frac{\nu \rho g y z}{E} \\
w &= \frac{\nu g z^2}{2E} + \frac{\nu \rho g}{2E}\left(x^2 + y^2\right) - \frac{\nu g l^2}{2E}
\end{aligned}
\tag{2.33}
$$

Evidently, because of the Poisson's effects, lateral displacements, u and v, have nonzero values except along the axis of the member ($x = y = 0$). Solution technique, based on stress function approach, as demonstrated with the elementary example of prismatic bar, is useful when the stress distribution inside a body can be described analytically based on *a priori* knowledge of the system response. Alternative analytical solution technique, based on displacement field assumption, is described in Section 2.6.

2.6   SOLUTION OF ELASTICITY PROBLEMS USING DISPLACEMENT VARIATION FUNCTIONS

In this analysis technique, displacement variation functions, $u(x, y, z)$, $v(x, y, z)$, $w(x, y, z)$, are introduced to satisfy all necessary conditions, i.e. boundary conditions, equilibrium equations, and compatibility conditions.

EXAMPLE: STRETCHING OF A PRISMATIC BAR UNDER AXIAL TENSION

We consider a prismatic bar subjected to an axial tensile force, P, at end (2), and is restrained at end (1) (Figure 2.8). Considering a linear variation of deformation over the element length, the following displacement response functions are assumed for the element (Poisson's effect ignored):

$$u = ax \quad v = 0 \quad w = 0 \tag{2.34}$$

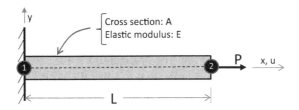

FIGURE 2.8   Stretching of a prismatic bar under axial tension.

Using the strain–displacement relationships of equation (2.4), we get the following expressions for strains inside the body:

$$\varepsilon_x = \frac{\partial u}{\partial x} = a \qquad \varepsilon_y = \frac{\partial v}{\partial y} = 0 \qquad \varepsilon_z = \frac{\partial w}{\partial z} = 0 \qquad \gamma_{xy} = \gamma_{yz} = \gamma_{zx} = 0 \qquad (2.35)$$

Strain components in equation (2.35) satisfy the compatibility conditions defined in equations (2.8)–(2.13). Displacement boundary condition of $u = 0$ at $x = 0$ is also satisfied by the displacement functions (2.34). Now considering the force boundary condition at $x = L$ (node 2):

$$Internal\ resistance = \left(A.E.\varepsilon_x\right)\Big|_{x=L} = \left(A.E.a\right) = Extenral\ force\ P \qquad (2.36)$$

Equation of equilibrium (2.36) yields, $a = P/AE$. Substituting it in equations (2.34) and (2.35), we get the following functions for displacements, strains, and stresses:

$$u = \frac{Px}{AE} \qquad v = 0 \qquad\qquad w = 0$$

$$\varepsilon_x = \frac{P}{AE} \qquad \varepsilon_y = \varepsilon_z = 0 \qquad \gamma_{xy} = \gamma_{yz} = \gamma_{zx} = 0 \qquad (2.37)$$

$$\sigma_x = \frac{P}{A} \qquad \sigma_y = \sigma_z = 0 \qquad \tau_{xy} = \tau_{yz} = \tau_{zx} = 0$$

Equations (2.37) represent the correct solution for the problem in Figure 2.8 as all necessary conditions (compatibility, equilibrium, and boundary conditions) have been satisfied. Solution method outlined above proceeded through a systematic procedure of displacement function definition, verification of necessary conditions, calculation of unknown coefficients in displacement functions from the boundary conditions, followed by calculation of strains and stresses.

Stress function method in Section 2.5 started with the assumption of stress distribution definition; and followed similar systematic procedure as the displacement method. However, *a priori* definition of either of the field variable, displacement or stress, is often not evident for general shaped bodies with complex loading and boundary conditions, such as the example shown schematically in Figure 1.12. The definition of displacement variation, however, provides a convenient tool to define the response mechanism of general shape smaller elements that can be inter-connected to define the overall response of a complex element assembly (Figure 1.13). This concept of discretizing the displacement field of a continuous solid body, to an assembly of interconnected elements, forms the basis of finite element simulation models. A basic requirement of finite element formulation is the derivation of element stiffness properties that can be used to solve the system equilibrium equations involving nodal DOF (equations 1.5). An objective formulation of the element stiffness property is desired to find a unique solution for the overall displacement response without ambiguity. Several theoretical alternatives, eventually leading to the same element stiffness definition, are often discussed in books on finite element methods. Among different techniques (virtual work, minimization of potential energy, weighted residual methods, etc.), the virtual work method is presented in Section 2.7 as it leads to the target formulation in a short easily interpretable way.

2.7 STIFFNESS METHOD (FINITE ELEMENT METHOD) OF STRUCTURAL ANALYSIS

System equilibrium equations based on stiffness properties of discrete spring elements have been discussed in Section 1.4. Following derivations present a generalized procedure for stiffness property calculations of a solid element (Figure 2.9), hereafter referred to as "finite element". Element is connected to a set of nodes that have displacement degrees of freedom, $\{u\}$. External forces acting on the nodes are denoted by vector $\{P\}$. Internal element resistances, producing equilibrium with external forces, generate a set of internal stresses $\{\sigma\}$. The corresponding internal strains are represented by the vector $\{\varepsilon\}$. Let us impose a virtual displacement, $\bar{U}$, on the system ($\bar{U}$ is consistent with the prescribed boundary conditions). Using the strain–displacement relationship of equation (2.6), virtual strains corresponding to the virtual displacement field $\bar{U}$ are given by

$$\{\bar{\varepsilon}\} = [B]\{\bar{U}\} \tag{2.38}$$

Internal virtual work done by the strains $\{\bar{\varepsilon}\}$:

$$Internal\ virtual\ work = \int\{\bar{\varepsilon}\}^T\{\sigma\}dV = \bar{U}^T\int[B]^T\{\sigma\}dV \tag{2.39}$$

External virtual work done on the forces $\{P\}$:

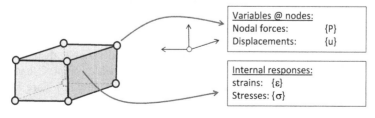

> Variables @ nodes:
> Nodal forces: $\{P\}$
> Displacements: $\{u\}$

> Internal responses:
> strains: $\{\varepsilon\}$
> Stresses: $\{\sigma\}$

FIGURE 2.9 Definition of a finite element.

$$\textit{External virtual work} = \{\bar{U}\}^{T}.\{P\} \tag{2.40}$$

Equating the internal virtual work with the external virtual work, we get

$$\bar{U}^{T} \int [B]^{T}\{\sigma\}dV = \{\bar{U}\}^{T}.\{P\} \tag{2.41}$$

As the relationship (2.41) is valid for any value of virtual displacement, the equality of multipliers must exist, i.e.:

$$\int [B]^{T}.\{\sigma\}.dV = \{P\} \tag{2.42}$$

Equation (2.42) represents the equilibrium between internal stresses and external forces. Substituting the relationships (2.26) and (2.6) into equation (2.42), and after re-arranging the terms, we get

$$\left[\int [B]^{T}.[C].[B].dV\right].\{u\} = \{P\} \tag{2.43}$$

Comparing equation (2.43) with the familiar system equilibrium equation (1.5), we get the following general definition of the finite element stiffness matrix:

$$[k] = \left[\int [B]^{T}.[C].[B].dV\right] \tag{2.44}$$

Integral in equation (2.44) is calculated with numerical integration – accuracy of which depends on element types and integration order chosen in specific element formulations. As discussed earlier, finite element analysis procedure starts with breaking down of a large complex system into an assembly of smaller "elements" – interconnected at "node" points (Figure 1.13). Deformation response of the body is represented by the vector of displacement values at discrete nodes. Analytical assumptions are used to describe local deformation inside finite elements, and to determine the derivatives of displacements that define the strain–displacement relationship matrix [B] in equation (2.6). Element-specific geometric property matrix [B], and the material stress–strain relationship matrix [C] (equation 2.26), are used to calculate the individual element stiffness matrix (equation 2.44). Calculated element stiffness matrix is mapped into the element-specific degrees of freedom in the NxN stiffness matrix in equation (1.5). Overall stiffness matrix of the larger body is assembled from the individual contributions of finite elements that make up the entire structure (Figure 1.13).

Load vector, {P}, on the right side of equation (2.43) is calculated as follows:

$$\{P\} = -\int B^{T}\sigma_{0}dV + \int f^{B}dV + \int f^{s}dS + \sum R_{c} \tag{2.45}$$

where σ_{o} is initial stress, f^{B} is internal body force per unit volume, f^{S} is surface force per unit surface area, and R_{c} represents concentrated forces applied directly at model nodes. Following the assembly of system stiffness matrix and load vector from the contributions of individual elements, the following steps are executed to determine the finite element model responses:

- Identify known forces and displacement constraints
- Modify equations of equilibrium for known boundary conditions and external loads
- Solve equations of equilibrium to determine unknown displacements: $[k] * \{u\} = \{P\}$
- Calculate strains inside all elements: $\{\varepsilon\} = [B]\{u\}$
- Calculate internal stresses: $\{\sigma\} = [C]\{\varepsilon\}$
- Total strain energy stored can be calculated from the summation of element contributions: $U_0 = \int \frac{1}{2}\{\varepsilon\}^T.\{\sigma\}.dV$

A critical step in developing the finite element model of a structure involves the assumption of displacement field inside individual elements – that in turn define the strain–displacement relationship matrix $[B]$ in the above calculations. The assumption of known displacement variation inside elements is analogous to the displacement method solution technique discussed in Section 2.6. However, finite element method does not impose an analytical displacement field equation over the entire geometry of solution domain. Piece-by-piece analytical description of discrete finite element responses reproduces the complex displacement response of a general structural body.

The finite element stiffness formulation, represented by equation (2.44), applies to deformation response field of any complexity (axial, shear, bending, torsion, etc.). The remainder of this chapter will focus on the demonstration of stiffness formulation method for the simplest element form that provides resistance to axial deformation only (truss, cable, etc.). Formulations for more complex 2D and 3D finite elements will be introduced in subsequent chapters.

2.8 STIFFNESS PROPERTIES OF 1-D TRUSS ELEMENT PROVIDING RESISTANCE TO AXIAL DEFORMATION ONLY

Figure 2.10 shows two example structural systems, (a) a cable supported bridge and (b) a truss tower, that are assemblies of members providing resistance to axial deformation only. Displacement degrees of freedom, at nodes 1 and 2 of element in Figure 2.11, are identified as u_1 and u_2. Positions of the two end nodes of element are defined by global coordinates x_1 and x_2. A local coordinate definition r is introduced (Figure 2.11) to define any point inside the element in the range of $r = -1$ to $r = +1$. For linear variation of coordinate over the element length, interpolation functions for internal coordinate locations are given by, $H_1 = (1 - r)/2$ and $H_2 = (1 + r)/2$. Coordinate position, x, at any point inside the element is defined as follows:

$$x = H_1.x_1 + H_2.x_2 \tag{2.46}$$

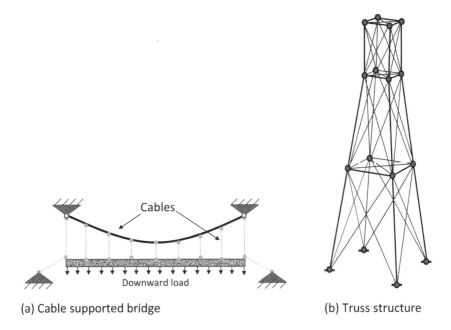

(a) Cable supported bridge   (b) Truss structure

FIGURE 2.10 Structural assemblies of elements providing resistance to axial deformation only.

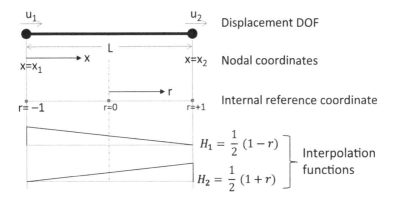

FIGURE 2.11 Interpolation functions for the nodal variables of a truss element.

After inserting expressions for H_1 and H_2 in equation (2.46), and re-arranging the terms:

$$r = \frac{1}{L/2}\left(x - \frac{x_1 + x_2}{2}\right) \tag{2.47}$$

where L is the length of the element ($L = x_2 - x_1$). Taking partial derivative of both sides of equation (2.47) with respect to x:

$$\frac{\partial r}{\partial x} = \frac{2}{L} \tag{2.48}$$

Assuming that the deformation varies linearly over the element length, internal deformation at any point inside the element can be expressed in terms of the nodal displacements by using the same coordinate interpolation functions of Figure 2.11:

$$u = H_1.u_1 + H_2.u_2 = \frac{1-r}{2}.u_1 + \frac{1+r}{2}.u_2 \tag{2.49}$$

Internal strain caused by the nodal displacements, u_1 and u_2 (Figure 2.11), can be defined as follows:

$$\varepsilon = \frac{\partial u}{\partial x} = \left(\frac{\partial u}{\partial r}\right).\left(\frac{\partial r}{\partial x}\right) \tag{2.50}$$

Substituting the expressions (2.48) and (2.49) into equation (2.50), we get the following definition for internal strain:

$$e = \frac{1}{L}\left(u_2 - u_1\right) \tag{2.51}$$

Relationship between internal strain and nodal displacements, given by equation (2.51), can be re-written in the following form:

$$\{\varepsilon\} = \begin{bmatrix} \dfrac{-1}{L} & \dfrac{1}{L} \end{bmatrix} \begin{Bmatrix} u_1 \\ u_2 \end{Bmatrix} \tag{2.52}$$

Comparing equation (2.52) with (2.6), strain–displacement relationship matrix [B] for two-node truss element of Figure 2.11 is defined as follows:

$$[B] = \begin{bmatrix} \dfrac{-1}{L} & \dfrac{1}{L} \end{bmatrix} \tag{2.53}$$

Stress–strain relation for one-dimensional axial deformation response of the truss element is simply defined by the elastic modulus, E, thus giving the following definition of elasticity matrix [C]:

$$[C] = [E] \tag{2.54}$$

Substituting the definitions of [B] and [C] from equations (2.53) and (2.54) into equation (2.44), stiffness matrix of two-node truss element is calculated as follows:

$$[k] = \int \begin{bmatrix} -1/L \\ 1/L \end{bmatrix} [E] \begin{bmatrix} \dfrac{-1}{L} & \dfrac{1}{L} \end{bmatrix} dV \tag{2.55}$$

Element geometry and material parameters inside equation (2.55) are constants that can be taken out of integral sign, thus, reducing the integral operation in equation (2.55) to a simple integral of the volume of the truss element. Writing $dV = A.dx$ (A being the cross-sectional area of prismatic bar), we get the following modified form of equation (2.55):

$$[k] = \begin{bmatrix} \dfrac{E}{L^2} & \dfrac{-E}{L^2} \\ \dfrac{-E}{L^2} & \dfrac{E}{L^2} \end{bmatrix} \cdot \int_{x_1}^{x_2} A.dx \tag{2.56}$$

Integral in equation (2.56) is simply the volume of the element $[= A(x_2 - x_1) = A.L]$. Stiffness matrix of the 2-DOF truss element, aligned with the coordinate direction x (Figure 2.11), is thus given by the following equation:

$$[k] = \begin{bmatrix} \dfrac{AE}{L} & \dfrac{-AE}{L} \\ \dfrac{-AE}{L} & \dfrac{AE}{L} \end{bmatrix} \tag{2.57}$$

Stiffness matrix in equation (2.57) has been derived based on axial resistance of truss element against the displacement DOF u_1 and u_2 that are aligned with member axis direction, x (Figure 2.11). A truss element in actual structural assembly can be oriented in any direction in 3D space. Figure 2.12(a) shows 3 global displacement DOF at each end node, i and j, of an arbitrarily oriented two-node truss element. Figure 2.12(b) shows local member axis, x, and relevant local DOF, u_1 and u_2, for internal element response calculations. Relationships between global and local DOF are defined as follows:

$$\begin{Bmatrix} u_1 \\ u_2 \end{Bmatrix} = \begin{bmatrix} l_{ij} & m_{ij} & n_{ij} & 0 & 0 & 0 \\ 0 & 0 & 0 & l_{ij} & m_{ij} & n_{ij} \end{bmatrix} \begin{Bmatrix} U_i \\ V_i \\ W_i \\ U_j \\ V_j \\ W_j \end{Bmatrix} \to \{u\} = [T]\{U\} \tag{2.58}$$

where l_{ij}, m_{ij}, and n_{ij} are direction cosines of member (i,j) with reference to the global coordinate system (X, Y, Z). Stiffness matrix of truss element, calculated with reference to local member axis direction (equation 2.57), can be transformed by using the coordinate transformation matrix $[T]$ (equation 2.58), to get the stiffness matrix of truss element having 6-DOF in global coordinate directions (Figure 2.12(a)), given by equation (2.59):

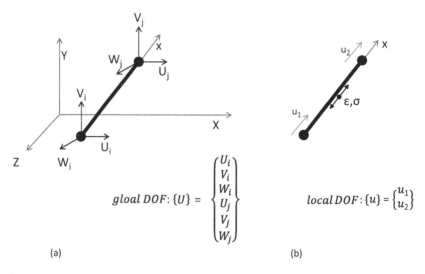

FIGURE 2.12 (a) 6-DOF truss element in 3D space with 3-DOF at each node; (b) local reference direction for axial response calculation of two-node truss element.

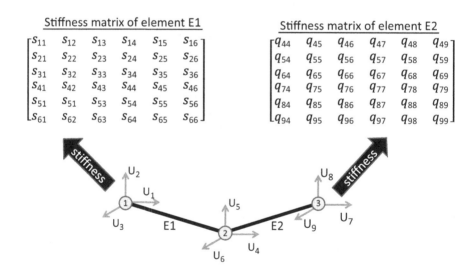

FIGURE 2.13 Stiffness matrices of two truss elements each having 6 displacement DOF.

$$\left[K\right]_{6x6} = \left[T\right]_{6x2}^{T}\left[k\right]_{2x2}\left[T\right]_{2x6} \tag{2.59}$$

where $[k]_{2x2}$ is the local stiffness matrix defined in equation (2.57) and $[T]$ is the coordinate transformation matrix defined in equation (2.58).

Figure 2.13 shows examples of 6×6 stiffness matrices for two truss elements having arbitrary orientation in 2D space. Following the matrix assembly procedure

$$
\begin{bmatrix}
s_{11} & s_{12} & s_{13} & s_{14} & s_{15} & s_{16} & 0 & 0 & 0 \\
s_{21} & s_{22} & s_{23} & s_{24} & s_{25} & s_{26} & 0 & 0 & 0 \\
s_{31} & s_{32} & s_{33} & s_{34} & s_{35} & s_{36} & 0 & 0 & 0 \\
s_{41} & s_{42} & s_{43} & K_{44} & K_{45} & K_{46} & q_{47} & q_{48} & q_{49} \\
s_{51} & s_{51} & s_{53} & K_{54} & K_{55} & K_{56} & q_{57} & q_{58} & q_{59} \\
s_{61} & s_{62} & s_{63} & K_{64} & K_{65} & K_{66} & q_{67} & q_{68} & q_{69} \\
0 & 0 & 0 & q_{74} & q_{75} & q_{76} & q_{77} & q_{78} & q_{79} \\
0 & 0 & 0 & q_{84} & q_{85} & q_{86} & q_{87} & q_{88} & q_{89} \\
0 & 0 & 0 & q_{94} & q_{95} & q_{96} & q_{97} & q_{98} & q_{99}
\end{bmatrix}
\begin{Bmatrix}
U_1 \\ U_2 \\ U_3 \\ U_4 \\ U_5 \\ U_6 \\ U_7 \\ U_8 \\ U_9
\end{Bmatrix}
=
\begin{Bmatrix}
P_1 \\ P_2 \\ P_3 \\ P_4 \\ P_5 \\ P_6 \\ P_7 \\ P_8 \\ P_9
\end{Bmatrix}
$$

where stiffness @ shared DOF:
$K_{ij} = s_{ij} + q_{ij}$

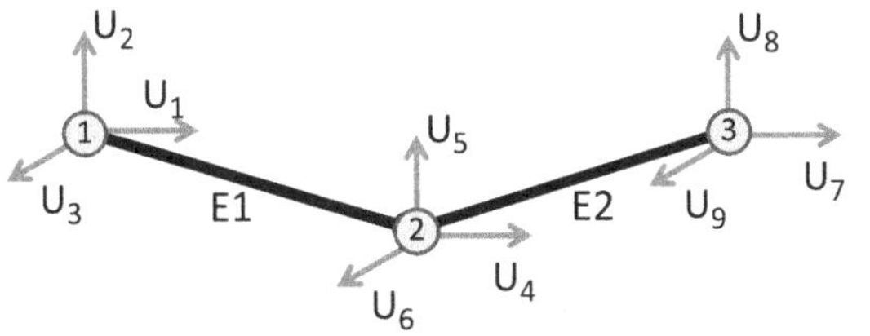

FIGURE 2.14 Overall stiffness matrix of two truss element assembly.

described in Section 1.4, stiffness matrix of a structure can be built from the contributions of as many elements as needed to completely describe the structure. Figure 2.14 shows an assembled stiffness matrix for two truss element systems. After introducing nodal loads and boundary conditions, system responses can be calculated following the same procedure described in Section 1.4 for two-spring system. The same procedure can also be used for structural systems comprising of a very large number of elements.

Analysis steps described in the above present the simplest form of finite element analysis of a structural assembly that is made up of members capable of providing resistance to axial deformation only. More complex structural response predictions require more advanced formulations for the calculation of element stiffness properties that will be introduced in Chapters 3–8.

2.9 HIGHER ORDER TRUSS ELEMENT AND MODEL REFINEMENT

Two-node truss element described in Section 2.8 allows linear variation of deformation over the element length (Figure 2.11). However, truss element can be re-formulated with one additional node in the middle to allow higher order displacement variation within the element (Figure 2.15). Interpolation functions H_1, H_2, and H_3 (Figure 2.15) can be used to derive the stiffness matrix of this three-node truss element by following the same procedure described for two-node element in Section 2.8. Three-node truss element is specifically suitable to model the curved profile of a cable as shown in Figure 2.16. A trivial attempt may lead to a model of the cable with multiple two-node truss elements – describing the curved geometric profile with piece-wise linear approximations. However, a cable made with multiple two-node elements will have

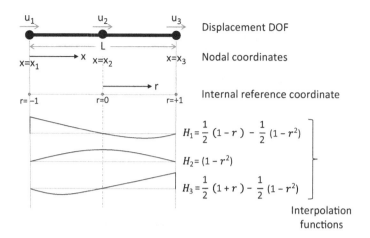

FIGURE 2.15 Interpolation functions for the nodal variables of a three-node truss element.

FIGURE 2.16 Cable modeled with a three-node truss element.

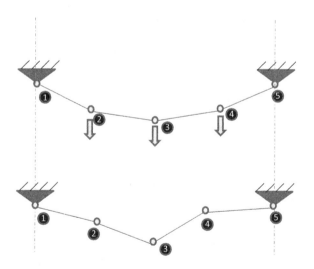

FIGURE 2.17 Instability in multi-element model of a cable.

inherent instability (as shown in Figure 2.17), thereby not meeting the essential requirement of a positive definite stiffness matrix formulation (Section 1.4). While a failed analysis is an indicator of structural instability, it is better to use upfront engineering judgment to determine the stability of a structural analysis model before embarking on the analysis task.

2.10 MODEL PREPARATION FOR COMPUTER-AIDED ANALYSIS OF STRUCTURES WITH FINITE ELEMENT SIMULATION SOFTWARE

Sections 2.7 and 2.8 have provided detailed descriptions of stiffness method analysis of structures – built as assembly of truss elements providing resistance to axial deformation only. In computer-aided analysis exercise, the actual model description tasks are undertaken by the engineers while all calculation tasks are left for finite element analysis software packages (group-3 in Figure 1.19). Each of those analysis software products has its own syntax rules for model description inputs although basic model input parameters (nodes, elements, loads, etc.) are similar. Example demonstrations in this book are presented for ABAQUS models (Dassault Systems 2020b). General-purpose model preprocessors, such as HyperMesh (Altair University 2020), possess the capability to import model input files of one finite element solver, and to export out the input data file for another solver. Computer-aided finite element model preparation process will be discussed in Chapters 3 and 4. Discussions in this chapter will remain limited to manual input data preparation for small-scale structural analysis problems. Descriptions of key data for simple truss analysis problem of Figure 2.18 are presented in the following. Data input syntax for ABAQUS software is presented

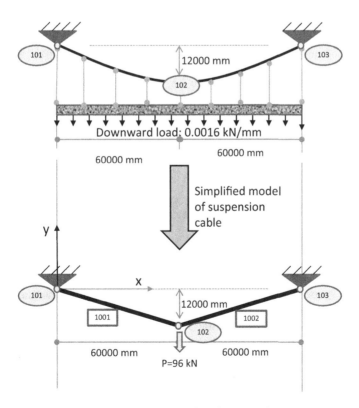

FIGURE 2.18 Simplified two truss element model of a suspension cable.

as a specific example. Specific syntax rules will need to be followed if a different analysis software is used for the practice problems.

Nodes: Each node in a model has a unique identification (ID) number and Cartesian coordinates. A set of nodes can be identified by assigning a name with the parameter "NSET". The following data block describes a set of three nodes with ID numbers, 101, 102, and 103, and their respective x-, y-, z-coordinates.

```
*NODE, NSET=setn1
101,0.0,0.0,0.0
102,60000.0, −12000.0,0.0
103,120000.0,0.0,0.0
```

Elements: Each element in a model has a unique ID number and a list of nodes that it attaches to. A set of elements can be grouped together with a name assigned by the parameter "ELSET". Element description also requires a "TYPE" deceleration with a standard type name selected from the ABAQUS element library. Based on selected type, ABAQUS chooses the specific calculation procedures for stresses, strains, and stiffness properties. For example, two-node truss element discussed in Section 2.8, is identified in ABAQUS library as "T2D2" – meaning a truss element in 2D space attached to 2 end nodes. Following is an example description of two truss elements in a set named "cable-1":

```
*ELEMENT, ELSET=cable-1, TYPE=T2D2
1001, 101, 102
1002, 102, 103
```

Element properties: Elements require type-specific property definitions. Truss element type discussed in Section 2.8 requires element cross-sectional area for stiffness calculation. Following example describes the member cross-section area ($A = 4560.4$ mm^2) for element set cable-1:

```
*SOLID SECTION, ELSET=cable-1, MATERIAL=mat1
4560.4
```

Material Properties: Material name identified in member property data must be defined with data block identifier *MATERIAL in ABAQUS input file. For temperature-independent elastic stress–strain response analysis, Young's modulus and Poisson's ratio are the only required material parameters. Example material data block for material named "mat1" is shown in the following assigning Young's modulus=210.0 GPa and Poisson's ratio=0.3:

```
*MATERIAL, NAME=mat1
*ELASTIC
210.0,0.3
```

<u>Boundary conditions</u>: External boundary conditions on nodes are defined by data block identifier *BOUNDARY. Boundary constraints can be imposed on selected DOF (1 to 6) to a node ID or to a set of nodes (identified by a set name). The following example imposes zero displacement constraint to two DOF (1–2) of nodes 101 and 103:

```
*BOUNDARY
101,1,2,0.0
103,1,2,0.0
```

<u>Loads</u>: *CLOAD identifies data block for concreted nodal loads. Each nodal load is described with node ID followed by associated DOF and the load magnitude. The following example specifies a load of 96 kN in negative y coordinate direction at node ID 102:

```
*CLOAD
102,2,-96.0
```

<u>Analysis step:</u> Data block for analysis step in ABAQUS starts with identifier *STEP; and it ends with identifier *END STEP. In between those two command lines, analysis types, loads, and output requests can be included. A small-displacement linear elastic analysis type is specified by including "PERTURBATION" key word with *STEP. Static load–deflection analysis is specified by *STATIC command. Following is an example analysis step including analysis type, applied load, and output requests:

```
*STEP, PERTURBATION
Load–deflection analysis for a concentrated load of –96 kN
*STATIC
*CLOAD
102, 2, –96.0
*NODE PRINT
U
RF
*EL PRINT
S
*END STEP
```

2.11 PRACTICE PROBLEMS – STRESS ANALYSIS OF A CABLE WITH FINITE ELEMENT ANALYSIS SOFTWARE ABAQUS

PROBLEM-1

Figure 2.18 shows a simplified two-element model of a bridge suspension cable. Cable is assumed to be solid circular section of diameter 76.2 mm (cross-sectional area = 4560.4 mm²). Cable material is steel with properties: E = 210 GPa and v=0.3. Distributed load from the bridge is represented by a "lumped" load of 96 kN at node #102. Input data for ABAQUS analysis are presented in the following:

```
*HEADING
Cable analysis with two truss elements
*NODE, NSET=setn1
101,0.0,0.0,0.0
102,60000,-12000
103,120000,0.0,0.0
*ELEMENT,TYPE=T2D2,ELSET=cable-1
1001,101,102
1002,102,103
*SOLID SECTION, MATERIAL=mat-1, ELSET=cable-1
4560.4
*MATERIAL, NAME=mat-1
*ELASTIC
210,0.3
*BOUNDARY
101,1,2,0.0
103,1,2,0.0
*STEP, PERTURBATION
96 kN vertical downward load at node 2
*STATIC
*CLOAD
102,2,-96.0
*NODE PRINT
U
RF
*EL PRINT
S
*END STEP
```

Problem to solve:

1. Type in the above input data in a text data file that can be named anything with an extension .inp (for example, cable.inp)
2. Run the model in ABAQUS and check the results in text output file (cable.dat)
3. Analyze the two-element truss assembly model by hand
   a. What is the stress in the cable? (hint: find member forces from Σforces=0 at node #102)
   b. Calculate the vertical deflection at node #102. Hint: external work $(1/2*P*u_{102})$ = total internal energy $\Sigma(1/2*\sigma*\varepsilon*vol)$
   c. How do hand calculation results compare with the ABAQUS results for two-element assembly model?

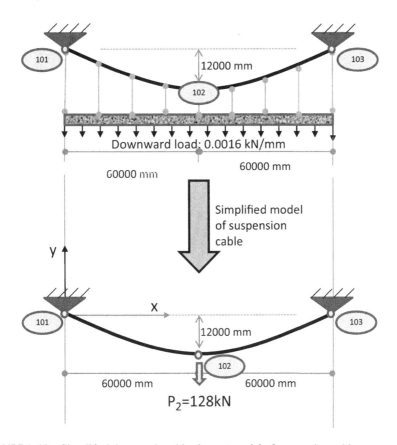

FIGURE 2.19 Simplified three-node cable element model of suspension cable.

PROBLEM-2

Figure 2.19 shows a three-node truss element model of suspension cable. Concentrated load at node #102 is calculated from distributed load by using the parabolic influence function H2 from Figure 2.15.

$$
\begin{aligned}
P_{102} &= \int \left(1 - r^2\right) * 0.0016 * dx \\
&= \int_{-1}^{+1} \left(1 - r^2\right)\left(0.0016 * \frac{L}{2}\right) * dr \\
&= 0.0016 * \frac{L}{2} * \int_{-1}^{+1} \left(1 - r^2\right) * dr \\
&= 0.0016 * \frac{120000}{2} * \frac{4}{3} \\
&= 128\,\text{kN}
\end{aligned}
$$

Modify the ABAQUS input file of Problem-1 to represent the three-node truss element model. Check the ABAQUS results of three-node truss model and compare that with those of Problem-1.

3 Analysis of Solids Represented by 2D Stress Fields

SUMMARY

Stress/deformation response of three-dimensional solids, in certain special cases, can be predicted with the analysis of simplified two-dimensional models. Theory of elasticity-based description of plane-strain and plane-stress problems falls in this category. Stress–strain relationships, specific to plane-strain state, are derived in Section 3.1. Additional essential conditions of elasticity problem solutions, compatibility condition, stress equilibrium equation, and stress boundary conditions, are also reviewed in this section. Section 3.2 presents key derivations for the special case of plane-stress elasticity problem. The fundamental elasticity formulations, discussed in Sections 3.1 and 3.2 for 2D planar solids, are combined in Section 3.3 to define a bi-harmonic compatibility condition that must be satisfied by any stress function solution of the 2D elasticity problems. Section 3.4 presents detailed analytical developments for the stress function-based solution of bending and shear deformation responses of a cantilever beam. Finding the desirable polynomial function for an engineering analysis problem, however, takes systematic execution of lengthy analytical process. The method has not seen numerical implementations for matrix method analysis of large structural systems. Although not an analytical ingredient, for the standard displacement-based finite element formulation technique, the stress function solution, however, provides a direct insight into the deformation behavior of elementary solids, particularly for the shear deformation behavior that is often difficult to simulate well with piecewise linearized displacement response of finite element models. The stress function-based solutions, presented for simple solid mechanics case studies, serve as important references for assessing the deformation behavior of solid and beam finite element formulations.

The stiffness formulation, for finite element simulation of general 2D solids, is introduced in Section 3.5 where element orientation is assumed parallel to the Cartesian x–y coordinate system. A more general-purpose formulation, based on isoparametric element definition, is introduced in Section 3.6 – followed by a discussion on numerical integration of element property matrices in Section 3.7. Higher order formulations, quadrilateral elements with mid-side nodes, are introduced in Section 3.8, which provide improved in-plane bending response compared to the simpler four-node elements. Relative risk-benefits of higher order versus lower order incompatible elements are also discussed in this section. Triangular elements are often used, in conjunction with four-node plane solid elements, in geometric discretization of curved boundaries. Numerical details for representing the constant stress–strain response of triangular elements, created by collapsing one side of four-node quadrilateral solid elements, are discussed in Section 3.9. Convergence characteristics of displacement and stress responses, predicted by finite element analysis technique, are discussed in Section 3.10 – with simulation results of an example case study.

Section 3.11 presents ABAQUS-specific element types that are available for building a simulation model for plane-stress or plane-strain conditions. Commonly used options to specify externally applied loads are described in Section 3.12. The primary focus of this chapter is to learn about the details of element formulations and their effects on predicted results. The study of convergence characteristics of finite element model, however, requires preparation of simulation models with large number of nodes and elements. Section 3.13 discusses about the modeling of a simple plane-stress square plate problem by using the general-purpose finite element model pre-processor: HyperMesh (Altair University 2020). More discussions on software-aided model pre-processing work will follow in Chapter 4. Finally, practice problems on 2D stress analysis of solids are presented in Section 3.14.

3.1 PLANE STRAIN – A SPECIAL FORM OF ELASTICITY PROBLEM

A long prismatic solid, with constant geometric and material properties along the member axis, subjected to a uniform lateral loading along the length, is shown in Figure 3.1. Member is held between two fixed-end conditions resulting in no deformation in the axis direction ($\varepsilon_z = \gamma_{yz} = \gamma_{zx} = 0$). End constraints, however, will lead to nonzero axial stress ($\sigma_z \neq 0$) on the cross-sectional plane. State of material deformation in the body can be described in a two-dimensional cross-sectional plane as shown in Figure 3.1 – with no variation occurring in axis direction. Strain–deformation relationships for this plane-strain condition are defined in the following equations (3.1 and 3.2):

$$\varepsilon_x = \frac{\partial u}{\partial x}, \quad \varepsilon_y = \frac{\partial v}{\partial y}, \quad \gamma_{xy} = \gamma_{yx} = \frac{\partial u}{\partial y} + \frac{\partial v}{\partial x} \tag{3.1}$$

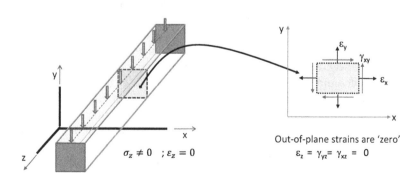

FIGURE 3.1 Plane-strain condition with material deformation occurring in 2D cross-sectional plane (no variation along the member axis).

$$\varepsilon_z = \frac{\partial w}{\partial z} = 0, \quad \gamma_{yz} = \gamma_{zy} = \frac{\partial v}{\partial z} + \frac{\partial w}{\partial y} = 0, \quad \gamma_{xz} = \gamma_{zx} = \frac{\partial u}{\partial z} + \frac{\partial w}{\partial x} = 0 \quad (3.2)$$

Three expressions for plane-strain condition (equation 3.1) are functions of only two in-plane displacement components (u, v). Compatibility condition among the strain components is represented by equation (2.8) alone. Now from the Hooke's law (equations 2.18), strain component in member axis direction gives the following relationship expressing the out-of-plane normal stress as a function of in-plane normal stresses:

$$\varepsilon_z = \left(\frac{\sigma_z}{E} - v \frac{\sigma_x}{E} - v \frac{\sigma_y}{E} \right) = 0 \rightarrow \sigma_z = v \left(\sigma_x + \sigma_y \right) \quad (3.3)$$

Combining equations (2.18) and (3.3), two in-plane normal strain components are given by the following equations:

$$\varepsilon_x = \left(1 - v^2 \right) \frac{\sigma_x}{E} - \left(v + v^2 \right) \frac{\sigma_y}{E}$$
$$\varepsilon_y = \left(1 - v^2 \right) \frac{\sigma_y}{E} - \left(v + v^2 \right) \frac{\sigma_x}{E} \quad (3.4)$$

Shear stress–strain relationship for plane-strain condition is given by the following :

$$\gamma_{xy} = \frac{\tau_{xy}}{G} = \frac{2(1+v)}{E} \tau_{xy} \quad (3.5)$$

Inverting equations (3.4) and (3.5), stress–strain relationships for plane-strain condition can be written as follows:

$$\begin{Bmatrix} \sigma_x \\ \sigma_y \\ \tau_{xy} \end{Bmatrix} = \frac{E}{(1+v)(1-2v)} \begin{bmatrix} 1-v & v & 0 \\ v & 1-v & 0 \\ 0 & 0 & \frac{1-2v}{2} \end{bmatrix} \begin{Bmatrix} \epsilon_x \\ \epsilon_y \\ \gamma_{xy} \end{Bmatrix} \quad (3.6)$$

Material property matrix on the right-hand side of equation (3.6) represents the elasticity matrix [C] to be used in finite element stiffness calculations (equation 2.44). Stress equilibrium equations for plane-strain state can be obtained from equations (1.10) by considering the stress components in x,y plane only:

$$\frac{\partial \sigma_x}{\partial x} + \frac{\partial \tau_{xy}}{\partial y} + F_x = 0$$
$$\frac{\partial \sigma_y}{\partial y} + \frac{\partial \tau_{xy}}{\partial x} + F_y = 0 \quad (3.7)$$

Using the stress–strain relationships (3.4 and 3.5), and equilibrium equations (3.7), compatibility equation (2.8) can be written in terms of stresses as in the following:

$$\left(\frac{\partial^2}{\partial x^2}+\frac{\partial^2}{\partial y^2}\right)\left(\sigma_x+\sigma_y\right)=-\frac{1}{1-v}\left(\frac{\partial F_x}{\partial x}+\frac{\partial F_y}{\partial y}\right) \tag{3.8}$$

Stress distribution function in a plane-strain condition must satisfy the compatibility condition (3.8) and the stress boundary conditions for a given problem. Three-dimensional stress boundary conditions of equations (1.14) can be simplified to describe the two-dimensional boundary conditions as in the following:

$$p_x = \sigma_x l + \tau_{xy} m \tag{3.9}$$
$$p_y = \tau_{xy} l + \sigma_y m$$

3.2   PLANE STRESS – A SPECIAL FORM OF ELASTICITY PROBLEM

Figure 3.2 shows a three-dimensional fabric form supported by a mast and cables. Fabric material provides resistance at any point only within the plane of differential element surrounding that point (as shown in the figure). Stress in the out-of-plane direction at any point is assumed zero. Strain–displacement relationships for this plane-stress condition are similar to those of equations (3.1) and (3.2) except the out of plane normal strain, ε_z, which is not zero:

$$\varepsilon_z = \frac{\partial w}{\partial z} \neq 0 \tag{3.10}$$

Stress–strain relationships for plane-stress state are given by the following (equation 3.11):

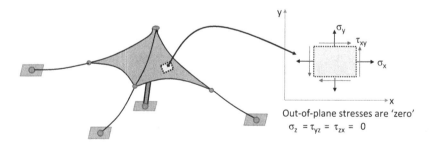

FIGURE 3.2   Cable-supported fabric – plane-stress condition in 3D space with material providing resistance to in-plane deformation only.

$$\varepsilon_x = \frac{\sigma_x}{E} - v\frac{\sigma_y}{E}$$

$$\varepsilon_y = \frac{\sigma_y}{E} - v\frac{\sigma_x}{E} \tag{3.11}$$

$$\varepsilon_z = -\frac{v}{E}\left(\sigma_x + \sigma_y\right) = \frac{-v}{1-v}\left(\varepsilon_x + \varepsilon_y\right)$$

$$\gamma_{xy} = \frac{\tau_{xy}}{G} = \frac{2(1+v)}{E}\tau_{xy}$$

Inverting equations (3.11), stress–strain relationships for plane-stress state are given by the following:

$$\begin{Bmatrix} \sigma_x \\ \sigma_y \\ \tau_{xy} \end{Bmatrix} = \frac{E}{1-v^2} \begin{bmatrix} 1 & v & 0 \\ v & 1 & 0 \\ 0 & 0 & \frac{1-v}{2} \end{bmatrix} \begin{Bmatrix} \epsilon_x \\ \epsilon_y \\ \gamma_{xy} \end{Bmatrix} \tag{3.12}$$

Material property matrix on the right-hand side of equation (3.12) represents the elasticity matrix [C] that can be used in finite element stiffness calculations (equation 2.44). Following the same steps of plane-strain problem description, compatibility equation for plane-stress problem can be written as follows:

$$\left(\frac{\partial^2}{\partial x^2} + \frac{\partial^2}{\partial y^2}\right)\left(\sigma_x + \sigma_y\right) = -\left(1+v\right)\left(\frac{\partial F_x}{\partial x} + \frac{\partial F_y}{\partial y}\right) \tag{3.13}$$

Stress distribution function in a plane-stress condition must satisfy the compatibility condition (3.13) and the stress boundary conditions similar to those given by equations (3.9).

3.3 STRESS FUNCTIONS FOR 2D PLANE STRAIN AND PLANE STRESS ELASTICITY PROBLEMS

When body forces are negligible, $F_x = F_y \approx 0$, stress equilibrium equations (3.7), and compatibility equations (3.8) and (3.13) for plane-strain and plane-stress conditions, take the following forms:

$$\frac{\partial \sigma_x}{\partial x} + \frac{\partial \tau_{xy}}{\partial y} = 0$$

$$\frac{\partial \sigma_y}{\partial y} + \frac{\partial \tau_{xy}}{\partial x} = 0 \qquad (3.14)$$

$$\left(\frac{\partial^2}{\partial x^2} + \frac{\partial^2}{\partial y^2}\right)(\sigma_x + \sigma_y) = 0$$

Above essential conditions for 2D stress-field solutions are readily satisfied when a stress function $\phi(x, y)$ is defined such that stress components are defined by the following derivative forms:

$$\sigma_x = \frac{\partial^2 \varphi}{\partial y^2}, \ \sigma_y = \frac{\partial^2 \varphi}{\partial x^2}, \ \tau_{xy} = -\frac{\partial^2 \varphi}{\partial x\, \partial y} \qquad (3.15)$$

Substituting the stress component definitions from equation (3.15) into compatibility equation (3.13) yields:

$$\frac{\partial^4 \varphi}{\partial x^4} + 2\frac{\partial^4 \varphi}{\partial x^2\, \partial y^2} + \frac{\partial^4 \varphi}{\partial y^4} = \nabla^4 \varphi = 0 \qquad (3.16)$$

In absence of body force effects, solution of a 2D plane-stress or plane-strain problem involves the solution of bi-harmonic equation (3.16) subject to boundary conditions of the problem given by equations (3.9). Despite the simplification of 3D problem to 2D condition, direct solution of equation (3.16) is still a challenging task. Common practice is to use indirect method (also known as inverse or semi-inverse method). In that approach, based on examination of body configuration and externally applied loads and constraints, a stress function ϕ is defined by a polynomial of known and unknown coefficients. Unknown coefficients are later determined by satisfying the stress boundary conditions (3.9). For example, a second-degree polynomial (equation 3.17) will clearly satisfy the bi-harmonic compatibility equation (3.16), and it will also provide non-zero stress values (equation 3.15):

$$\varphi = ax^2 + bxy + cy^2 + \ldots. \qquad (3.17)$$

A higher order polynomial can also be used provided that compatibility condition $\nabla^4\phi = 0$ is satisfied with the adjustment of non-zero coefficients. Few elementary stress functions are presented in the following.

The second-degree polynomial of equation (3.18) satisfies the compatibility condition (equation 3.16). Associated stress components, defined by relations (3.15), are shown in equation (3.19):

$$\varphi_2 = \frac{a_2}{2}x^2 + b_2 xy + \frac{c_2}{2}y^2 \qquad (3.18)$$

$$\sigma_x = c_2, \quad \sigma_y = a_2, \quad \tau_{xy} = -b_2 \tag{3.19}$$

Figure 3.3 shows graphically the constant stress components (σ_x, σ_y and τ_{xy}) represented by function (3.18). One or more of the polynomial coefficients can be adjusted to represent special stress states of uniaxial (only c_2 or a_2 is non-zero), biaxial (both c_2 and a_2 are non-zero; $b_2 = 0$), or pure shear condition (only b_2 is non-zero).

The third-degree polynomial of equation (3.20) satisfies the compatibility condition (equation 3.16); the associated stress components are given by equation (3.21):

$$\varphi_3 = \frac{a_3}{6} x^3 + \frac{b_3}{2} x^2 y + \frac{c_3}{2} xy^2 - \frac{d_3}{6} y^3 \tag{3.20}$$

$$\sigma_x = c_3 x - d_3 y, \quad \sigma_y = a_3 x + b_3 y, \quad \tau_{xy} = -b_3 x - c_3 y \tag{3.21}$$

For the special case of $a_3 = b_3 = c_3 = 0$, stress definitions of equation (3.21) reduce to the following form (equation 3.22) – representing the special case of in-plane pure bending as shown with the example of a differential element in Figure 3.4.

$$\sigma_x = -d_3 y, \quad \sigma_y = \tau_{xy} = 0 \tag{3.22}$$

The fourth-degree polynomial of equation (3.23) satisfies the compatibility condition (3.16) when $e_4 = -(2c_4 + a_4)$. Stress components associated with this stress function are defined by equations (3.24):

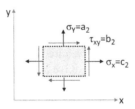

FIGURE 3.3 Stress field described by second-degree polynomial function $\varphi_2 = \frac{a_2}{2} x^2 + b_2 xy + \frac{c_2}{2} y^2$

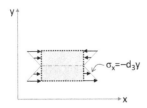

FIGURE 3.4 Stress field of pure bending case described by a third-degree polynomial function $\varphi_3 = -\frac{d_3}{6} y^3$.

$$\varphi_4 = \frac{a_4}{12} x^4 + \frac{b_4}{6} x^3 y + \frac{c_4}{2} x^2 y^2 + \frac{d_4}{6} xy^3 + \frac{e_4}{12} y^4 \qquad (3.23)$$

$$\sigma_x = c_4 x^2 + d_4 xy - \left(2c_4 + a_4\right) y^2$$
$$\sigma_y = a_4 x^2 + b_4 xy + c_4 y^2 \qquad (3.24)$$
$$\tau_{xy} = -\frac{b_4}{2} x^2 - 2c_4 xy - \frac{d_4}{2} y^2$$

The fifth-degree polynomial of equation (3.25) satisfies the compatibility condition (3.16) when $(3a_5 + 2c_5 + e_5).x + (b_5 + 2d_5 + 3f_5).y = 0$. Since this condition needs to be satisfied for any coordinate position (x, y), polynomial coefficients are further reduced by $e_5 = -(3a_5 + 2c_5)$ and $b_5 = -(2d_5 + 3f_5)$.

$$\varphi_5 = \frac{a_5}{20} x^5 + \frac{b_5}{12} x^4 y + \frac{c_5}{6} x^3 y^2 + \frac{d_5}{6} x^2 y^3 + \frac{e_5}{12} xy^4 + \frac{f_5}{20} y^5 \qquad (3.25)$$

Stress components corresponding to the stress function of equation (3.25) are defined by the following equations (3.26):

$$\sigma_x = \frac{c_5}{3} x^3 + d_5 x^2 y - \left(3a_5 + 2c_5\right) xy^2 + f_5 y^3$$

$$\sigma_y = a_5 x^3 - \left(3f_5 + 2d_5\right) x^2 y + c_5 xy^2 + \frac{d_5}{3} y^3 m \qquad (3.26)$$

$$\tau_{xy} = \frac{1}{3}\left(3f_5 + 2d_5\right) x^3 - c_5 x^2 y - d_5 xy^2 - \frac{1}{3}\left(3d_5 + 2c_5\right) y^3$$

Polynomial stress functions described above can be combined to define a stress field that is relatively more complex in nature. The following example demonstrates the use of analytical stress functions to solve problems of elementary solid mechanics. It can be mentioned here that stress function method is not directly implemented in the standard finite element analysis software products that are coded based on displacement field assumptions. Nonetheless, an example problem analysis, presented in Section 3.4, highlights the capability of stress function-based analysis technique, and it also highlights certain aspects of stress-deformation response that must be respected by the finite element simulation models.

3.4   EXAMPLE USE OF STRESS FUNCTION APPROACH TO SOLVE ELASTICITY PROBLEMS

A thin cantilever beam, thickness $t \ll$ beam depth 2h, is subjected to a shear load P at the free end (Figure 3.5). Magnitude of load P is considered large enough to make the effects of body weight negligible ($F_x = F_y \approx 0$ in equations 3.7). Stress state in this thin body is of plane-stress type ($\sigma_z = \tau_{yz} = \tau_{xz} = 0$). The top and bottom edges of the

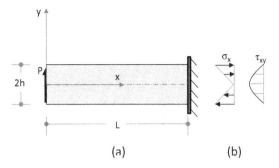

FIGURE 3.5 (a) A cantilever beam in x–y plane (with thickness $= t$ in the z-direction) and subjected to a shear load P at the free end; and (b) internal beam stresses as given by elementary solid mechanics principles.

beam are stress free. The left end of cantilever beam is free of normal stresses, and the resultant of shear stresses on that edge must be equal to the externally applied force P. Equations (3.27) summarize the stress boundary conditions on the member.

$$\left(\tau_{xy}\right)_{y=\pm h} = 0°, \quad \left(\sigma_y\right)_{y=\pm h} = 0, \left(\sigma_x\right)_{x=0} = 0$$

$$t * \int_{(-h)}^{(+h)} \tau_{xy} dy = P \tag{3.27}$$

Given that the bending moment on any section in the beam varies linearly with coordinate distance x, and normal stress at any point on the section varies linearly with y coordinate, the normal stress σ_x, which is a second derivative of yet-to-be-defined stress function ϕ in equation (3.15), can be expressed by the following equation (3.28), where c_1 is a constant:

$$\sigma_x = \frac{\partial^2 \varphi}{\partial y^2} = c_1 xy \tag{3.28}$$

Double integration of equation (3.28) with respect to y gives the following general expression for the stress function ϕ where $f_1(x)$ and $f_2(x)$ are functions of x to be determined:

$$\varphi = \frac{1}{6} c_1 xy^3 + y f_1(x) + f_2(x) \tag{3.29}$$

Substitution of stress function ϕ in compatibility condition (3.16) gives the following expression:

$$y \frac{\partial^4 f_1}{\partial x^4} + \frac{\partial^4 f_2}{\partial x^4} = 0 \tag{3.30}$$

Since the condition of equation (3.30) must hold for any value of y, two fourth-order derivative terms must be independently "zero", i.e.

$$\frac{\partial^4 f_1}{\partial x^4} = 0 \text{ and } \frac{\partial^4 f_2}{\partial x^4} = 0 \tag{3.31}$$

Integration of the expressions in equation (3.31) four times with respect to x leads to the following, where c_2, c_3, etc., are constants of integration:

$$\begin{aligned} f_1(x) &= c_2 x^3 + c_3 x^2 + c_4 x + c_5 \\ f_2(x) &= c_6 x^3 + c_7 x^2 + c_8 x + c_9 \end{aligned} \tag{3.32}$$

Inserting the function definitions of equations (3.32) into equation (3.29), the general definition of the stress functions is obtained as follows:

$$\varphi = \frac{1}{6} c_1 x y^3 + \left(c_2 x^3 + c_3 x^2 + c_4 x + c_5 \right) y + c_6 x^3 + c_7 x^2 + c_8 x + c_9 \tag{3.33}$$

Stress components corresponding to the stress function of equation (3.33) follow from equation (3.15):

$$\sigma_y = \frac{\partial^2 \varphi}{\partial x^2} = 6 \left(c_2 y + c_6 \right) x + 2 \left(c_3 y + c_7 \right) \tag{3.34}$$

$$\tau_{xy} = -\frac{\partial^2 \varphi}{\partial x \, \partial y} = -\frac{1}{2} c_1 y^2 - 3 c_2 x^2 - 2 c_3 x - c_4 \tag{3.35}$$

Using the stress boundary condition equations (3.27) into equations (3.34) and (3.35), the following values of the polynomial constants are obtained, where $I = 2/3(th^3)$ is the moment of inertia of rectangular beam cross-section about the neutral axis:

$$c_2 = c_3 = c_6 = c_7 = 0$$

$$c_4 = -\frac{1}{2} c_1 h^2 \tag{3.36}$$

$$c_1 = -\frac{3P}{2th^3} = -\frac{P}{I}$$

Using the polynomial coefficient values in equations (3.28), (3.34), and (3.35), the following expressions are obtained for the stress components inside the cantilever beam:

$$\sigma_x = -\frac{Pxy}{I}, \qquad \sigma_y = 0, \qquad \tau_{xy} = -\frac{P}{2I}\left(h^2 - y^2\right) \tag{3.37}$$

Stress responses given by equations (3.37) are exactly same as the ones given by beam bending theory in solid mechanics. Displacement response of the beam can now be calculated by using the strain–displacement and stress–strain relationships:

$$\varepsilon_x = \frac{\partial u}{\partial x} = \frac{\sigma_x}{E} = -\frac{Pxy}{EI}, \qquad \varepsilon_y = \frac{\partial v}{\partial y} = -v\frac{\sigma_x}{E} = v\frac{Pxy}{EI} \tag{3.38}$$

$$\gamma_{xy} = \frac{\partial u}{\partial y} + \frac{\partial v}{\partial x} = \frac{\tau_{xy}}{G} = -\frac{P}{2GI}\left(h^2 - y^2\right) \tag{3.39}$$

Integrating equations (3.38), we get the following expressions (equations 3.40) for displacements u and v, where $f_3(y)$ and $f_4(x)$ are unknown functions of y and x, respectively.

$$u = -\frac{Px^2 y}{2EI} + f_3\left(y\right) \qquad\qquad v = v\frac{Pxy^2}{2EI} + f_4\left(x\right) \tag{3.40}$$

Now substituting the expressions for u and v from equation (3.40) into equation (3.39), and after rearranging the terms we get the following relationship:

$$\left[-\frac{Px^2}{2EI} + \frac{\partial f_4}{\partial x}\right] + \left[v\frac{Py^2}{2EI} - \frac{Py^2}{2GI} + \frac{\partial f_3}{\partial y}\right] = -\frac{Ph^2}{2GI} \tag{3.41}$$

The first part inside the parenthesis on the left side of equation (3.41) is a function of x, the second part is a function of y, and the right-hand side is a constant for a given prismatic beam element. In order for the equality condition to hold true for any coordinate position (x, y), two parts inside the parenthesis on the left-hand side must be separately constant leading to the following relationships (where d and e are constants):

$$\left[-\frac{Px^2}{2EI} + \frac{\partial f_4}{\partial x}\right] = a\,constant\left(d\right),$$

$$and\left[v\frac{Py^2}{2EI} - \frac{Py^2}{2GI} + \frac{\partial f_3}{\partial y}\right] = a\,constant\left(e\right) \tag{3.42}$$

Constants d and e, however, are related to each other by equation (3.43):

$$d + e = -\frac{Ph^2}{2GI} \tag{3.43}$$

Re-arrangement of the terms in equation (3.42) leads to the following:

$$\frac{\partial f_4}{\partial x} = \frac{Px^2}{2EI} + d, \text{ and } \frac{\partial f_3}{\partial y} = -v\frac{Py^2}{2EI} + \frac{Py^2}{2GI} + e \tag{3.44}$$

Integration of both sides of equations (3.44) gives the following expressions for unknown functions f_3 and f_4 where p and q are integration constants:

$$f_3(y) = -v\frac{Py^3}{6EI} + \frac{Py^3}{6GI} + ey + p, \quad f_4(x) = \frac{Px^3}{6EI} + dx + q \tag{3.45}$$

Combining equations (3.45) and (3.40), we get the following equations for displacement response of the beam element:

$$u = -\frac{Px^2y}{2EI} - v\frac{Py^3}{6EI} + \frac{Py^3}{6GI} + ey + p \tag{3.46}$$

$$v = v\frac{Pxy^2}{2EI} + \frac{Px^3}{6EI} + dx + q \tag{3.47}$$

Constants d, e, p, and q can be determined by using the boundary constraints and the relationship (3.43). The lateral deflection equation for the beam axis can be obtained from equation (3.47) by substituting $y = 0$:

$$v_{y=0} = \frac{Px^3}{6EI} + dx + q \tag{3.48}$$

Now taking the derivative of expression (3.48) with respect to x, and setting it to zero at $x = L$ ("zero" rotation condition at the fixed end of beam), we get the following value for constant d:

$$d = -\frac{PL^2}{2EI} \tag{3.49}$$

Combining equations (3.43) and (3.49), we get the following value for constant e:

$$e = \frac{PL^2}{2EI} - \frac{Ph^2}{2GI} \tag{3.50}$$

Inserting zero lateral displacement boundary condition at the fixed end, $v = 0$ at $x = L$, in equation (3.47), and using the expression for constant e from equation (3.50), we get the following value for constant q:

$$q = \frac{PL^3}{3EI} \tag{3.51}$$

Inserting zero axial deformation boundary constraint at the fixed end, $u = 0$ at $x = L$ and $y = 0$ in equation (3.46) yields $p = 0$. Finally, inserting the values of constants d, e, p, and q in equations (3.46) and (3.47), we get the following equations for displacement response of the cantilever beam when subjected to a lateral load P at the free end:

$$u = -\frac{Px^2y}{2EI} - \nu\frac{Py^3}{6EI} + \frac{Py^3}{6GI} + \left(\frac{PL^2}{2EI} - \frac{Ph^2}{2GI}\right)y \tag{3.52}$$

$$v = \nu\frac{Pxy^2}{2EI} + \frac{Px^3}{6EI} - \frac{PL^2}{2EI}x + \frac{PL^3}{3EI} \tag{3.53}$$

And the expression for deflection of the beam axis, equation (3.48), takes the following form:

$$v_{y=0} = \frac{Px^3}{6EI} - \frac{PL^2}{2EI}x + \frac{PL^3}{3EI} \tag{3.54}$$

From equation (3.54), deflection of the cantilever beam at the free end (x=0) comes out to be $PL^3/3EI$ which is exactly equal to the value we get from beam theory analysis in solid mechanics. Stress function method, however, provides additional insights into the deformation response of the cantilever beam. We can take the derivative of displacement function u in equation (3.46) with respect to y, and set it to zero at $x = L$ (zero rotation at the fixed end of the beam), thus giving the following expression for constants (Timoshenko and Goodier 1982):

$$e = \frac{PL^2}{2EI}, d = -\frac{PL^2}{2EI} - \frac{Ph^2}{2GI} \tag{3.55}$$

Substituting these in equation (3.47), we get the following expression for deflection of the beam axis:

$$v_{y=0} = \frac{Px^3}{6EI} - \frac{PL^2}{2EI}x + \frac{PL^3}{3EI} + \frac{Ph^2}{2GI}(L - x) \tag{3.56}$$

Compared to equation (3.54), equation (3.56) gives an additional beam axis deflection term of $Ph^2(L - x)/2GI$ which is attributed to the shear deformation of the solid (not accounted for in the Euler Bernoulli beam theory analysis).

The cantilever beam example, analyzed by using the stress function-based analysis technique, has simple geometry, loading, and boundary conditions. Two more plane-stress examples of similar degree of analysis complexity are shown in Figure 3.6; solutions for these problems are available in Timoshenko and Goodier

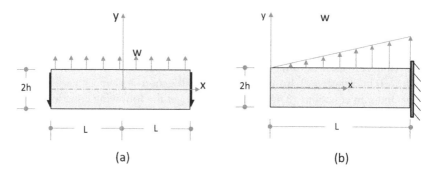

FIGURE 3.6   (a) Thin beam subjected to a uniformly distributed load w per unit length; (b) thin beam subjected to a linearly varying load over the member length.

(1982). As evident from the detailed analysis presentation of the cantilever beam example, finding the desirable polynomial function for an engineering analysis problem takes systematic execution of lengthy analytical process. Moreover, the method has not seen numerical implementations for matrix method analysis of large structural systems. The stress function method, however, provides insightful results about the deformation behavior of elementary solids, particularly for the shear deformation behavior that is often difficult to simulate well with finite element simulation models that rely on linear displacement field assumptions within finite element domains. The stress function-based analysis results, derived in the above, will be used as reference for quality assessment of displacement-based finite element simulation models. The stiffness formulation, for finite element simulation models of general 2D solids, is introduced in Section 3.5 where element orientation is assumed parallel to the Cartesian x–y coordinate system. A more general-purpose formulation, based on iso-parametric element definition, will be introduced in Section 3.6 – followed by a discussion on numerical integration of element property matrices in Section 3.7.

3.5   STIFFNESS METHOD ANALYSIS OF SOLIDS REPRESENTED BY 2D STRESS-DEFORMATION FIELDS

Figure 3.7(a) shows a solid element in 2D (x–y) plane connected to four corner nodes ($n_1...n_4$) with each node having two in-plane translational DOF (u,v). Figure 3.7(b) shows assumed displacement variation over the element domain for unit magnitude at node n_1 and zero values at the other three nodes. Analytical function of that displacement field, often referred to as shape function or interpolation function, is defined by equation (3.57):

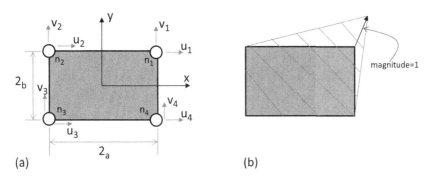

(a)   (b)

FIGURE 3.7 (a) A solid element in 2D *x–y* plane, connected to four corner nodes ($n_1...n_4$), with each node having 2 displacement DOF (u_i, v_i); (b) interpolation function over the 2D plane for a unit magnitude at node n_1.

$$H_1 = \frac{1}{4}\left(1+\frac{x}{a}\right)\left(1+\frac{y}{b}\right)$$ (3.57)

Similarly, shape functions for other nodes can be defined as follows:

$$H_2 = \frac{1}{4}\left(1-\frac{x}{a}\right)\left(1+\frac{y}{b}\right)$$

$$H_3 = \frac{1}{4}\left(1-\frac{x}{a}\right)\left(1-\frac{y}{b}\right)$$ (3.58)

$$H_4 = \frac{1}{4}\left(1+\frac{x}{a}\right)\left(1-\frac{y}{b}\right)$$

Displacement responses at any point (x,y) inside the element can be related to the nodal displacements (u_i, v_i), with the use of interpolation functions (3.57 and 3.58), as follows:

$$u(x,y) = H_1u_1 + H_2u_2 + H_3u_3 + H_4u_4$$
$$v(x,y) = H_1v_1 + H_2v_2 + H_3v_3 + H_4v_4$$ (3.59)

Using the above displacement functions in the strain–displacement relationships of equations (2.4), we get the following expressions for in-plane strains in the 2D solid:

$$\varepsilon_x = \frac{\partial u}{\partial x} = \frac{1}{4a}\left(1+\frac{y}{b}\right).u_1 + \frac{-1}{4a}\left(1+\frac{y}{b}\right).u_2$$

$$+ \frac{-1}{4a}\left(1-\frac{y}{b}\right).u_3 + \frac{1}{4a}\left(1-\frac{y}{b}\right).u_4$$

$$\varepsilon_y = \frac{\partial v}{\partial y} = \frac{1}{4b}\left(1+\frac{x}{a}\right).v_1 + \frac{1}{4b}\left(1-\frac{x}{a}\right).v_2$$

$$+ \frac{-1}{4b}\left(1-\frac{x}{a}\right).v_3 + \frac{-1}{4b}\left(1+\frac{x}{a}\right).v_4$$

$$\gamma_{xy} = \frac{\partial u}{\partial y} + \frac{\partial v}{\partial x} = \frac{1}{4b}\left(1+\frac{x}{a}\right).u_1 + \frac{1}{4b}\left(1-\frac{x}{a}\right).u_2$$

$$+ \frac{-1}{4b}\left(1-\frac{x}{a}\right).u_3 + \frac{-1}{4b}\left(1+\frac{x}{a}\right).u_4$$

$$+ \frac{1}{4a}\left(1+\frac{y}{b}\right).v_1 + \frac{-1}{4a}\left(1+\frac{y}{b}\right).v_2$$

$$+ \frac{-1}{4a}\left(1-\frac{y}{b}\right).v_3 + \frac{1}{4a}\left(1-\frac{y}{b}\right).v_4$$

(3.60)

Compatibility condition in equation (2.8) will be satisfied by the above strain functions – a basic requirement of finite element shape function definitions. Re-writing the equations (3.60) in vector and matrix forms, we get the following equation (3.61), where matrix [B] is the strain–displacement relationship matrix defined by equation (3.62):

$$\begin{Bmatrix} \varepsilon_x \\ \varepsilon_y \\ \gamma_{xy} \end{Bmatrix} = \begin{bmatrix} B \end{bmatrix} \begin{Bmatrix} u_1 \\ u_2 \\ u_3 \\ u_4 \\ v_1 \\ v_2 \\ v_3 \\ v_4 \end{Bmatrix}$$

(3.61)

$$[B] = \frac{1}{4ab}\begin{bmatrix} (b+y) & -(b+y) & -(b-y) & (b-y) & 0 & 0 & 0 & 0 \\ 0 & 0 & 0 & 0 & (a+x) & (a-x) & -(a-x) & -(a+x) \\ (a+x) & (a-x) & -(a-x) & -(a+x) & (b+y) & -(b+y) & -(b-y) & (b-y) \end{bmatrix}$$ (3.62)

In addition to the strain–displacement relationship matrix, [B], equation (2.44) for element stiffness calculation also requires the material stress–strain relationship matrix [C]. From equations (3.6) and (3.12), [C] matrix can be defined by equations (3.63) and (3.64), respectively, for plane-strain and plane-stress conditions in a solid:

$$\left[C\right]_{plane-strain} = \frac{E}{\left(1+v\right)\left(1-2v\right)} \begin{bmatrix} 1-v & v & 0 \\ v & 1-v & 0 \\ 0 & 0 & \dfrac{1-2v}{2} \end{bmatrix} \qquad (3.63)$$

$$\left[C\right]_{plane-stress} = \frac{E}{1-v^2} \begin{bmatrix} 1 & v & 0 \\ v & 1 & 0 \\ 0 & 0 & \dfrac{1-v}{2} \end{bmatrix} \qquad (3.64)$$

Assuming that material properties are constant over an element domain, matrix [C] defined by equation (3.63) or (3.64), as applicable, is a constant in the right side integral of equation (2.44). Stiffness property of the element, thus becomes, a second-order polynomial function based on the definition of [B] matrix given by equation (3.62). Integral of stiffness function can be determined analytically for the simple rectangular solid element that aligns with the x–y coordinate directions. Rest of the analysis method will follow the same general steps outlined in Section 2.7 for stiffness-based matrix method analysis of structures.

Unlike the simple rectangular finite element definition of Figure 3.7, not all finite elements in a general structural analysis model will be aligned perfectly with the user-defined global coordinate system. Strain–displacement relationship functions (equations 3.60), therefore, need to be modified to represent an arbitrary orientation of the element in the user-defined x–y coordinate plane. Sections 3.6 and 3.7 present the general formulations for a solid element having arbitrary orientation in the 2D global reference coordinate system.

3.6 ISO-PARAMETRIC DEFINITION OF 2D SOLID FINITE ELEMENTS

Figure 3.8(a) shows a solid element having arbitrary orientation with respect to the global (x,y) coordinate system in 2D plane. Nodal coordinates and displacement DOF are defined in global directions, while a local coordinate system (r, s) is introduced bounding the element domain in the ranges of $r = -1$ to $r = +1$, and

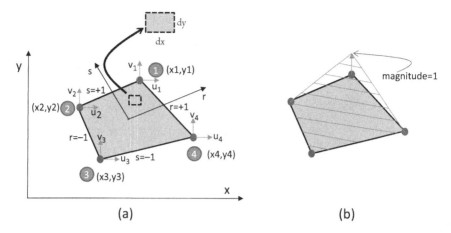

FIGURE 3.8 Iso-parametric formulation for 2D solid element: (a) global coordinates (x_i, y_i), and local coordinate system (r, s); (b) element shape function for node (1).

$s = -1$ to $s = +1$. Figure 3.8(b) shows the element shape function for unit magnitude at node #1. Similar shape functions for the other three connecting nodes of the element can also be drawn (not shown graphically). Analytical definitions of shape functions (H_i, $i = 1...4$), in terms of local (r, s) coordinates, are given by equations (3.65):

$$H_i = \frac{1}{4}(1 \pm r)(1 \pm s), \text{where}$$

$$H_1 = \frac{1}{4}(1 + r)(1 + s)$$

$$H_2 = \frac{1}{4}(1 - r)(1 + s) \qquad (3.65)$$

$$H_3 = \frac{1}{4}(1 - r)(1 - s)$$

$$H_4 = \frac{1}{4}(1 + r)(1 - s)$$

Interpolating the nodal coordinates, global coordinates of a point inside the element can be defined by the following functions:

$$x = H_1 x_1 + H_2 x_2 + H_3 x_3 + H_4 x_4$$
$$y = H_1 y_1 + H_2 y_2 + H_3 y_3 + H_4 y_4 \qquad (3.66)$$

Taking partial derivates of x and y with respect to r and s:

$$\frac{\partial x}{\partial r} = \sum \frac{\partial H_i}{\partial r}.x_i; \quad \frac{\partial x}{\partial s} = \sum \frac{\partial H_i}{\partial s}.x_i$$

$$\frac{\partial y}{\partial r} = \sum \frac{\partial H_i}{\partial r}.y_i; \quad \frac{\partial y}{\partial s} = \sum \frac{\partial H_i}{\partial s}.y_i \ i = 1...4 \tag{3.67}$$

Using the shape function definitions from equations (3.65), partial derivative terms in equations (3.67) are found to be:

$$\frac{\partial H_i}{\partial r} = \pm \frac{1}{4}(1 \pm s) \ ; \quad \frac{\partial H_i}{\partial s} = \pm \frac{1}{4}(1 \pm r) \tag{3.68}$$

Following the rules of partial differentiation:

$$\frac{\partial H_i}{\partial r} = \frac{\partial H_i}{\partial x}.\frac{\partial x}{\partial r} + \frac{\partial H_i}{\partial y}.\frac{\partial y}{\partial r}$$

$$\frac{\partial H_i}{\partial s} = \frac{\partial H_i}{\partial x}.\frac{\partial x}{\partial s} + \frac{\partial H_i}{\partial y}.\frac{\partial y}{\partial s} \tag{3.69}$$

Writing the relations (3.69) in vector and matrix forms,

$$\begin{Bmatrix} \dfrac{\partial H_i}{\partial r} \\ \dfrac{\partial H_i}{\partial s} \end{Bmatrix} = \begin{bmatrix} J \end{bmatrix} * \begin{Bmatrix} \dfrac{\partial H_i}{\partial x} \\ \dfrac{\partial H_i}{\partial y} \end{Bmatrix}, where \begin{bmatrix} J \end{bmatrix} = \begin{bmatrix} \dfrac{\partial x}{\partial r} & \dfrac{\partial y}{\partial r} \\ \dfrac{\partial x}{\partial s} & \dfrac{\partial y}{\partial s} \end{bmatrix} \tag{3.70}$$

Matrix [J] in equation (3.70), known as Jacobian, expresses the relationship between local element coordinates (r,s) and the global model coordinates (x, y). Terms inside the Jacobian matrix are given by equations (3.67) and (3.68). Physical interpretation of Jacobian can be provided by considering a small differential element of dimensions $dx*dy$ inside the finite element in Figure 3.8. Differential terms dx and dy can be expressed in terms of r and s variables as in the following:

$$dx = \frac{\partial x}{\partial r}.dr + \frac{\partial x}{\partial s}.ds; \quad dy = \frac{\partial y}{\partial r}.dr + \frac{\partial y}{\partial s}.ds \tag{3.71}$$

Area of the differential element is defined as follows:

$$dA = dx * dy = \left(\frac{\partial x}{\partial r}.\frac{\partial y}{\partial r} \right).(dr * dr) + \left(\frac{\partial x}{\partial r}.\frac{\partial y}{\partial s} \right).(dr * ds)$$

$$+ \left(\frac{\partial x}{\partial s}.\frac{\partial y}{\partial r} \right).(dr * ds) + \left(\frac{\partial x}{\partial s}.\frac{\partial y}{\partial s} \right).(ds * ds) \tag{3.72}$$

Setting the vector product terms $(dr*dr)$ and $(ds*ds)$ to "zero", equation (3.72) takes the form:

$$dA = \left(\frac{\partial x}{\partial r}\cdot\frac{\partial y}{\partial s}\right)\cdot(dr*ds) + \left(\frac{\partial x}{\partial s}\cdot\frac{\partial y}{\partial r}\right)\cdot(ds*dr) \tag{3.73}$$

Substituting $ds*dr = -dr*ds$ in equation (3.73):

$$dA = \left(\frac{\partial x}{\partial r}\cdot\frac{\partial y}{\partial s} - \frac{\partial x}{\partial s}\cdot\frac{\partial y}{\partial r}\right)\cdot(dr*ds) \tag{3.74}$$

Using the definition of Jacobian from equation (3.70), equation (3.74) can be re-written as follows:

$$dA = dx*dy = (\det J).(dr*ds) \tag{3.75}$$

A positive definite value of "J" defines a one-to-one relationship between global $(x\text{–}y)$ and local $(r\text{–}s)$ coordinate systems. Using the equations (3.65), (3.66) and (3.67), we get the following relations for the special case of rectangular element in Figure 3.7(a);

$$\frac{\partial x}{\partial r} = a, \quad \frac{\partial y}{\partial r} = 0, \quad \frac{\partial x}{\partial s} = 0, \quad \frac{\partial y}{\partial s} = b \tag{3.76}$$

Jacobian of the rectangular element is thus defined by the following:

$$[J] = \begin{bmatrix} a & 0 \\ 0 & b \end{bmatrix} \tag{3.77}$$

Determinant of Jacobian given in equation (3.77) represents a quarter of the 2a x 2b rectangular element of Figure 3.7(a). This positive definite value thereby defines a one-to-one coordinate transformation relationship in equation (3.75). Ratio between the two diagonal terms of Jacobian, J_{11} and J_{22}, defines the aspect ratio of element. Evidently, for a square shape element, two diagonal terms will have equal values. A distorted element shape, like the one in Figure 3.9, leads to a variable Jacobian definition over the element domain (equation 3.78):

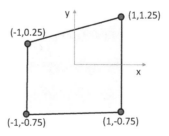

FIGURE 3.9   A distorted 2D solid element.

$$[J] = \frac{1}{4}\begin{bmatrix} 4 & (1+s) \\ 0 & (3+r) \end{bmatrix} \tag{3.78}$$

For this distorted element example, coordinate transformation relationship of equation (3.75) becomes a function of (r, s) (because of the location-dependent definition of J in equation 3.78). In finite element modeling, it is preferable to have square or rectangular shape elements that have constant Jacobian value within a given element domain, thus, facilitating a robust calculation procedure for the stiffness properties as discussed in Section 3.7.

3.7 NUMERICAL CALCULATION OF STIFFNESS MATRIX FOR ISO-PARAMETRIC 2D SOLID FINITE ELEMENTS

Similar to the coordinate interpolation exercise of equation (3.66), displacement responses inside a 2D solid element (u, v) can be related to the nodal displacements (u_i, v_i) with the use of shape functions H_i given in equations (3.65):

$$u = \sum_{i=1}^{4} H_i u_i \quad v = \sum_{i=1}^{4} H_i v_i \tag{3.79}$$

Strains corresponding to the displacement responses (u, v) are given by the following expressions:

$$\varepsilon_x = \frac{\partial u}{\partial x} = \sum \frac{\partial H_i}{\partial x}.u_i$$

$$\varepsilon_y = \frac{\partial v}{\partial y} = \sum \frac{\partial H_i}{\partial y}.v_i \tag{3.80}$$

$$\gamma_{xy} = \left[\frac{\partial u}{\partial y} + \frac{\partial v}{\partial x}\right] = \sum \frac{\partial H_i}{\partial y}.u_i + \sum \frac{\partial H_i}{\partial x}.v_i$$

Writing the equations (3.80) with vector and matrix forms, we get the following strain–displacement relationships:

$$
\begin{Bmatrix} \varepsilon_x \\ \varepsilon_y \\ \gamma_{xy} \end{Bmatrix} =
\begin{bmatrix}
\dfrac{\partial H_1}{\partial x} & 0 & \dfrac{\partial H_2}{\partial x} & 0 & \dfrac{\partial H_3}{\partial x} & 0 & \dfrac{\partial H_4}{\partial x} & 0 \\
0 & \dfrac{\partial H_1}{\partial y} & 0 & \dfrac{\partial H_2}{\partial y} & 0 & \dfrac{\partial H_3}{\partial y} & 0 & \dfrac{\partial H_4}{\partial y} \\
\dfrac{\partial H_1}{\partial y} & \dfrac{\partial H_1}{\partial x} & \dfrac{\partial H_2}{\partial y} & \dfrac{\partial H_2}{\partial x} & \dfrac{\partial H_3}{\partial y} & \dfrac{\partial H_3}{\partial x} & \dfrac{\partial H_4}{\partial y} & \dfrac{\partial H_4}{\partial x}
\end{bmatrix}
\begin{Bmatrix} u_1 \\ v_1 \\ u_2 \\ v_2 \\ u_3 \\ v_3 \\ u_4 \\ v_4 \end{Bmatrix} \tag{3.81}
$$

Strain–displacement relationship matrix [B], as used in stiffness matrix calculation of equation (2.44), is defined from equations (3.81) as follows:

$$
[B] = \begin{bmatrix}
\dfrac{\partial H_1}{\partial x} & 0 & \dfrac{\partial H_2}{\partial x} & 0 & \dfrac{\partial H_3}{\partial x} & 0 & \dfrac{\partial H_4}{\partial x} & 0 \\[2mm]
0 & \dfrac{\partial H_1}{\partial y} & 0 & \dfrac{\partial H_2}{\partial y} & 0 & \dfrac{\partial H_3}{\partial y} & 0 & \dfrac{\partial H_4}{\partial y} \\[2mm]
\dfrac{\partial H_1}{\partial y} & \dfrac{\partial H_1}{\partial x} & \dfrac{\partial H_2}{\partial y} & \dfrac{\partial H_2}{\partial x} & \dfrac{\partial H_3}{\partial y} & \dfrac{\partial H_3}{\partial x} & \dfrac{\partial H_4}{\partial y} & \dfrac{\partial H_4}{\partial x}
\end{bmatrix} \quad (3.82)
$$

Terms inside the [B] matrix, comprising of x and y derivatives of shape functions H_i, can be determined by inverting the relationships (3.70) as in the following:

$$
\begin{Bmatrix} \dfrac{\partial H_i}{\partial x} \\[2mm] \dfrac{\partial H_i}{\partial y} \end{Bmatrix} = [J]^{-1} * \begin{Bmatrix} \dfrac{\partial H_i}{\partial r} \\[2mm] \dfrac{\partial H_i}{\partial s} \end{Bmatrix} \quad (3.83)
$$

The inverse of Jacobian [J] exists when it is positive definite. As discussed earlier, for rectangular and square shape elements, [J] is a diagonal matrix of constant values over an element domain. The right-side terms in equation (3.82), defined by equations (3.68), are linear functions of r and s only, thus making the [B] matrix in equation (3.82) a linear function of r and s. Considering the above definition of [B] matrix, and the constant stress–strain relationship matrix [C], given by equation (3.63) or (3.64) for a 2D problem, stiffness terms inside the integral of equation (2.44) will be a second-order function of local element coordinates r and s. These stiffness terms for a given finite element are generally calculated by using numerical integration methods.

Figure 3.10 shows numerical integration examples for area calculations under one-dimensional functions that are defined with single variable x. For linear function of Figure 3.10(a), area under the line is precisely calculated with single-point calculation – function value at the domain center (f_0) multiplied by the domain length (α_0). For a second-order function in one dimension, area under the curve can be accurately calculated with judicious selection of two integration points as outlined in Figure 3.10(b). Extending this numerical integration procedure of one-dimensional second-order function to a two-dimensional domain, volume under a second-order surface can be calculated by evaluating the function values at 2×2 integration points (Figure 3.11(a)). Exact value of volume under the surface can be calculated by using the Gauss integration rule (Press et al. 2007) that specifies integration points at (± 0.57735, ± 0.57735); and weight factor values as $\alpha_{ij} = 1.0$ (Figure 3.11(b)). The second-order variation of stiffness function for a four-node iso-parametric finite element, as

(a) 1st order variation (straight line)

$$y = ax + b$$

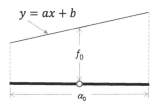

f_0

α_0

Area under the line

$$= \int (ax + b)\, dx \;=\; \alpha_0 * f_0$$

(b) 2nd order variation (curve)

$$y = ax^2 + bx + c$$

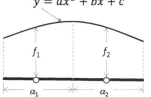

f_1 f_2

α_1 α_2

Area under the curve

$$= \int (ax^2 + bx + c)\, dx$$

$$= \alpha_1 * f_1 \;+\; \alpha_2 * f_2 \;=\; \sum_{i=1}^{2} \alpha_i * f_i$$

FIGURE 3.10 Numerical integration: (a) one-point integration for linear variation; (b) two-point integration for second-order variation.

(a) 2nd order variation (surface)

$$z = f(x^2, y^2, xy, x, y, ..)$$

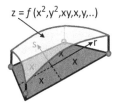

Volume under the surface

$$= \int z.\,dx.\,dy$$

$$= \alpha_{11} * z_{11} \;+\; \alpha_{12} * z_{12} + \alpha_{21} * z_{21} \;+\; \alpha_{22} * z_{22}$$

$$= \sum_{i=1}^{2}\sum_{j=1}^{2} \alpha_{ij} * z_{ij}$$

(b) Gauss integration parameters

$$(\alpha_{ij} = 1.0)$$

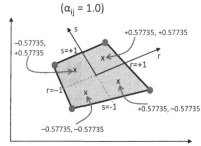

FIGURE 3.11 (a) Numerical integration of volume under a second-order surface over (x, y) plane, and (b) corresponding Gauss integration parameters.

discussed earlier, can be schematically compared with the second-order general function over the two-dimensional plane of Figure 3.11 (a). Equation (3.84) shows the numerical expressions for the stiffness matrix terms of four-node quadrilateral element, where f_{ij} are the stiffness function values at Gauss quadrature points ($i = 1,2$ and $j = 1,2$), and α_{ij} are weight factors ($= 1.0$):

$$[k] = \int [B]^T \cdot [C] \cdot [B].dV = \begin{bmatrix} \sum \alpha_{ij} f_{ij} .. & .. & .. & .. & .. & .. & .. & .. \\ .. & .. & .. & .. & .. & .. & .. & .. \\ .. & .. & .. & .. & .. & .. & .. & .. \\ .. & .. & .. & .. & .. & .. & .. & .. \\ .. & .. & .. & .. & .. & .. & .. & .. \\ .. & .. & .. & .. & .. & .. & .. & .. \\ .. & .. & .. & .. & .. & .. & .. & .. \\ .. & .. & .. & .. & .. & .. & .. & .. \end{bmatrix} \quad (3.84)$$

2×2 Gauss integration rule provides exact calculation of stiffness function in equation (3.84) for four-node square or rectangular shape element where Jacobian J remains constant over the element domain. Jacobian of a distorted element (example in Figure 3.9) becomes a function of element local coordinates (r, s) (equation 3.78), thus, introducing higher order terms in the strain–displacement relationship matrix defined by equations (3.82) and (3.83). Stiffness matrix terms in equation (3.84), thus, also contain higher order terms. 2×2 Gauss quadrature rule, used in many standard finite element software products for stiffness calculation of four-node quadrilateral elements (equation 3.84), provides accurate values for second-order functions of stiffness variation. Higher order stiffness variation terms for distorted elements will be approximately evaluated by the default 2×2 Gauss integration rule, thus leading to some calculation errors. It is therefore important to keep four-node element shapes close to square or rectangular for accuracy of calculations that use default 2×2 Gauss integration rule. Figure 3.12 demonstrates the negative effect of element shape distortion on predicted stress response of an elementary example.

An element is called "fully integrated" when the stiffness function is accurately/fully integrated with the appropriate order of integration over the element domain. For computational speed, finite element software packages may adopt fewer number

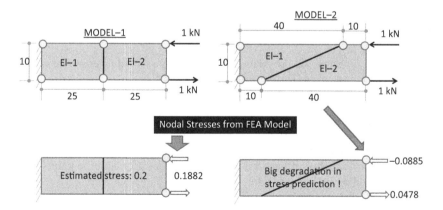

FIGURE 3.12 Negative effects of distorted element shape (MODEL-2) on predicted stress response.

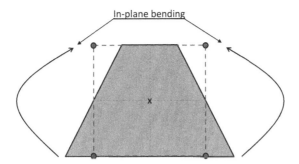

FIGURE 3.13 Single-point reduced integration (x) of four-node quadrilateral element – producing zero resistance to in-plane bending.

of integration points compared to what is required for exact integration of the stiffness function. Such reduced integration of element properties may affect the deformation and resistance response mechanisms of finite elements. Single center-point calculations for a four-node quadrilateral element (Figure 3.13) capture no strains and stresses under in-plane bending action. This lack of stress resistance may lead to an unrealistic zero energy hourglass deformation mode in an assembly of similar under-integrated finite elements.

Linear shape functions for four-node quadrilateral elements (Figures 3.7 and 3.8) imply piecewise linear approximations of curved deformation profile. This piece-by-piece linear approximations of curved deformation profile often make the system response stiff. An alternative to piecewise linearization of displacement field is to use higher order finite elements that are capable of producing higher order deformation response over an element domain (Section 3.8).

3.8   HIGHER ORDER PLANE-STRESS/PLANE-STRAIN ELEMENTS

Figure 3.14 shows an eight-node solid element for analysis of 2D plane-stress or plane-strain response field. Shape functions for corner node (1), and for mid-side node (2) are written in terms of local (r, s) coordinate system as follows:

$$H_1 = \frac{1}{4}(1+r)(1+s)(r+s-1)$$
$$H_2 = \frac{1}{2}(1-r^2)(1+s)$$

(3.85)

Similar third-degree shape functions for other nodal DOF of the element can be derived by considering one DOF at a time. These higher order shape functions can be used to calculate the element stiffness matrix by following the same steps presented in Section 3.7 for four-node quadrilateral element. Third-degree polynomial equations for eight-node element shape functions, however, will lead to second-degree functions for strain–displacement relationships in equations (3.82), and to fourth-degree functions in element stiffness terms of equation (3.84). Accurate calculation

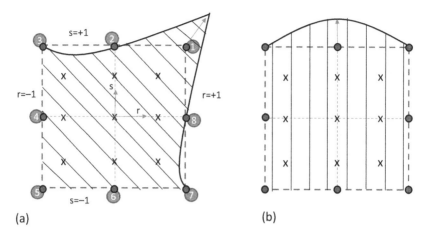

FIGURE 3.14 (a) Eight-node element for 2D stress-field analysis – with shape function shown for one corner node, and (b) shape function for a mid-side node.

of these stiffness terms can be achieved with 3×3 numerical integration points (x) as marked in Figure 3.14. Reduced 2×2 integration rule for eight-node elements is used occasionally for higher computation efficiency. Elements having mid-side boundary nodes, such as the 8-node element, are called "serendipity" elements. Addition of an internal node, possibly at the center ($r = s = 0$) makes the element a "Lagrangian" quadratic element (Figure 3.15). The name "Lagrangian" is used because the element shape functions can be obtained by taking the products of one-dimensional Lagrange interpolants. The element behavior is at its best when the interior node is located at the element center. The geometry of Lagrangian element is completely defined by the coordinates of eight boundary nodes in 2D space; the ninth node (if considered in finite element formulations) is added during element calculations. The shape function associated with the ninth node of quadratic Lagrangian element (Figure 3.15) is defined by equation (3.86):

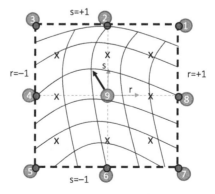

FIGURE 3.15 Nine-node Lagrangian element for 2D stress-field analysis – with shape function shown for ninth node at the element center.

$$H_9 = \left(1 - r^2\right)\left(1 - s^2\right) \tag{3.86}$$

All iso-parametric elements lose accuracy when distorted from rectangular shape. The nine-node element is much less sensitive than the eight-node element to non-rectangularity, to curvature of sides, and to placing of side nodes away from mid-points of the sides. Compared to the response of four-node quadrilateral elements, eight- and nine-node elements provide much better simulation of in-plane bending behavior because of the inherent flexibility introduced by the mid-side nodes. In the past, special form of four-node element has been developed by using higher order deformation function in between the corner nodes to replicate the edge bending behavior (Cook et al. 1989). Formulations of higher order deformation mode, without the use of mid-side nodes, commonly known as in-compatible finite elements, have been implemented in some finite element software packages. These elements use higher order polynomial functions to define parabolic deformation modes between nodes, thus, making the element behavior more flexible compared to that of a standard four-node element. However, absence of the mid-side nodes provides the opportunity for in-compatible deformation between adjacent elements (discussed further in Chapter 4). With the present availability of abundant computing power, it is better to use eight- or nine-node elements that provide computationally stable response without the risk of unstable numerical response that can emerge from the use of lower order incompatible elements.

3.9 CONSTANT STRESS/STRAIN TRIANGULAR ELEMENT

Triangular elements are often used, in conjunction with four-node plane solid elements, in geometric discretization of curved boundaries. A triangular solid element can be constructed by collapsing one side of a four-node solid element. Figure 3.16(a) shows a four-node square element, and Figure 3.16(b) shows a triangular element created by collapsing the edge 1–2 of the four-node element. Combining equations

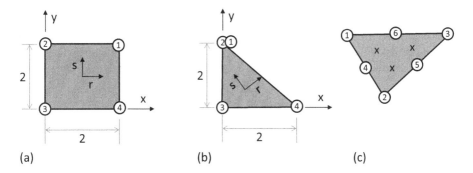

(a)   (b)   (c)

FIGURE 3.16 (a) Four-node element, and (b) triangular element created by collapsing the side (1–2) with coincident nodes (1) and (2); and (c) six-node triangular element with three integration points (x).

(3.65) and (3.66), and assigning same coordinate values for coincident nodes 1 and 2, i.e. $x_1 = x_2$ and $y_1 = y_2$, the following equations are obtained for coordinate definitions at any point inside the triangular element:

$$x = \frac{1}{2}(1+s).x_2 + \frac{1}{4}(1-r)(1-s).x_3 + \frac{1}{4}(1+r)(1-s).x_4$$
$$y = \frac{1}{2}(1+s).y_2 + \frac{1}{4}(1-r)(1-s).y_3 + \frac{1}{4}(1+r)(1-s).y_4$$

$$(3.87)$$

For the example case in Figure 3.16(b), using the actual nodal coordinates into equations (3.87), we get the following expressions for coordinates (x,y):

$$x = \frac{1}{2}(1+r)(1+s)$$
$$y = (1+s)$$

$$(3.88)$$

Now using these expressions for x and y, the Jacobian definition of equation (3.70) takes the following form:

$$[J] = \begin{bmatrix} \dfrac{\partial x}{\partial r} & \dfrac{\partial y}{\partial r} \\ \dfrac{\partial x}{\partial s} & \dfrac{\partial y}{\partial s} \end{bmatrix} = \begin{bmatrix} \dfrac{1-s}{2} & 0 \\ -\dfrac{1+r}{2} & 1 \end{bmatrix}$$

$$(3.89)$$

Inverting the Jacobian matrix of equation (3.89), we get:

$$[J]^{-1} = \begin{bmatrix} \dfrac{2}{1-s} & 0 \\ \dfrac{1+r}{1-s} & 1 \end{bmatrix}$$

$$(3.90)$$

Analogous to the coordinate definitions of equations (3.87), displacement responses inside the triangular element can be written in terms of the nodal displacements as follows:

$$u = \frac{1}{2}(1+s).u_2 + \frac{1}{4}(1-r)(1-s).u_3 + \frac{1}{4}(1+r)(1-s).u_4$$
$$v = \frac{1}{2}(1+s).v_2 + \frac{1}{4}(1-r)(1-s).v_3 + \frac{1}{4}(1+r)(1-s).v_4$$

$$(3.91)$$

Taking partial derivatives of u and v with respect to element local variables r and s:

$$\frac{\partial u}{\partial r} = -\frac{1}{4}(1-s).u_3 + \frac{1}{4}(1-s)u_4$$

$$\frac{\partial u}{\partial s} = \frac{1}{2}u_2 - \frac{1}{4}(1-r).u_3 - \frac{1}{4}(1+r)u_4$$

$$\frac{\partial v}{\partial r} = -\frac{1}{4}(1-s).v_3 + \frac{1}{4}(1-s)v_4 \qquad (3.92)$$

$$\frac{\partial v}{\partial s} = \frac{1}{2}v_2 - \frac{1}{4}(1-r).v_3 - \frac{1}{4}(1+r)v_4$$

Now using the coordinate transformation rule of equation (3.83),

$$\begin{Bmatrix} \dfrac{\partial u}{\partial x} \\[2mm] \dfrac{\partial u}{\partial y} \end{Bmatrix} = [J]^{-1} \begin{Bmatrix} \dfrac{\partial u}{\partial r} \\[2mm] \dfrac{\partial u}{\partial s} \end{Bmatrix} = \begin{bmatrix} \dfrac{2}{1-s} & 0 \\[2mm] \dfrac{1+r}{1-s} & 1 \end{bmatrix} \cdot \begin{bmatrix} 0 & 0 & -\dfrac{1-s}{4} & 0 & \dfrac{1-s}{4} & 0 \\[2mm] \dfrac{1}{2} & 0 & -\dfrac{1-r}{4} & 0 & -\dfrac{1+r}{4} & 0 \end{bmatrix} \cdot \begin{Bmatrix} u_2 \\ v_2 \\ u_3 \\ v_3 \\ u_4 \\ v_4 \end{Bmatrix}$$

which eventually leads to,

$$\begin{Bmatrix} \dfrac{\partial u}{\partial x} \\[2mm] \dfrac{\partial u}{\partial y} \end{Bmatrix} = \begin{bmatrix} 0 & 0 & -\dfrac{1}{2} & 0 & \dfrac{1}{2} & 0 \\[2mm] \dfrac{1}{2} & 0 & -\dfrac{1}{2} & 0 & 0 & 0 \end{bmatrix} \begin{Bmatrix} u_2 \\ v_2 \\ u_3 \\ v_3 \\ u_4 \\ v_4 \end{Bmatrix} \qquad (3.93)$$

Similarly,

$$\begin{Bmatrix} \dfrac{\partial v}{\partial x} \\[2mm] \dfrac{\partial v}{\partial y} \end{Bmatrix} = \begin{bmatrix} 0 & 0 & 0 & -\dfrac{1}{2} & 0 & \dfrac{1}{2} \\[2mm] 0 & \dfrac{1}{2} & 0 & -\dfrac{1}{2} & 0 & 0 \end{bmatrix} \begin{Bmatrix} u_2 \\ v_2 \\ u_3 \\ v_3 \\ u_4 \\ v_4 \end{Bmatrix} \qquad (3.94)$$

From equations (3.93) and (3.94), strains inside the triangular element at any point (x, y) are given by:

$$\{\varepsilon\} = \left\{ \begin{array}{c} \dfrac{\partial u}{\partial x} \\[2ex] \dfrac{\partial v}{\partial y} \\[2ex] \dfrac{\partial u}{\partial y} + \dfrac{\partial v}{\partial x} \end{array} \right\} = \begin{bmatrix} 0 & 0 & -\dfrac{1}{2} & 0 & \dfrac{1}{2} & 0 \\[2ex] 0 & \dfrac{1}{2} & 0 & -\dfrac{1}{2} & 0 & 0 \\[2ex] \dfrac{1}{2} & 0 & -\dfrac{1}{2} & -\dfrac{1}{2} & 0 & \dfrac{1}{2} \end{bmatrix} \cdot \left\{ \begin{array}{c} u_2 \\ v_2 \\ u_3 \\ v_3 \\ u_4 \\ v_4 \end{array} \right\} = [B] \cdot \left\{ \begin{array}{c} u_2 \\ v_2 \\ u_3 \\ v_3 \\ u_4 \\ v_4 \end{array} \right\} \quad (3.95)$$

Evidently, strain–displacement relationship matrix [B] in equation (3.95) is independent of the element internal coordinates (r, s). This means that strain values are constant throughout the element. Triangular elements therefore represent constant strain (and constant stress) state internally. Constant strain triangular element in the above has been derived by degenerating a four-node quadrilateral element to a triangular one. Property matrices of a triangular element can also be derived directly by considering the three-node geometric configuration. Nonetheless, the stress/strain will remain constant irrespective of how the element stiffness properties are derived. Some finite element software packages use explicit triangular element formulations while some other packages use degenerated formulation from four-node element description. Either way, actual use of constant-strain elements should be kept to minimum when a significant gradient is expected in the stress field (Figure 3.4). Alternatively, higher order triangular element (six-node element in Figure 3.16(c)) can be used to model stress variations in a field. Further discussion on the selection of element shape functions and integration order is presented in Chapter 4.

3.10 CONVERGENCE OF FINITE ELEMENT MODEL SOLUTIONS

Convergence of finite element simulation models depends on the quality and accuracy of finite element formulations that are used for piecewise discretization of a solid. Developers of finite element software packages often provide analysis examples including a commonly used form of convergence test known as "patch test" (Irons and Ahmad 1980). The basic idea of patch test is to demonstrate that finite element model solutions converge to the exact theoretical solution as the mesh is refined. It has been argued that patch test is neither sufficient nor necessary for convergence (Bathe 1996). Nonetheless, the study of finite element mesh refinement effect on predicted results is important to engineers for deciding what degree of finite element mesh refinement is sufficient to produce good quality results. Figure 3.17 shows a 50×50 mm plate of 1 mm thickness-subjected to in-plane stresses and boundary conditions. Four finite element mesh models with different degrees of mesh refinement, and the contours of predicted vertical displacements are shown in Figure 3.18(a). An important characteristic of finite element displacement solution is its continuity across element boundaries. As expected from the problem description given in Figure 3.17, vertical displacement response is zero at the left boundary, and it increases gradually towards the right with the maximum response occurring at the

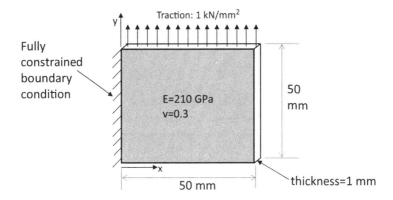

FIGURE 3.17 A 50 mm × 50 mm plate of unit thickness – subjected to plane-stress loading and boundary conditions.

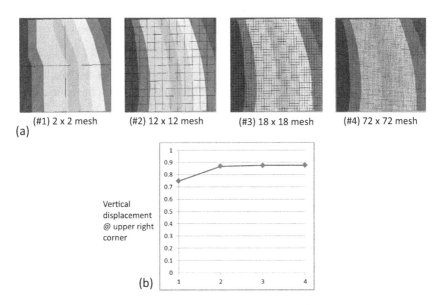

FIGURE 3.18 (a) Smooth contours of vertical displacements (in the *y*-direction) predicted by finite element mesh models of four different refinements; (b) convergence of predicted displacement response at upper right corner.

upper right corner. Figure 3.18(b) plots the maximum displacement responses obtained from 4 different mesh models. The coarse (2 × 2) mesh model, being the stiffest, predicts the lowest displacement value. The predicted maximum displacement response, however, does not change with increasing refinement beyond model #2 (12 × 12). In general, the overall displacement response in finite element analysis tends to converge quickly with a reasonable degree of mesh refinement. In the problem presented above, mesh refinement with 12 elements was good enough for piece-by-piece linearized simulation of displacement variation along the *x*-direction. In practical finite element analysis, two models can be prepared initially – one model

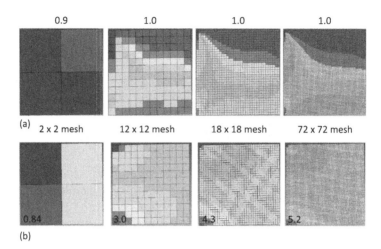

FIGURE 3.19   (a) Smooth contours of stresses in vertical (*y*) direction predicted by finite element mesh models of four different refinements; (b) contours of stresses in horizontal (*x*) direction – with non-convergence at the bottom left corner

having double the refinement level compared to the other one. If the displacement response of the refined model is substantially different from that of the course model, further refinement of the model can be undertaken until a satisfactory state of convergence is achieved.

Figure 3.19(a) shows contours of vertical direction stresses (σ_{yy}) obtained from 4 finite element mesh models discussed earlier. Coarse mesh model (2×2) provides a poor prediction of the stress boundary condition on the upper edge. Predicted stress profiles become smoother as the mesh refinement progresses beyond 12×12 level. More refined finite element models provide better predictions of stress profile in the body. Similar general conclusion can also be drawn for the contours of horizontal direction stresses (σ_{xx}) shown in Figure 3.19(b). However, the predicted stress value at the bottom left corner becomes increasingly higher as the mesh refinement becomes higher – implying that the predicted local stress response, near the geometric discontinuity at the left bottom corner, fails to converge with mesh refinement. This is a classical problem inherited from theory of elasticity solutions for stress responses at the points of stress singularity. Further discussions on this issue will follow in Chapter 9. Stress variations near smooth geometric profiles, however, can be well predicted by selective refinement of finite element mesh as discussed in Chapter 4.

3.11 SELECTION OF ELEMENT TYPES FOR 2D STRESS ANALYSIS WITH FEA SOFTWARE PACKAGES

As described in Section 2.10, a critical input in finite element analysis models is the element type – appropriate for the analysis of a given problem. Finite element simulation software packages provide coded names to be used by the analysts for selecting desired element formulation types. For example, four-node plane-stress element, with 2×2 Gauss integration rule, is identified in ABAQUS library as

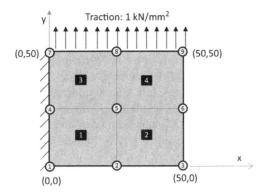

FIGURE 3.20 A four-element model of 50 mm × 50 mm plate– subjected to plane-stress loading and boundary conditions.

"CPS4" – meaning a continuum ("solid") element in 2D space attached to four corner nodes. Considering the four-element model of example problem discussed in Section 3.10 (Figure 3.20), plane-stress elements in ABAQUS input file will look as follows (where the element corner nodes are listed following a counter-clockwise sequence (as per the data input syntax for 2D elements in ABAQUS):

*ELEMENT, ELSET=Plate50×50, TYPE=CPS4
1, 1, 2, 5, 4
2, 2, 3, 6, 5
3, 4, 5, 8, 7
4, 5, 6, 9, 8

Coordinates of nodes (1,2,...9), used to describe the elements in the above model, can be defined by ABAQUS data block under "*NODE" keyword as described in Section 2.10. Reduced one-point integration rule for four-node quadrilateral element (Figure 3.13) can be selected by choosing "TYPE=CPS4R" in the above element description. Element type "CPS4" can be used to create "degenerated" triangular plane-stress elements by assigning the same node ID number twice in the description of a quadrilateral element. Example of degenerated triangular element in Figure 3.16(b) can be described in ABAQUS as follows:

*ELEMENT, TYPE=CPS4, ELSET=set2
element ID, 3, 4, 2,2

Above element can also be described by choosing the standard three-node plane-stress element type "CPS3" from ABAQUS element library:
*ELEMENT, TYPE=CPS3, ELSET=set3
element ID, 3, 4, 2

Fully integrated six-node triangular element of Figure 3.16(c) can be described in ABAQUS input file, starting with the corner nodes first, followed by the listing of mid-side nodes as follows:

```
*ELEMENT, TYPE=CPS6, ELSET=set4
element ID, 1,2, 3, 4,5,6
```

Reduced one-point integrated version of six-node triangular element can be selected by specifying "TYPE=CPS6R" in the above data block. Higher order eight-node plane-stress element, using 3×3 full Gauss integration rule, is identified as "TYPE=CPS8" in ABAQUS element library. Element example shown in Figure 3.14(a) can be defined in ABAQUS input file by defining the corner nodes first, followed by the identification of mid-side nodes, as in the following:

```
*ELEMENT, TYPE=CPS8, ELSET=set5
element ID, 1, 3, 5,7,2,4,6,8
```

Eight-node plane-stress element, using reduced 2×2 Gauss integration rule, can be selected by specifying "TYPE=CPS8R" in the above data description. ABAQUS elements for modeling plane-strain problems follow the exact same data format described in the above for plane-stress elements, but with unique element type code names, such as CPE4 instead of CPS4, CPE3 instead of CPS3, etc.

As discussed earlier, higher order eight- and six-node elements are relatively more flexible compared to the corresponding lower order four- and three-node elements of the same geometric shape. Lower order elements obviously facilitate faster computational efficiency, but with stiffer response prediction. An artificial improvement to the flexibility of four-node quadrilateral element, as also discussed earlier, is achieved by using the in-compatible elements. ABAQUS formulation for that element is defined by "TYPE=CPS4I" and "TYPE=CPE4I", respectively, for plane-stress and plane-strain conditions. In ideal model setup with regular element shapes, incompatible elements provide results that are comparable to those of higher order elements, but with fewer degrees of freedom. Parabolic deformation profile over the element boundary, however, can naturally lead to incompatible deformation profiles across element boundaries (discussed further in Chapter 4). Response of incompatible elements tends to be very sensitive to geometric distortion. Eight-node elements provide a much better and robust alternative to these incompatible elements. Nine-node Lagrangian elements, with a center node added to the eight-node element, provide more flexible and robust behavior when element shape is distorted from regular square or rectangular shape. However, not all finite element software packages include Lagrangian element option.

Each set of ABAQUS solid elements requires a property data block with material name and part thickness information as described in the following:

```
*SOLID SECTION, ELSET= Plate50×50, MATERIAL=mat1
1.0
```

where the value "1.0" is an example for part thickness (Figure 3.17). Material properties, for the given material name "*mat1*" in the above, are described in the ABAQUS data block under "*MATERIAL" keyword as described in Section 2.10. These material properties are used by ABAQUS to calculate stress–strain relationship matrix [C], using equation (3.63) or (3.64), depending on the stress-field assumption of plane strain or plane stress, respectively. Nodal coordinates, element connectivity to nodes, element type selection, part thickness information, and material property data provide essential and sufficient information to ABAQUS for the calculation of stiffness matrix using equation (3.84). Assignment of boundary conditions on nodes follows the same procedure outlined in Section 2.10. In addition to concentrated nodal loads (described in Section 2.10), solid elements can be subjected to distributed loads along the edges. Figure 3.20 shows distributed traction applied on the upper edges of elements 3 and 4. ABAQUS syntax rule is to define pressure data by identifying element ID followed by the element face or edge, and the magnitude of applied pressure (+ve sign is used for pressure on the edge). Element face ID of an element is defined by following the order of element connectivity definitions. Face between the first and second nodes, listed with the element ID in *ELEMENT definition, is referred to as face ID "1"; face ID "2" is defined as that defined by second and third nodes of an element, and so on. ABAQUS input data, for the applied distributed load on example elements of Figure 3.20, is given in the following:

*DLOAD
3, P3, –1.0
4, P3, –1.0

In the above, "3" and "4" are element ID numbers, "P3" refers to pressure load on face #3 of the elements, and the negative value of 1.0 indicates a distributed traction load (a pressure load is associated with +ve sign in ABAQUS). Analysis step definition in ABAQUS for the given load follows the same format with "*STEP" and "*END STEP" keywords as discussed in Section 2.10. A complete ABAQUS input description of the four-element model, shown in Figure 3.20, is provided in the following:

*HEADING
Plane-stress analysis model using four-node quadrilateral elements
*NODE, NSET=setn1
1,0.0,0.0,0.0
2,25,0
3,50,0
4,0,25
5,25,25
6,50,25
7,0,50
8,25,50
9,50,50
*ELEMENTS, ELSET=Plate50×50, TYPE=CPS4
1, 1, 2, 5, 4
2, 2, 3, 6, 5
3, 4, 5, 8, 7

4, 5, 6, 9, 8
*SOLID SECTION, MATERIAL=mat-1, ELSET= Plate50×50
1.0
*MATERIAL, NAME=mat-1
*ELASTIC
210,0.3
*BOUNDARY
1,1,2,0.0
4,1,2,0.0
7,1,2,0.0
*STEP, PERTURBATION
Distributed pressure load on elements
*STATIC
*DLOAD
3, P3, –1.0
4, P3, –1.0
*NODE PRINT
U
RF
*EL PRINT
S
*END STEP

3.12   DESCRIPTION OF LOAD TYPES IN GENERAL STRESS ANALYSIS PROBLEMS

In actual solution of matrix equilibrium equations, loads are represented at the nodes aligned with the associated DOF. General-purpose FEA software packages accept input data describing surface and or body loads (e.g. *DLOAD specification in ABAQUS). These non-nodal loads are internally substituted by the software packages with equivalent nodal loads for actual analysis purpose. Element shape functions are generally used to determine the influence functions in nodal load calculations. For example, influence function for node #1 in eight-node solid element of Figure 3.14(a) is defined by substituting $s = 1$ in the 1st expression of equation (3.85):

$$H_1 = \frac{1}{2}r.(1+r) \qquad (3.96)$$

Equivalent load on node #1, for a distributed load $p(r)$ acting on the edge 1–2–3 of that element, will be calculated as follows:

$$P_1 = \int_{-1}^{+1} H_1.p(r).dr \qquad (3.97)$$

Similarly, equivalent loads on nodes 2 and 3 can be calculated by using the influence functions for nodes 2 and 3, respectively. Similar procedure is also applied to substitute surface pressure and body loads with equivalent nodal loads in 3D finite element simulation models. Distributed loads on element edge, surface, and body may also be simply lumped to the affected nodes during model build process without using the software provided option of distributed load description.

Pre-existing stresses and strains can be considered in finite element analysis of structures. General-purpose finite element software packages allow the initial stress inputs at finite element integration points. Equivalent nodal loads are calculated by the software packages internally:

$$\{P\} = -\int [B]^T \{\sigma_0\}.dV \qquad (3.98)$$

where $\{\sigma_0\}$ is the vector of initial stresses at element integration points. Pore fluid pressure, in porous solids, can also be represented by $\{\sigma_0\}$ in equation (3.98). Temperature change effects are represented by equivalent thermal strains in the material:

$$\varepsilon_T = \alpha.(T - T_0) \qquad (3.99)$$

where T is current temperature at element integration point, T_0 is the stress-free reference temperature, and α is the thermal expansion coefficient of the material. Thermal strains from equation (3.99) are converted to equivalent thermal stresses by using Hooke's laws, and the equivalent nodal loads for temperature change effects are calculated by using equation (3.98). Theoretical manuals of software packages generally contain descriptions of how distributed surface and body loads are substituted with equivalent nodal loads for internal calculations.

3.13   REFINED FINITE ELEMENT MODEL PREPARATION WITH A PRE-PROCESSOR (HYPERMESH)

Manual model preparation, as described in Section 3.11 for four-element 2D finite element model, is not suitable to prepare refined models with large number of elements and nodes (such as the 12 × 12, 18 × 18, and 72 × 72 models shown in Figure 3.19). Altair's HyperMesh (Altair University 2020) is a general-purpose finite element model preparation software with in-built templates to prepare models for several finite element analysis software packages including ABAQUS. A step-by-step video demonstration of preparing a fine mesh model of the plane-stress problem (of Figure 3.17) is provided in the support data website (File name: *HyperMesh_ Preparation_ of_ABAQUS_InputFile_for_a_2D_Plane_Stress Problem.mp4*). Key data blocks saved by HyperMesh in the ABAQUS input format are listed in the following:

*NODE
5,0.0,0.0,0.0
…..
…
*ELEMENTS, ELSET=auto1, TYPE=CPS4
…
…

*SOLID SECTION, MATERIAL=mat-1, ELSET= auto1
1.0
*MATERIAL, NAME=mat1
*ELASTIC
210,0.3
*STEP, PERTURBATION
*STATIC
*BOUNDARY
…..
…….
*CLOAD
…
……
*END STEP

ABQUS model input file (Plate 25x25.inp), saved at the end of HyperMesh session, can be run by ABAQUS and the results of that run can be post-processed either in ABAQUS/CAE graphical post-processing tool or be processed with Altair's HyperView post-processing tool. Similar model building exercise can be undertaken with other pre-processor tools if desired.

3.14   PRACTICE PROBLEMS – STRESS ANALYSIS OF A PLATE SUBJECTED TO IN-PLANE STRESSES

PROBLEM 1
Four-element model of the 50 mm × 50 mm plane-stress plate shown in Figure 3.20 underestimated the displacement response (Figure 3.18), and produced crude profiles of stress distributions in the plate (Figure 3.19). Re-analyze the problem using four eight-node plane-stress elements (CPS8) and compare the results with those presented in Figure 3.19.

PROBLEM 2
An approximate stress function for the problem in Figure 3.17 is given by the following equation (where $c = 25$ mm, and $p = 1$ kN/mm):

$$\varphi = \frac{p}{40c^3}\left(10.c^3.(2c-x)^2 + 15.c^2.(2c-x)^2.(y-c)-2.c^2.(y-c)^3 - 5.(2c-x)^2.(y-c)^3 + (y-c)^5\right)$$

(3.100)

Check that the stress function satisfies the compatibility equation and stress boundary conditions. Predict the vertical displacement at upper right corner of the plate; and compare the result with those reported in Figure 3.18 from finite element analysis models.

PROBLEM 3
Prepare a refined finite element model for the problem in Figure 3.17, using eight-node plane-stress elements, and compare the results with those obtained in Problem-1.

4 FEA Model Preparation and Quality Checks

SUMMARY

This chapter focuses on how to produce good-quality finite element models of two-dimensional solids by using general-purpose model pre-processing software. Although specific references are made to HyperMesh software features for model build operations, and to ABAQUS software for actual FEA solutions, the discussions on key aspects of quality model preparation are equally applicable to other software products for model preparation and validation of results. Introductory ideas for adaptation of finite element grid refinement to local variations of stress field are presented in Section 4.1. The important step of choosing finite element formulation type for appropriate representation of the stress field problem is discussed in Section 4.2. A brief discussion on the processing of geometric design data for stress analysis exercise is presented in Section 4.3. Section 4.4 presents (a)-to-(g) critical details of finite element model preparation and quality checks. Section 4.5 focuses on basic quality checks of analysis results that are essential for validating a model prior to drawing engineering conclusions from the model predictions. Practice problems on model preparation and quality checks are presented in Section 4.6. Discussions and examples in this chapter have remained primarily focused on analysis of 2D stress field problems. However, many of the same metrics for model and result quality checks will apply to finite element simulation of 3D problems that will be introduced in next chapters.

4.1 ADAPTATION OF FINITE ELEMENT MESH TO STRESS FIELD VARIATIONS

Finite element mesh refinement examples shown in Figures 3.18 and 3.19 have used uniform grid patterns. However, mesh grid refinement may need to be varied within the same analysis model depending on the complexity of variation in a stress field. Stress variation within a homogeneous solid body, in presence of body forces, is described by equilibrium equations (1.10). In absence of body force, one-dimensional stress field in a homogeneous body (Figure 4.1) shows no variation of stress:

$$\frac{\partial \sigma_x}{\partial x} = 0 \tag{4.1}$$

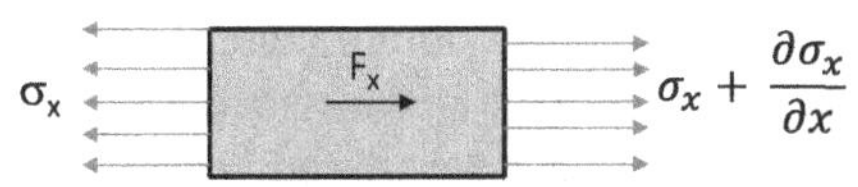

FIGURE 4.1 Stress variation in a one-dimensional field.

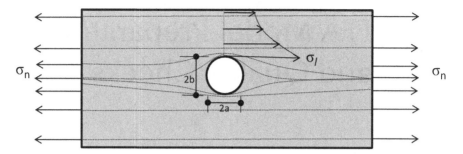

FIGURE 4.2   Stress flow around a hole in an axially loaded flat plate.

However, a geometric discontinuity, such as a hole in a plate (Figure 4.2), can cause a perturbation of the stress flow. Maximum local tangential stress on the upper edge of the hole can be defined as follows:

$$\sigma_l = k_l . \sigma_n \tag{4.2}$$

where σ_n is the nominal axial stress acting on the left and right boundary edges of the plate; and k_ℓ is the stress concentration factor defining the amplification of stress at the local measurement point. For a general elliptical hole of dimensions $2a$ x $2b$ at the center of a very large plate, the maximum stress concentration on the upper edge of hole is defined by equation (4.3) (Ugural and Fenster 2012):

$$k_l = 1 + 2. \left(\frac{b}{a} \right) \tag{4.3}$$

For a circular shape hole ($a = b$), equation (4.3) gives a stress concentration factor $k_\ell = 3$, thus defining the local stress in equation (4.2) as three times the magnitude of applied nominal stress (σ_n). Axially loaded plate with hole (Figure 4.2) is a good example with known theoretical solution that is studied in this chapter to verify the convergence property of finite element stress analysis models. It also provides the opportunity to verify Saint Venant's principle (Saint-Venant 1797–1886) that the local disturbance in a stress field disappears at a distance away from the point of geometric imperfection.

   Figure 4.3 shows approximate subdivisions based on the gradient of variation in the stress field. Finite element modeling can be adapted to this variation by using more refined mesh in the higher gradient area, and less refined mesh in the relatively smoother stressed areas (Figure 4.4). Adaptation of mesh refinement with high number of lower order elements (i.e. linear solid elements in this particular example) is known as "h-adaptivity" where "h" refers to the element size that is reduced in higher stress gradient area. An alternative to h-adaptivity is to use higher order polynomial functions (shape functions), with fewer number of elements, to capture the gradient of stress variation in a model. Figure 4.5 shows a model of the plate with hole – using eight-node quadratic elements around the curved perimeter of the hole. This mesh adaptation is known as "p-adaptivity", where p refers to the degree of element shape

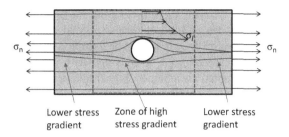

FIGURE 4.3   Subdivisions based on the gradient of stress variation in the axially loaded plate with a circular hole at the center.

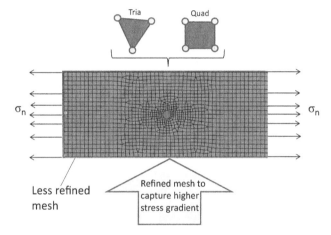

FIGURE 4.4   Finite element model with fine mesh of lower order solid elements (Tria and Quad) in the area of high stress gradient.

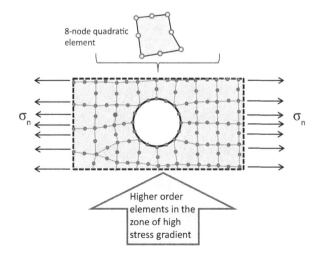

FIGURE 4.5   Finite element model of higher order elements (eight-node quadratic) in the area of high stress gradient.

function. Quadratic solid elements in Figure 4.5 use mid-side nodes approximately in-between two corner nodes of solid elements. Special element formulations by moving middle nodes closer to one corner can be used to reproduce high gradient of stress–strain variations near points of stress singularity (Zienkiewicz and Taylor 1989). Choice of mesh adaptation technique, higher number of lower order elements versus higher order elements, depends on the analyst's experience with specific analysis problems. Specific nature of the deformation field of some problems may make the use of higher order elements more suitable than the lower order ones as discussed in the following section.

4.2 ELEMENT TYPE SELECTION FOR A GIVEN STRESS ANALYSIS PROBLEM

Figure 4.6(a) shows bending deformation profile of beam type structure where planes AB and CD, initially perpendicular to beam longitudinal axis, remain plane after bending, and rotate with the beam axis. Edges AC and BD, shown in Figure 4.6(b), follow the curved profiles of free boundaries. Planar response of sections AB and CD implies linear variation of normal strains along the beam section. Any finite element, used to model the beam bending response, should ideally be capable of representing this linear variation of strain in direction y (on cross-section perpendicular to the beam axis). Triangular elements (Section 3.9), which can represent constant stress/strain response within element domain, make the overall response stiff. Bi-linear shape functions (equations 3.65), representing the deformation profile of quadrilateral solid elements, can naturally reproduce the linear rotation behavior of beam sections AB and CD. However, use of the linear function, to represent node-to-node deformation profile along element boundary, does not naturally reproduce the

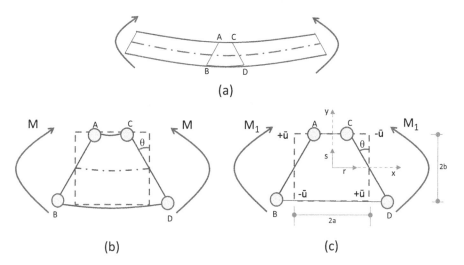

FIGURE 4.6   (a) Bending deformation of beam, (b) idealized deformation of an element under in-plane bending, and (c) linear deformation profiles along the edges of a four-node 2D element.

bending-induced curved profiles on edges AC and BD in Figure 4.6 (c). Using parametric coordinate system definitions from Section (3.6), relationships between local (r,s) and global (x,y) coordinates for the rectangular element of Figure 4.6(c) are defined by

$$r = \frac{x}{a}; \; s = \frac{y}{b} \tag{4.4}$$

Strain components inside the 2D solid element are given by the following equations (4.5):

$$\begin{aligned}
\varepsilon_x &= \frac{\partial u}{\partial x} = \frac{\partial u}{\partial r} \cdot \frac{\partial r}{\partial x} + \frac{\partial u}{\partial s} \cdot \frac{\partial s}{\partial x} = \frac{1}{a} \cdot \frac{\partial u}{\partial r} \\
\varepsilon_y &= \frac{\partial v}{\partial y} = \frac{\partial v}{\partial r} \cdot \frac{\partial r}{\partial y} + \frac{\partial v}{\partial s} \cdot \frac{\partial s}{\partial y} = \frac{1}{b} \cdot \frac{\partial v}{\partial s} \\
\gamma_{xy} &= \frac{\partial u}{\partial y} + \frac{\partial v}{\partial x} = \cdots = \frac{1}{b} \cdot \frac{\partial u}{\partial s} + \frac{1}{a} \cdot \frac{\partial v}{\partial r}
\end{aligned} \tag{4.5}$$

Pure bending-induced element boundary rotation, "θ", produces element nodal displacements, $\pm\bar{u}$, in the x-coordinate direction (Figure 4.6(c)). Displacement field inside the element can, thus, be defined by the following equations (4.6):

$$u = (r.s)\bar{u}, \qquad v = 0 \tag{4.6}$$

Substituting the displacement functions from equation (4.6) into equation (4.5), we get the following expressions for internal strains inside the quadrilateral element of Figure 4.6(c):

$$\varepsilon_x = \frac{s}{a}.\bar{u}, \qquad \varepsilon_y = 0, \qquad \gamma_{xy} = \frac{r}{b}.\bar{u} \tag{4.7}$$

Linear variation of normal strain on beam section, caused by applied bending moment, is re-produced by the first part of equation (4.7). The third part in equation represents an artificial shear strain (often called "parasitic" shear) as there should be no shear deformation under pure bending load. This artificial shear strain is introduced by the assumption of linear deformation profile along element boundaries – preventing the reproduction of curvilinear deformation profiles along edges AC and BD (Figure 4.6(c)). This parasitic shear strain response produces parasitic shear stress inside the material, thus producing a different bending resistance (M_1) compared to the bending resistance (M) of the ideal element behavior shown in Figure 4.6(b). Displacement field for true bending deformation of element in Figure 4.6(b) is described by the following equation (4.8):

$$u = (r.s)\bar{u}, \qquad v = \left[(1-r^2)\frac{a}{2b} + v.(1-s^2)\frac{b}{2a} \right].\bar{u} \tag{4.8}$$

Element strain responses corresponding to that displacement field are given by equation (4.9), where natural condition of zero shear strain for pure bending effect is reproduced:

$$\varepsilon_x = \frac{s}{a}.\bar{u}, \quad \varepsilon_y = -v\frac{s}{a}.\bar{u}, \quad \gamma_{xy} = 0 \tag{4.9}$$

By comparing the strain energies corresponding to the two deformation conditions of Figures 4.6(b) and 4.6(c), the following expression can be derived for the ratio between artificial element bending resistance and true beam bending resistance (Cook et al. 1989):

$$\frac{M_1}{M} = \frac{1}{1+v}\left[\frac{1}{1-v} + \frac{1}{2}\left(\frac{a}{b}\right)^2\right] \tag{4.10}$$

For square shape elements ($a = b$), and material Poisson's ratio $v = 0.3$, equation (4.10) gives $M_1 = 1.48 \times M$. In-plane bending resistance (M_1) of quadrilateral solid element is, thus, very high compared to "true" theoretical bending resistance (M). And this artificial resistance becomes too big as the element aspect ratio a/b increases. The use of fully integrated 4-node quadrilateral solid elements should be minimized in the analysis of in-plane bending problems.

A numerical remedy for parasitic shear problem of quadrilateral elements is to conduct internal calculations at element centroid ($r = s = 0$), thus getting zero value for parasitic shear strain ($\gamma_{xy} = 0$) in equation (4.7). This action, however, also produces zero normal strain ($\varepsilon_x = 0$) at element centroid, making the element vulnerable to zero energy hourglass deformation mode as discussed in Figure 3.13 and Section 3.7. Software implementation of reduced integrated solid elements often includes numerical countermeasures to reduce the risk of developing uncontrolled hourglass deformation mode. ABAQUS element library, for example, includes reduced integration forms of quadrilateral solid elements CPS4R and CPE4R, respectively, for plane-stress and plane-strain conditions. An artificial internal resistance is added to element formulation for producing some resistance to hourglass deformation mode. Use of reduced integration elements works well when many elements are used to capture the bending profile of a solid member. Higher order solid elements, discussed in Section 3.8, provide more stable and better choice for simulating the bending deformation problems. Selection of desired element type is a critical step in preparation of finite element analysis models by using model pre-processing software products. Section 4.3 in the following introduces a plane stress analysis problem having geometric discontinuity. Section 4.4 describes the critical model preparation steps, with selective mesh refinement (h-adaptivity) in areas of expected stress concentration, by using the general-purpose finite element model pre-processor HyperMesh (Altair University 2020).

4.3 INITIAL GEOMETRIC DESIGN OF STRUCTURAL COMPONENTS

Initial geometric attributes of component design can be defined by using point, line, surface, and volume attribute definition features available inside a finite element model preparation software package (group 2 in Figure 1.19). However, in complex engineering projects, it is a standard practice for one team to develop the CAD model by using a specialized design software such as the ones listed in group 1 of Figure 1.19, while another team performs analysis work on the same design by using different software packages from groups 2 and 3. And at the same time, a third team responsible for manufacturing the product may use yet another different software to process the initial design data. In-built database formats of CAD design software packages are generally not compatible with those of the analysis software packages used by the downstream users and suppliers. Transfer of CAD data from initial design software to downstream user platform is achieved through the use of standard data formats. Two of the widely used data exchange formats are IGES and STEP. IGES (Initial Graphics Exchange Specification), last published by the American National Standards Institute (ANSI) in 1996, was developed primarily for the exchange of pure geometric data between computer-aided engineering platforms. This data exchange format is supported by many general-purpose CAE software packages. However, STEP (STandard for the Exchange of Product model data), developed later based on ISO 10303-11 standard (ISO 2020), is currently the default choice for engineering data exchanges. ISO 10303 employs the neutral ASCII format as the medium for data exchanges between different CAE platforms. Data file "Plate_with_hole.step", provided as support electronic data with this book, contains the geometry data of the plate with hole shown in Figure 4.7. Data inside a STEP file is described by using the EXPRESS modeling language specified in Part 11 of the standard ISO 10303-11. EXPRESS is not a programming language; it is a standard for the computer-interpretable representation and exchange of product data through

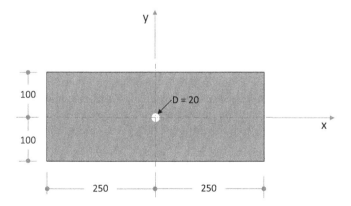

FIGURE 4.7   Geometric description of a plate with hole.

schemas and constraints. Example syntax of data description in STEP file is shown in the following:

ISO-10303-21;
HEADER;
...
DATA;
....
#161=CARTESIAN_POINT(' ',(2.5E+02,10.E+01,0.));
...
....
#176=LINE(",#161,#171);.
....
END-ISO-10303-21;

General-purpose finite element model processing software products, listed in group-2 of Figure 1.19, can be used to import the geometry data file ("Plate_with_hole.step"), and a finite element model can be prepared for analysis with one of the target FEA software packages listed in group 3 of Figure 1.19. Section 4.4 in the following describes the key steps for developing an ABAQUS finite element analysis model of the plate with hole example by using Altair's HyperMesh software (Altair University 2020).

4.4   FEA MODEL PREPARATION FOR CASE-STUDY: PLATE WITH HOLE (BY USING HYPERMESH)

HyperMesh can prepare finite element model input file for several different FEA packages. Choosing the target FEA software at the launch of HyperMesh session, through a dialog box as shown in Figure 4.8, helps to build an error-free model in subsequent steps. Key model preparation steps are described in this section. A video session file of this specific HyperMesh model preparation example is provided in data download site as specified in the preface of this book (HyperMesh_Session_for_ ABAQUS_Model_Preparation_of_2D_Plane_Stress_Plate_with_Hole.mp4). A blended discussion of quality model preparation steps, with HyperMesh specific menu options, is presented in the following. Users planning to try out HyperMesh model pre-processor may want to review the following text simultaneously with the above-mentioned video session file. Key model preparation steps described here are equally applicable if a user prefers to use a different FEA model preparation software (from group 2 in Figure 1.19).

4.4.1   CAD DATA PREPARATION FOR FEA MODELING

First step in the model preparation task is to import the CAD data into the pre-processor database, for example by using "File-Import-Geometry" steps from HyperMesh menu. Upon successful completion of the data import step, model pre-processing software will show the imported geometry data in its display window.

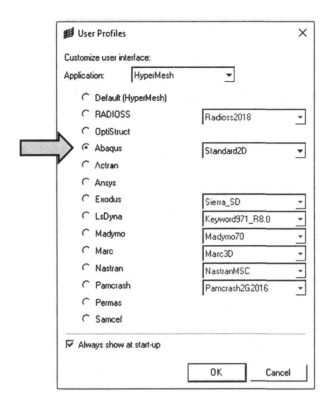

FIGURE 4.8   Target FEA solver pre-selection in a model pre-processor.

FIGURE 4.9   Geometry data of plate with hole.

Figure 4.9 shows a shaded view of imported geometry data from file Plate-with-hole. step (solid surface area is shown filled by selecting "Shaded Geometry and Surface Edges" from HyperMesh Panel menu). At this stage of model pre-processing, substantial effort can be required for cleaning of the imported CAD data -specifically for complex part geometries. Current example case study of plate with hole presents very clear CAD data (as shown in Figure 4.9) without the need for further data cleaning steps.

4.4.2 FINITE ELEMENT MESHING OF THE CAD GEOMETRY

As discussed in Section 4.2, a critical step before actual mesh modeling of part geometry is to identify the desired element types for appropriate simulation of the target stress field. Menu panel "2D" in HyperMesh provides a pick option of "elem types". Activation of that task bar brings up ABAQUS element type choices for 2D problems discussed in Section 3.11. Figure 4.10 shows the selection of CPS3 and CPS4 for tria3 and quad4 element types, respectively. Higher order CPS6 and CPS8 element types can be selected against tria6 and quad8 categories if these are desired to be used in subsequent meshing step. These element type names are specific to ABAQUS element library available for plane stress modeling of part geometry lying in a 2D plane. Element type names will vary if a different FEA software name is chosen for target model preparation. Pre-selection of target FEA software, with appropriate element type names, minimizes the downstream manual work for model corrections. The next important step in software-aided finite element model preparation from CAD geometry data is to choose a suitable algorithm for discretization of geometry to finite elements and nodes. HyperMesh, for example, provides many different options under "2D" menu panel for meshing of both two- and three-dimensional surfaces. Choosing the "automesh" option brings up the panel menu for user-guided interactive meshing of the CAD data. Parametric controls for interactive mesh generation operation can be defined at this step such as minimum or maximum element size, element internal angle, degree of polynomial (linear vs quadratic), etc. A preliminary mesh of first-order quad and tria elements, with approximate element size not exceeding 10 mm, shows a refined enough mesh at outer boundaries of 200 mm × 500 mm plate, but a very crude discretization around the hole of 20 mm diameter. Problem-specific prior knowledge is critical at this stage to accept the mesh or reject it if the discretization is not expected to meet target engineering analysis results. For example, plate with hole is expected to have high stress-gradient in the immediate vicinity of the hole boundary. HyperMesh provides the desired opportunity for adapting mesh refinement to specific areas by immediately executing the "reject" option, which leads to interactive window showing geometric boundary lines with expected number of grid divisions in the display window (Figure 4.11). Number of grid division on each CAD feature line can be increased or decreased with mouse click on the respective displayed number. After adjusting the grid density to desired

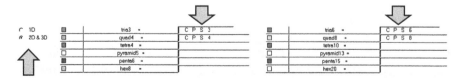

FIGURE 4.10   Pre-selection of element types to be used in finite element model.

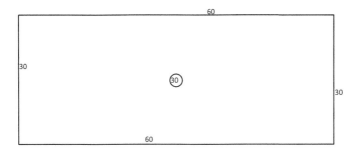

FIGURE 4.11   Discretization of geometric edges for selective adjustment of finite element grid refinement.

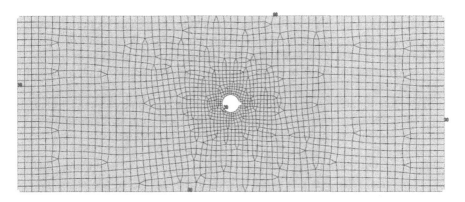

FIGURE 4.12   Automatic mesh generation by a model pre-processor based on user inputs for mesh grid refinement.

number of divisions on each feature line, a new mesh is generated with the execution of "mesh" command. Figure 4.12 shows a mesh generated with 30 grid points on the perimeter of hole, giving an element size of approximately 2.1 mm around the hole, while element sizes along the boundaries of the plate remain at about 6–8 mm. A revised version of the mesh can be generated again by rejecting the current displayed version and adjusting the number of grid points on geometric features lines as desired. Evidently, *a priori* knowledge of the analysis problem is essential to guide the interactive meshing exercise, with visual assessment of the finite element mesh to meet the eventual analysis objectives.

4.4.3   FINITE ELEMENT MESH QUALITY CHECKS

As evident from Figure 4.12, element shapes do not remain at ideal square or equilateral triangular shapes as the mesh transitions from refined density at the vicinity of the circular hole to coarser density along boundaries of the rectangular plate. Negative effects of element distortion on predicted result quality have been discussed with an elementary stress analysis example in Section 3.7 and Figure 3.12. It is important to check element quality after developing the mesh model of a component. HyperMesh provides several element quality metrics like skew, warpage, aspect ratio, Jacobian, interior angles, etc., to report the quality of a generated mesh. Metrics for mesh quality checks vary from software to software. Few of the commonly used element quality measurements are discussed in the following.

> Warpage: A quad element is temporarily split into two adjacent triangles (for quality check only), and warp angle is defined as the angle between the normals to two triangular planes. A quad element can be split into triangles in two different ways by using either of the two diagonal lines. HyperMesh reports the maximum value of the two possible angles. Ideal value of warpage is "zero" while an acceptable limit may be set at $\leq 10°$. Warpage check is important in 3D surface models. Here in the current example of 2D plane stress problem, warpage angle is zero.
>
> Aspect ratio: Aspect ratio of an element is defined as the ratio between maximum element edge length and minimum edge length. Ideal value of aspect ratio is "1" while an acceptable limit can be considered up to <5 for the current analysis problem subjected to in-plane traction mode of deformation only.
>
> Skew: Interior angle at each element corner node of a quad element is calculated, and the skewness of element is defined as 90° minus minimum internal angle of the quad element. Ideal value for skewness is "0" while an acceptable limit is generally considered to be $\leq 45°$.
>
> Jacobian: The determinant of Jacobian matrix (equation 3.70) measures the numerical relationship between element local and global coordinate systems. HyperMesh evaluates the determinant of the Jacobian matrix at each integration point, and it reports the ratio between the smallest and the largest values. An ideal square shape element has that ratio at 1. As the element becomes distorted, that ratio approaches a zero value, and a concave shape element reaches a value of -1. Acceptable limit of Jacobian ratio is set at ≥ 0.7.
>
> Min/max angle: Minimum and maximum values of internal angles of triangular and quad elements are reported to check the general distortion of elements. Ideal and acceptable values for quad elements are, respectively, 90° and $\geq 45°$. Corresponding values for triangular elements are 60° and $\geq 20°$.

HyperMesh reports of element quality checks can be generated by going through "Tool" menu panel and choosing "check elements" submenu item. Figure 4.13 shows a graphical report generated by HyperMesh displaying elements with Jacobian ratio values less than 0.7. Quality issues of computer-generated meshes may need to be corrected manually by modifying the nodes and elements. A completely new mesh may need to be created occasionally when there are too many mesh quality issues in the vicinity of critical engineering interest. Other than element shape quality issues discussed above, computer-aided finite element mesh generation may also experience other quality glitches. For example, free edges of finite elements should always match with free boundaries of CAD geometry. Discovery of internal free edges along

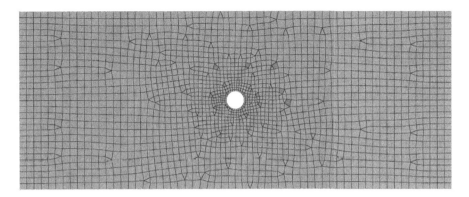

FIGURE 4.13 Graphical display highlighting the elements that fail to meet specific quality check criterion.

inter-element boundaries indicates absence of inter-element connectivity along those free edges. Under the "Tool" menu panel, HyperMesh provides the submenu option of "edges" that provides command option of "find edges" in selected components. After deleting the free edges, nodes within immediate vicinity of one another can be displayed by using the command "preview equiv" with a very small value of tolerance (0.1 mm for example). Displayed nodes from "preview equiv" step can be merged together by choosing the command "equivalence" at that stage. Extra caution must be exercised to check the physical validity of such merger operation as it may lead to unacceptable representation of the physical geometric features.

Mistakes made during meshing operations can result in duplication or multiplication of nodes and elements overlaid on the same geometric entity. Once again, modern software packages come handy with built-in tools to correct such meshing errors. Under HyperMesh menu item "Tool", "check elems" submenu lists a command named "duplicates" to activate duplicate element check. Unnecessary duplicate elements, found through that check process, can be deleted at that point. Multi-step model creation and cleaning operations may also leave extra nodes floating in the database without connectivity to element mesh. Under the "Geom" menu panel, HyperMesh lists "temp nodes" submenu option that can be used to visualize all nodes; and to eliminate all floating nodes by using the command button "clear all". As discussed in Section 3.11, ABAQUS expects that 2D solid elements be defined by listing the corner nodes following counter-clockwise sequencing, which means that normal to the elements in the x–y plane will be directed in the z-direction toward the observer facing the model on the x–y plane. Automatic mesh generation may occasionally lead to opposite orientation of element normal, which must be checked and corrected if needed. In the "Tool" menu panel of HyperMesh, "normals" submenu provides the capability to display element normals by putting a positive numerical value under "vector display", and by executing the menu mutton "display". Finite element model may need to be rotated in 3D space for proper visualization of the element normal direction. If the normals appear in the opposite direction of ABAQUS specification, one click "reverse" button will set the element normals to opposite direction. Normal definition, assumed correct for ABAQUS, may not be correct for other finite element analysis packages as there is no universal standard for how nodal connectivity should be sequenced in element definition. Users should consult the manuals of specific FEA solver to identify the correct element definition format.

Above discussions include common mesh quality checks while preparing a finite element analysis model by using the general-purpose model preparation software HyperMesh. An audio-visual presentation on mesh quality checks with HyperMesh is available online at Altiar University (2014). Some of these same items may apply to model quality checks with other pre-processors as well. Upon verification of the mesh quality, user may proceed with the following key steps, 4.4.4 through 4.4.7, for the completion of an ABSQUS finite element analysis model by using HyperMesh pre-processing software.

4.4.4 MATERIAL AND PART PROPERTY ASSIGNMENT

Essential material property data for the calculation of elasticity matrix (equation 3.64) are provided through *MATERIAL data block in ABAQUS input files (discussed in Section 2.10). Through the "Materials" menu panel in HyperMesh, a user can proceed to specify a material name (for example, Mat-1), and can choose "ABAQUS_MATERIAL" for "card image" option. Execution of "create/edit" task bar opens the menu for preparation of the ABAQUS material model inputs (equivalent of manual input description under *MATERIAL data block). Checking of "Elastic" option will open input cells to define isotropic material properties of Young's modulus (E) and Poisson's ratio (v). For the case study example of plate with hole, $E = 70$ and $v = 0.333$ are provided as inputs for the elastic behavior of aluminum material. Execution of "return" button saves the data in HyperMesh database, and returns the user to main menu panel. Selection of "Property" attribute icon from main meu panel will open the dialogue subpanel for creating property data for a part. Required inputs are property name (for example Prop-1), "SOLID SECTION" as input for "card image", and selection of previously defined "Mat-1" under the selection window of "material=". Execution of "create/edit" will lead to new submenu for defining the section properties of the plane-stress plate. Checking the "DataLine" option provides an input box under "Attribute_Value" where user can specify part thickness value ("1.0" for the example plane-stress plate problem). Execution of the "return" button will take the user back to property creation subpanel. Selection of "assign" button on that subpanel leads the user to choose elements "all" or "displayed". Previously created property data card (Prop-1) can be selected in the "property" input box. Execution of the "assign" option at that point assigns the selected property data (including the associated material data with it) to all selected elements. An essential step of model quality check is to select "Card Edit" from main menu panel, and then verify that the material and property data have been correctly assigned to the finite element model in display window. Assignment of material and section property data to the elements completes the physical description of the part to be analyzed.

4.4.5 ANALYSIS PARAMETERS – BOUNDARY CONSTRAINTS, EXTERNAL LOADS AND MODEL OUTPUTS

Specification of proper boundary conditions is an essential step for building a stable analysis model. Figure 4.14 shows a schematic description of loads and boundary constraints to be applied on the example problem of plate with hole. Problem-specific descriptions of constraints and loads can be described through graphical interfaces of HyperMesh without the need for downstream manual modifications of the final model input file. "Analysis" option under HyperMesh menu panel brings up selection menu for assigning boundary conditions and loads on a finite element model. The "constraints" selection bar from the menu subpanel leads to the main operation stage for defining nodal boundary constraints. Several selection options are available under the "nodes" selection panel. A tedious but visually verifiable way is to click on the "nodes" selection button, and choose the nodes directly on the model display window for boundary condition assignment. Figure 4.15 shows selected nodes on the left-side boundary where constraints on x-displacement can be assigned by checking the "dof1=0.0" in the menu selection panel, and subsequently pressing the button "create". Center node skipped in the selection on Figure 4.15 can be selected separately in the next step, and boundary constraints in both x and y directions can be assigned by checking the options "dof1=0.0" and "dof2=0.0" in the selection panel. Pressing of "create" task bar will add this boundary constraint definition to the model database. Boundary conditions defined in above two steps prevents any rigid body motion possibility of the model in x–y plane. No constraint definitions are needed in z-direction since this particular example problem has been setup as a 2D analysis model in x–y plane.

Loading conditions on structures can be described as discrete or distributed forces. Figure 4.14 shows a uniformly distributed traction force of 0.1 GPa acting on the right-side boundary of the plate-with-hole. However, in actual finite element analysis, all forces are represented as nodal forces in the solution of equilibrium equations (1.5). Assuming that the nodes along the right-side boundary are

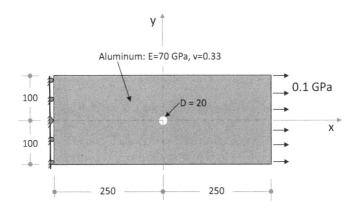

FIGURE 4.14   External loading and boundary constraints acting on the plate in the x–y plane.

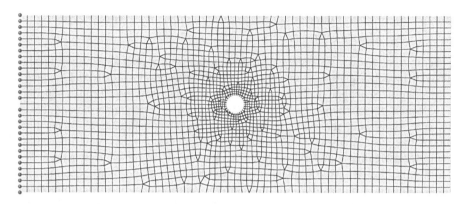

FIGURE 4.15   Node selections to specify *x*-constraints on the left-boundary of plate.

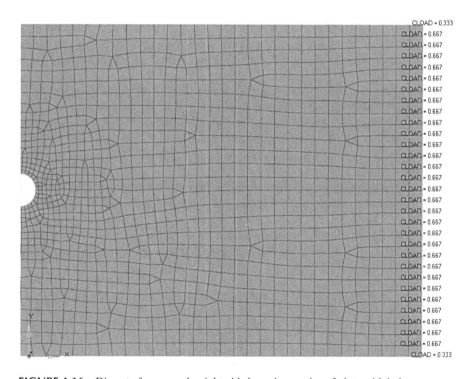

FIGURE 4.16   Discrete forces on the right-side boundary nodes of plate with hole.

uniformly spaced @ 6.67 mm, effective load on each boundary node of the unit-
thickness plate will be 0.667 kN except at the two corner nodes that will see an
effective load of 0.333 kN (Figure 4.16). ABAQUS expects discrete nodal loads to
be defined with "*CLOAD" keyword as discussed in Section 2.10. Similar to the
definition process for nodal constraints, HyperMesh provides the option to generate
nodal loads with the item "Forces" listed under "Analysis" menu panel. Target nodes
subjected to 0.667 kN discrete loads can be picked interactively from the display

window and the relevant load amplitude can be applied in the x-direction. A similar second step of 0.333 kN load application can be defined by choosing the two corner nodes on the right side boundary. Boundary constraints and loads, described in HyperMesh in the above two steps, need to be included in the definition of ABAQUS analysis step as described in the following Section 4.4.6. Solution of system equilibrium equations (1.5) produces results for nodal displacements and reaction forces. Strains and stresses inside finite elements are calculated by using equations (2.6) and (2.26). ABAQUS stores all calculated nodal and element response results in a binary database file (*.ODB). Selective results for nodes, elements, and global model (such as energy balance report) can be requested to be saved in the text format results file named with an extension ".DAT". Specific requests for such results can be defined through "output bock" that is listed under the "Analysis" menu panel in HyperMesh.

4.4.6   DEFINITION OF ABAQUS ANALYSIS STEP

Linear elastic load–deflection analysis of plate with hole, subjected to boundary constraints and loads described above, can be executed by specifying static perturbation analysis option of ABAQUS as discussed in Section 2.10. The "load steps" task bar listed in "Analysis" menu panel of HyperMesh provides the relevant menu selection items. The "edit" task bar opens the selection menu to choose "PERTURBATION" analysis with "Static" load type. Boundary conditions and loads described earlier need to be selected under "loadcols" selection button. Output requests defined earlier can be included under the "outputblocks" option.

4.4.7   EXPORTING THE MODEL INPUT FILE FOR ABAQUS ANALYSIS

Finite element model entities described thus far can be saved as internal database of HyperMesh. However, for ABAQUS analysis, these data need to be exported out of HyperMesh by using the "File-Export" menu choice. This can be achieved by choosing ABAQUS under "solver deck" with "standard 2D" option, and by opting to export all data to an external data file named with extension ".INP". Model data for ABAQUS is saved in the ASCII format, and the saved file can be edited with a standard text editor if needed.

Experienced model builders can generally produce a usable finite element model by going through the multi-step model preparation process – described in the above, and also summarized with (a)-to-(g) steps in Figure 4.17. However, less-than-perfect execution steps often produce incomplete description of the model, thus leading to failed analysis by the downstream FEA software package. So, it is important to visually check the exported model data file before sending the model to analysis execution. Obviously, checking the mesh quality of a large complex model is beyond the scope of manual data check at that point. However, visual checks often identify common mistakes that occur in key model assumptions such as element types, loads, boundary conditions and analysis type selection. Essential key words in ABQUS input file of plate with hole problem are shown in the following, where dot point clusters refer to bulk datasets that are omitted from this presentation:

```
*HEADING
…………
*NODE
………..
*ELEMENT, TYPE=CPS3, ELSET=Plate-with-hole
…………
*ELEMENT, TYPE=CPS4, ELSET=Prate-with-hole
…………
*SOLID SECTION, MATERIAL=Mat-1, ELSET=Plate-with-hole
1.0
*MATERIAL, NAME=Mat-1
*ELASTIC
70.0, 0.33
*STEP, PERTURBATION
*STATIC
*BOUNDARY
……………
*CLOAD
……………
*NODE PRINT
RF
…..
*ENRGY PRINT
*END STEP
```

Submission of analysis model file to ABAQUS, and extraction of result files follow the same procedures described for practice problems presented in Chapters 1–3.

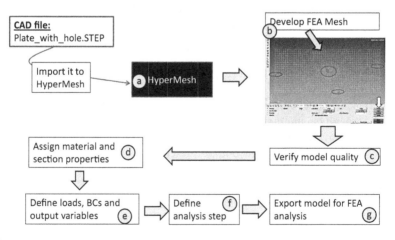

FIGURE 4.17 Key-process steps (a)–to-(g) in computer-aided model preparation.

4.5 POST-PROCESSING OF FEA RESULTS

The very first step in post-processing of finite element analysis results is the basic quality check of results to validate the reliability of analysis model. Few of the routine result quality checks include: (i) energy balance check, (ii) verification of system equilibrium, and (iii) compatibility checks of displacement and stress analysis results. In static load analysis cases, velocity and acceleration responses at nodal degrees-of-freedom are "zero". Internal energy stored in the system must be equal to the external work done, as defined in the following equation (4.11):

$$\text{Internal strain energy}, \int \left[\frac{1}{2} \cdot \sigma^T \cdot \varepsilon \right] dV = \text{External work done}, \frac{1}{2} \cdot \{P\}^T \cdot \{u\} \quad (4.11)$$

Energy error in finite element analysis of linear elastic static problems should be:

$$Energy\ error = \left(External\ work\ done - Internal\ strain\ energy \right) \approx 0 \quad (4.12)$$

Energy balance report from ABAQUS can be checked in ".DAT" output file when "*ENERGY PRINT" command is included in model preparation step (e) of Figure 4.17. Verification of energy balance in an example analysis is shown in Figure 4.18. Verification of the system equilibrium is another check that should be performed at the beginning of result post-processing step. In the plate with hole analysis example, applied load on the right-side boundary is 200*1*0.1 = 20 kN in the x-direction and "zero" is the y-direction. Summation of reaction forces saved in the "*.DAT" file produces a value of "20.0" in the x-direction, and "0.0" in the y-direction, thus passing the equilibrium quality check of results (Figure 4.19). Other simple checks of result quality can be achieved by verifying the displacement and stress boundary conditions, as shown in Figures 3.18 and 3.19 for plane-stress analysis of a plate. A special check of the stress analysis results, obtained for plate with hole, can be achieved through the contour plot of the x-directional stresses in the plate (Figure 4.20). Maximum stress value predicted at the crest of hole by the finite element analysis model is found to be 0.2974 GPa, which is very close to the theoretically expected value of 0.3 GPa (three times the applied boundary stress as per equation 4.3). Model refinement around the hole at the center of plate analysis example has, thus, adequately captured the high stress concentration at the vicinity of hole in the stressed plate.

4.6 PRACTICE PROBLEMS – STRESS ANALYSIS OF PLATE WITH HOLE

PROBLEM 1

Re-analyze the plate with hole example problem described in Figure 4.14 by using six- and eight-node quadratic plane-stress elements. Compare the results with those obtained and discussed in this chapter by using linear tria and quad elements. Prepare

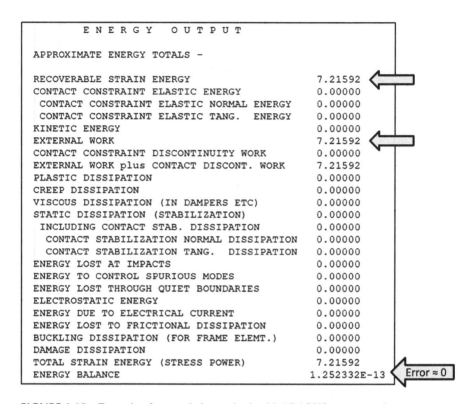

FIGURE 4.18   Example of energy balance check with ABAQUS output results.

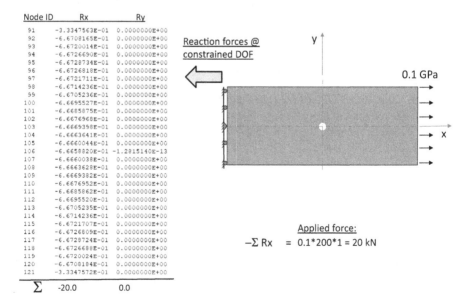

FIGURE 4.19   Equilibrium check with simulation model results.

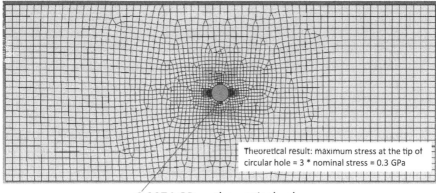

$$\sigma_x = 0.2974 \text{ GPa} \approx \text{theoretical value}$$

FIGURE 4.20   Contour plot of the x-directional stress in plate with hole.

a report with the information on: (i) FEA mesh picture with boundary conditions and applied loads; (ii) results of mesh quality checks; (iii) verification of FEA model results; and (iv) contour of stresses σ_{xx}. Determine at what distance the Saint Venant's principle is applicable to this stress analysis problem of plate with hole, i.e. the local disturbance in stress field disappears at a distance away from the point of geometric imperfection.

PROBLEM 2
Figure 4.21 shows a plane-stress plate similar to that of Figure 4.14, with one exception related to the shape of hole that is hexagonal in Figure 4.21. Conduct finite element stress analysis of this problem with two meshes of different refinement at the vicinity of the hole. Compare the results and comment on the convergence characteristic of finite element stress analysis results for this example problem.

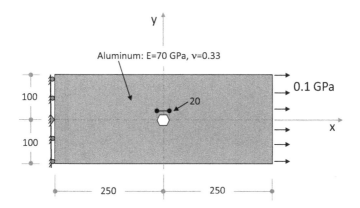

FIGURE 4.21   Plate with a hexagonal hole at the center.

5 Stress Analysis of Axisymmetric and General 3D Solids

SUMMARY

Stress analysis techniques, both analytical and numerical (finite element), have been presented in Chapter 3 for elastic solids that are amenable to two-dimensional idealization of plane strain or plane stress states. Three-dimensional elasticity problems, having an axis of symmetry, can also be simplified to 2D analysis models on the plane of axial symmetry. Section 5.1 presents the basic elasticity formulations for axisymmetric stress field by using the polar coordinate system that appears specifically suitable for this class of problems. Detailed analytical steps deriving the stress and deformation responses of uniformly pressurized thick cylinder example are presented in Section 5.2. Understandably, these detailed analytical steps are hard to replicate for solutions of practical engineering problems. The key ingredients of analytical stress–strain–displacement relationships, as well the analytical solutions derived for standard examples, serve as references for the finite element solution technique. Section 5.3 introduces the finite element method as general-purpose stress analysis tool for more complex axisymmetric problems that cannot be readily solved by using the analytical tools. An extension of the polar coordinate analytical framework is used in Section 5.4 to predict stresses in solids for the very specific case study of concentrated load effects. Once again, more complex 3D stress analysis cases require the use of finite element method that is introduced in Section 5.5 by simply adding a third dimension to the two-dimensional formulations of Chapter 3. Section 5.6 specifically focuses on the topic of software-aided model preparation and quality checks for 3D simulations. Finally, practice problems for both axisymmetric and general 3D solids are presented in Section 5.7.

5.1 AXISYMMETRIC – A SPECIAL FORM OF 3D ELASTICITY PROBLEMS

Figure 5.1 shows a cylindrical body having an axis of symmetry with respect to loading, geometry, material properties, boundary conditions, etc. Theory of elasticity-based analytical solutions for such problems is efficiently described by using polar coordinates (r, θ) (Figure 5.2). Relationships between polar and Cartesian coordinates in 2D cross-sectional plane of the member are given in the following:

$$x = r.\cos\theta; \;\; y = r.\sin\theta; \;\; r^2 = x^2 + y^2; \;\; \theta = \tan^{-1}\left(\frac{y}{x}\right) \tag{5.1}$$

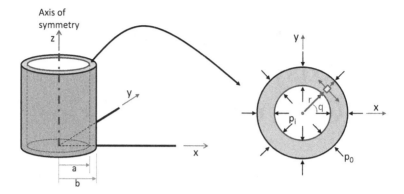

FIGURE 5.1   Three-dimensional problem with axis of symmetry (no variation in circumferential direction).

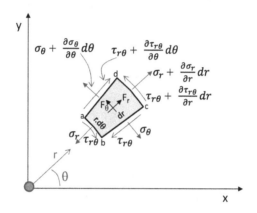

FIGURE 5.2   Expanded view of stress element from Figure 5.1 – shown in polar coordinates (r, θ).

Partial derivatives of equations (5.1) yield:

$$\frac{\partial r}{\partial x} \equiv \frac{x}{r} = \cos\theta, \quad \frac{\partial r}{\partial y} = \frac{y}{r} = \sin\theta$$

$$\frac{\partial r}{\partial x} \equiv \frac{x}{r} = -\frac{\sin\theta}{r}, \quad \frac{\partial \theta}{\partial y} = \frac{x}{r^2} = \frac{\cos\theta}{r} \tag{5.2}$$

Using the chain rule of partial differentiation, the following relations are obtained between partial derivatives of Cartesian and polar coordinate systems:

$$\frac{\partial}{\partial x} = \frac{\partial r}{\partial x} \cdot \frac{\partial}{\partial r} + \frac{\partial \theta}{\partial x} \cdot \frac{\partial}{\partial \theta} = \cos\theta \cdot \frac{\partial}{\partial r} - \frac{\sin\theta}{r} \cdot \frac{\partial}{\partial \theta}$$

$$\frac{\partial}{\partial y} = \frac{\partial r}{\partial y} \cdot \frac{\partial}{\partial r} + \frac{\partial \theta}{\partial y} \cdot \frac{\partial}{\partial \theta} = \sin\theta \cdot \frac{\partial}{\partial r} + \frac{\cos\theta}{r} \cdot \frac{\partial}{\partial \theta} \tag{5.3}$$

5.1.1 EQUATIONS OF EQUILIBRIUM IN POLAR COORDINATES

Considering unit thickness for the differential element in Figure 5.2, the equilibrium of forces in radial direction gives the following expression:

$$
\begin{aligned}
& \left(\sigma_r + \frac{\partial \sigma_r}{\partial r}.dr\right)\left(r+dr\right)d\theta - \sigma_r.r.d\theta \\
& - \left(\sigma_\theta + \frac{\partial \sigma_\theta}{\partial \theta}.d\theta\right).dr.\sin\frac{d\theta}{2} - \sigma_r.dr.\sin\frac{d\theta}{2} \\
& + \left(\tau_{r\theta} + \frac{\partial \tau_{r\theta}}{\partial \theta}.d\theta\right).dr.\cos\frac{d\theta}{2} - \tau_{r\theta}.dr.\cos\frac{d\theta}{2} + F_r.r.dr.d\theta = 0
\end{aligned}
\tag{5.4}
$$

where F_r is the body force per unit volume. Equation (5.4) can be simplified by substituting $d\theta/2$ and 1, respectively, for $\sin(d\theta/2)$ and $\cos(d\theta/2)$ when $(d\theta)$ is very small; and also by ignoring the higher order terms of the differential quantities dr and $d\theta$. A similar relationship can also be derived by considering the equilibrium of forces in tangential (θ) direction. Simplified expressions for equilibrium of stresses in r and θ are thus obtained as follows:

$$
\begin{aligned}
& \frac{\partial \sigma r}{\partial r} + \frac{1}{r}\frac{\partial \tau_{r\theta}}{\partial \theta} + \frac{\sigma_r - \sigma_\theta}{r} + F_r = 0 \\
& \frac{1}{r}.\frac{\partial \sigma_\theta}{\partial \theta} + \frac{\partial \tau_{r\theta}}{\partial r} + \frac{2\tau_{r\theta}}{r} + F_\theta = 0
\end{aligned}
\tag{5.5}
$$

In absence of body forces ($F_r = F_\theta = 0$), equilibrium equations (5.5) are satisfied by a stress function $\phi(r, \theta)$ when the stress components in radial and tangential directions are defined in terms of the stress function as follows:

$$
\begin{aligned}
\sigma_r &= \frac{1}{r}\frac{\partial \varphi}{\partial r} + \frac{1}{r^2}.\frac{\partial^2 \varphi}{\partial \theta^2} \\
\sigma_\theta &= \frac{\partial^2 \varphi}{\partial r^2} \\
\tau_{r\theta} &= \frac{1}{r^2}\frac{\partial \varphi}{\partial \theta} - \frac{1}{r}.\frac{\partial^2 \varphi}{\partial r \partial \theta} = -\frac{\partial}{\partial r}\left(\frac{1}{r}\frac{\partial \varphi}{\partial \theta}\right)
\end{aligned}
\tag{5.6}
$$

5.1.2 STRAIN–DISPLACEMENT RELATIONSHIPS

Denoting u for the displacement in radial direction of the differential element "abcd" in Figure 5.2, the radial strain can be defined as follows:

$$
\varepsilon_r = \frac{\partial u}{\partial r}
\tag{5.7}
$$

The tangential strain owing to radial deformation u can be derived as follows:

$$\varepsilon_\theta = \frac{\left[2\pi\left(r+u\right)-2\pi r\right]}{2\pi r} = \frac{u}{r} \tag{5.8}$$

Denoting the tangential deformation in element "abcd" of Figure 5.2 by v, the corresponding tangential strain can be expressed as follows:

$$\varepsilon_\theta = \frac{\left(\partial v/\partial\theta\right).d\theta}{r.d\theta} = \frac{1}{r}\cdot\frac{\partial v}{\partial\theta} \tag{5.9}$$

Combining the contributions of deformations u and v, total tangential strain is given by

$$\varepsilon_\theta = \frac{1}{r}\cdot\frac{\partial v}{\partial\theta} + \frac{u}{r} \tag{5.10}$$

Shear strain caused by the radial deformation u can be expressed as follows:

$$\gamma_{r\theta} = \frac{\left(\partial u/\partial\theta\right).d\theta}{r.d\theta} = \frac{1}{r}\cdot\frac{\partial u}{\partial\theta} \tag{5.11}$$

And the shear strain caused by the tangential deformation v is given by:

$$\gamma_{r\theta} = \frac{\partial v}{\partial r} - \frac{v}{r} \tag{5.12}$$

Combining equations (5.11) and (5.12), resultant shear strain is given by the following equation:

$$\gamma_{r\theta} = \frac{\partial v}{\partial r} + \frac{1}{r}\cdot\frac{\partial u}{\partial\theta} - \frac{v}{r} \tag{5.13}$$

Equations (5.11), (5.12), and (5.13) provide the strain–displacement relations in polar coordinate system for axisymmetric solid.

5.1.3 STRESS–STRAIN RELATIONSHIPS

Substituting r and θ for x and y in equations (3.4) and (3.5), Hooke's laws for axisymmetric plane strain condition are given by

$$\begin{aligned}
\varepsilon_r &= \left(1-v^2\right)\frac{\sigma_r}{E} - \left(v+v^2\right)\frac{\sigma_\theta}{E} \\
\varepsilon_\theta &= \left(1-v^2\right)\frac{\sigma_\theta}{E} - \left(v+v^2\right)\frac{\sigma_r}{E} \\
\gamma_{r\theta} &= \frac{\tau_{r\theta}}{G} = \frac{2\left(1+v\right)}{E}\tau_{r\theta}
\end{aligned} \tag{5.14}$$

Similarly, Hooke's laws for plane stress axisymmetric condition are obtained from equations (3.11) as follows:

$$\varepsilon_r = \frac{\sigma_r}{E} - v\frac{\sigma_\theta}{E}$$

$$\varepsilon_\theta = \frac{\sigma_\theta}{E} - v\frac{\sigma_r}{E} \tag{5.15}$$

$$\gamma_{r\theta} = \frac{\tau_{r\theta}}{G} = \frac{2(1+v)}{E}\tau_{r\theta}$$

5.1.4 COMPATIBILITY CONDITION

Following the steps used for Cartesian coordinate system in Section 2.3, strain compatibility condition in the polar coordinate system can be obtained from equations (5.7), (5.10), and (5.13):

$$\frac{\partial^2\varepsilon_\theta}{\partial r^2} + \frac{1}{r^2}\cdot\frac{\partial^2\varepsilon_r}{\partial\theta^2} + \frac{2}{r}\frac{\partial\varepsilon_\theta}{\partial r} - \frac{1}{r}\frac{\partial\varepsilon_r}{\partial r} = \frac{1}{r}\cdot\frac{\partial^2\gamma_{r\theta}}{\partial r.\partial\theta} + \frac{1}{r^2}\cdot\frac{\partial\gamma_{r\theta}}{\partial\theta} \tag{5.16}$$

To describe the compatibility condition in term of stress function ϕ, the Laplacian operator of Cartesian coordinate system (equation 2.30) can be transformed into the polar coordinate system by using equations (5.2) and (5.3):

$$\nabla^2\varphi = \frac{\partial^2\varphi}{\partial x^2} + \frac{\partial^2\varphi}{\partial y^2} = \frac{\partial^2\varphi}{\partial r^2} + \frac{1}{r}\cdot\frac{\partial\varphi}{\partial r} + \frac{1}{r^2}\cdot\frac{\partial^2\varphi}{\partial\theta^2} \tag{5.17}$$

Equation of compatibility (3.16) can thus be written in the polar coordinate system as follows:

$$\nabla^4\varphi = \left(\frac{\partial^2}{\partial r^2} + \frac{1}{r}\cdot\frac{\partial}{\partial r} + \frac{1}{r^2}\cdot\frac{\partial^2}{\partial\theta^2}\right)(\nabla^2\varphi) = 0 \tag{5.18}$$

Theory of elasticity developments in polar coordinate systems, described in the above, provide efficient solutions for simple axisymmetric engineering problems. Section 5.2 provides application examples of thick-wall cylinder problems. However, similar to the discussions in earlier chapters, on relative efficiency and limitations of analytical versus numerical analysis methods, solutions of general axisymmetric problems tend to become more manageable with the finite element technique which will be introduced in Section 5.3.

5.2 STRESS ANALYSIS OF AXISYMMETRIC EXAMPLE – THICK WALL CYLINDER

Stress and deformation responses inside the thick wall cylinder of Figure 5.1, subjected to uniform internal and external pressure, can be solved by using the basic equations of Section 5.1 – without using the stress-function-based semi-inverse method. Symmetric deformation response of the cylinder about z-axis implies that the shearing stress is zero, $\tau_{r\theta} = 0$. Assuming that the ends of the cylinder are open and unconstrained, stress in the z-direction is, $\sigma_z = 0$. Assuming z independence of the unit length ring of Figure 5.1, and assuming no body forces are acting on the system, the stress equilibrium equations (5.5) reduce to the following form:

$$\frac{\partial \sigma_r}{\partial r} + \frac{\sigma_r - \sigma_\theta}{r} = 0 \tag{5.19}$$

Radial and tangential stresses, σ_r and σ_θ, are related to radial and tangential strains, ε_r and ε_θ, that in turn are related to in-plane radial and tangential deformations u and v. However, symmetric deformation field about z- axis leads to the fact that tangential deformation $v = 0$. Strain–displacement relations of equations (5.7), (5.10), and (5.14), thus, take the following forms:

$$\varepsilon_r = \frac{\partial u}{\partial r}, \ \varepsilon_\theta = \frac{u}{r}, \ \gamma_{r\theta} = 0 \tag{5.20}$$

Combining the first two expressions of equation (5.20), we get a simple definition of compatibility condition as follows:

$$\frac{\partial u}{\partial r} - \varepsilon_r = \frac{\partial (r\varepsilon_\theta)}{\partial r} - \varepsilon_r \ \rightarrow r\frac{\partial \varepsilon_\theta}{\partial r} + \varepsilon_\theta - \varepsilon_r = 0 \tag{5.21}$$

Combining the Hooke's laws for plane-stress condition (equations 5.15) and the strain–displacement relations (equation 5.20), we get the following set of equations for axisymmetric ring of Figure 5.1:

$$\varepsilon_r = \frac{\partial u}{\partial r} = \frac{1}{E}(\sigma_r - v\sigma_\theta)$$
$$\varepsilon_\theta = \frac{u}{r} = \frac{1}{E}(\sigma_\theta - v\sigma_r) \tag{5.22}$$

Re-arranging the terms in equations (5.22):

$$\sigma_r = \frac{E}{1-v^2}(\varepsilon_r + v\varepsilon_\theta) = \frac{E}{1-v^2}\left(\frac{\partial u}{\partial r} + v\frac{u}{r}\right)$$
$$\sigma_\theta = \frac{E}{1-v^2}(\varepsilon_\theta + v\varepsilon_r) = \frac{E}{1-v^2}\left(\frac{u}{r} + v\frac{\partial u}{\partial r}\right) \tag{5.23}$$

Substituting equations (5.23) into equilibrium equation (5.19) yields the following differential equation for radial displacement u:

$$\frac{\partial^2 u}{\partial r^2} + \frac{1}{r}\frac{\partial u}{\partial r} - \frac{u}{r^2} = 0 \tag{5.24}$$

The solution of equation (5.24) is given by the following general expression with constants c_1 and c_2:

$$u = c_1 r + \frac{c_2}{r} \tag{5.25}$$

Substituting equation (5.25) into equation (5.23), we get the following expressions for radial and tangential stresses:

$$\sigma_r = \frac{E}{1-v^2}\left[c_1(1+v) - c_2\left(\frac{1-v}{r^2}\right) \right]$$

$$\sigma_\theta = \frac{E}{1-v^2}\left[c_1(1+v) + c_2\left(\frac{1-v}{r^2}\right) \right] \tag{5.26}$$

Constants c_1 and c_2 in equations (5.26) can be determined by using the relevant stress boundary conditions of a given problem. For the cylinder in Figure 5.1, stress boundary conditions on inner and outer surfaces can be described by equation (5.27), where negative sign indicates compressive stress on a surface:

$$\left(\sigma_r\right)_{r=a} = -p_i, \ \left(\sigma_r\right)_{r=b} = -p_0 \tag{5.27}$$

Using the stress boundary conditions of equation (5.27) into the first expression of equations (5.26), we get the following expressions for constants c_1 and c_2:

$$c_1 = \frac{1-v}{E} \cdot \frac{a^2 p_i - b^2 p_0}{b^2 - a^2}, \quad c_2 = \frac{1+v}{E} \cdot \frac{a^2 b^2 (p_i - p_0)}{b^2 - a^2} \tag{5.28}$$

Using the definitions of constants c_1 and c_2 from equations (5.28), stresses and deformation inside the thick cylinder of Figure 5.1 are finally given by the following equations (5.29) (known as Lame's equations):

$$\sigma_r = \frac{a^2 p_i - b^2 p_0}{b^2 - a^2} - \frac{a^2 b^2 (p_i - p_0)}{(b^2 - a^2) r^2}$$

$$\sigma_\theta = \frac{a^2 p_i - b^2 p_0}{b^2 - a^2} + \frac{a^2 b^2 (p_i - p_0)}{(b^2 - a^2) r^2} \tag{5.29}$$

$$u = \frac{1-v}{E} \cdot \frac{(a^2 p_i - b^2 p_0) r}{b^2 - a^2} + \frac{1+v}{E} \cdot \frac{a^2 b^2 (p_i - p_0)}{(b^2 - a^2) r}$$

For the special case of internal pressure only (external pressure: $p_0 = 0$), expressions in equations (5.29) take the following form:

$$\sigma_r = \frac{a^2 p_i}{b^2 - a^2}\left(1 - \frac{b^2}{r^2}\right)$$

$$\sigma_\theta = \frac{a^2 p_i}{b^2 - a^2}\left(1 + \frac{b^2}{r^2}\right) \tag{5.30}$$

$$u = \frac{a^2 p_i \cdot r}{E\left(b^2 - a^2\right)}\left[\left(1 - v\right) + \left(1 + v\right)\frac{b^2}{r^2}\right]$$

Numerical value of maximum radial stress occurs on inner surface of the cylinder ($r = a$). Since $b^2/r^2 > 1$, σ_r is negative (compressive) for all values of r except at $r = b$, where $\sigma_r = 0$. Tangential stress σ_θ is positive (tensile) for all values of r for this special case of internally pressured cylinder. When the cylinder wall is thin, $(b - a) < a/10$, expression for tangential stress in equations (5.30) takes the following familiar form, where t = wall thickness = $(b - a)$:

$$\sigma_\theta = \frac{p_i \cdot a}{t} \tag{5.31}$$

Next, for the special case of external pressure only (i.e. internal pressure: $p_i = 0$), equations (5.29) take the following form describing the internal stresses and deformation in the thick cylinder:

$$\sigma_r = -\frac{b^2 p_o}{b^2 - a^2}\left(1 - \frac{a^2}{r^2}\right)$$

$$\sigma_\theta = -\frac{b^2 p_o}{b^2 - a^2}\left(1 + \frac{a^2}{r^2}\right) \tag{5.32}$$

$$u = -\frac{b^2 p_o \cdot r}{E\left(b^2 - a^2\right)}\left[\left(1 - v\right) + \left(1 + v\right)\frac{a^2}{r^2}\right]$$

Numerical value of maximum radial stress occurs at $r = b$, and it is negative (compressive) throughout the thick wall. Tangential stress σ_θ is negative (compressive) when the cylinder is subjected to external pressure only. Applications of these analytical stress field solutions in designing shrink-fit multilayer compound cylinders and flywheels are available in Ugural and Fenster (2012). Evidently, foregoing analytical developments provide useful insights into the stress and deformation responses of axisymmetric problems having constant attributes (geometry, material, loading, boundary conditions, etc.) along the member axis. Application of this solution becomes challenging when one or more of these attributes vary along the member axis "z" such as the axisymmetric problem in Figure 5.3 where geometry, loading, and boundary conditions change along the axis of symmetry "z". Section 5.3 in the following presents stiffness-based finite element analysis method for solving general axisymmetric problems.

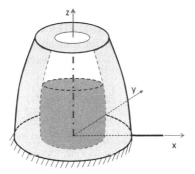

FIGURE 5.3 A partially filled axisymmetric cylinder of variable cross-sectional property along axis of symmetry.

5.3 FINITE ELEMENT ANALYSIS OF AXISYMMETRIC PROBLEMS

Figure 5.4 shows x–z cross-sectional view of the 3D solid from Figure 5.3. Member geometry, boundary conditions, and loading shown in the cross-sectional view are symmetric about the member axis "z". Axial symmetry of member deformation implies that any material point on the x–z cross section can have two translational degrees-of-freedom (u,w), but no deformation in out-of-plane direction $(v = 0)$. Three-dimensional problem of Figure 5.3, thus, essentially reduces to a two-dimensional analysis problem in x–z plane (Figure 5.4). Following the analytical developments for axisymmetric problem in Section 5.2, strain–displacement relations in two-dimensional x–z Cartesian coordinate system can be described as follows:

$$\varepsilon_x = \frac{\delta u}{\partial x}$$

$$\varepsilon_z = \frac{\delta w}{\partial z}$$

$$\gamma_{xz} = \left(\frac{\delta u}{\delta z} + \frac{\delta w}{\delta x} \right) \tag{5.33}$$

$$\varepsilon_\theta = \frac{u}{x}$$

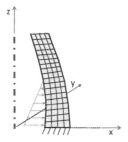

FIGURE 5.4 Finite element discretization of x–z cross-sectional plane of axisymmetric 3D body from Figure 5.3.

Strain components in Cartesian system are, thus, fully described by in-plane deformations (u, w). Following the iso-parametric 2D element formulations of Sections 3.6 and 3.7, strain–deformation relations for four-node axisymmetric finite element can be written by expanding the definitions from equations (3.81):

$$\begin{Bmatrix} \varepsilon_x \\ \varepsilon_z \\ \gamma_{xz} \\ \varepsilon_\theta \end{Bmatrix} = \begin{bmatrix} \dfrac{\partial H_1}{\partial x} & 0 & \dfrac{\partial H_2}{\partial x} & 0 & \dfrac{\partial H_3}{\partial x} & 0 & \dfrac{\partial H_4}{\partial x} & 0 \\ 0 & \dfrac{\partial H_1}{\partial z} & 0 & \dfrac{\partial H_2}{\partial z} & 0 & \dfrac{\partial H_3}{\partial z} & 0 & \dfrac{\partial H_4}{\partial z} \\ \dfrac{\partial H_1}{\partial z} & \dfrac{\partial H_1}{\partial x} & \dfrac{\partial H_2}{\partial z} & \dfrac{\partial H_2}{\partial x} & \dfrac{\partial H_3}{\partial z} & \dfrac{\partial H_3}{\partial x} & \dfrac{\partial H_4}{\partial z} & \dfrac{\partial H_4}{\partial x} \\ \dfrac{H_1}{x} & 0 & \dfrac{H_2}{x} & 0 & \dfrac{H_3}{x} & 0 & \dfrac{H_4}{x} & 0 \end{bmatrix} \cdot \begin{Bmatrix} u_1 \\ w_1 \\ u_2 \\ w_2 \\ u_3 \\ w_3 \\ u_4 \\ w_4 \end{Bmatrix} \qquad (5.34)$$

The matrix on the right-hand side of equation (5.34) describes the standard strain–displacement relationship matrix [B]. Terms inside [B] matrix can be calculated by following the same procedure described in Section 3.7. Stress–strain relationship matrix for the axisymmetric case can be written as follows:

$$\begin{Bmatrix} \sigma_x \\ \sigma_y \\ \tau_{xy} \\ \sigma_\theta \end{Bmatrix} = \frac{E(1-v)}{(1+v)(1-2v)} \begin{bmatrix} 1 & \dfrac{v}{1-v} & 0 & \dfrac{v}{1-v} \\ \dfrac{v}{1-v} & 1 & 0 & \dfrac{v}{1-v} \\ 0 & 0 & \dfrac{1-2v}{2(1-v)} & 0 \\ \dfrac{v}{1-v} & \dfrac{v}{1-v} & 0 & 1 \end{bmatrix} \begin{Bmatrix} \epsilon_x \\ \epsilon_y \\ \gamma_{xy} \\ \epsilon_\theta \end{Bmatrix} \qquad (5.35)$$

Strain–displacement relationship matrix [B], defined in equation (5.34), and stress–strain relationship matrix [C], defined in equation (5.35), are used in numerical integration scheme of equation (3.84) to calculate the element stiffness matrix [k]. Discussions presented in Sections 3.7 and 4.2, regarding the influences of element shape and integration rule on the accuracy of numerically calculated stiffness matrix, equally apply to the axisymmetric formulations as well. Higher order 2D element formulations, discussed in Section 3.8, and lower order constant strain triangular element formulation, discussed in Section 3.9, can also be used for axisymmetric problem analysis. Convergence properties of finite element models, as discussed in Section 3.10, are also applicable to the axisymmetric problem analysis.

Model preparation, for axisymmetric problem analysis with general-purpose finite element software packages, is very similar to that of plane-strain and plane-stress analysis models described in Chapters 3 and 4. ABAQUS, for example, requires that cross-sectional plane of axisymmetric geometry be defined by using global x–y coordinates, where x is the radial direction and y is the member axis direction. Finite element analysis model of an axisymmetric body can be developed from 2D cross-sectional geometry by following the (a)–(g) model preparation steps described in Section 4.4 for two-dimensional problems. Axisymmetric model description will require selection of appropriate element types, such as CAX3 for triangular shape constant stress/strain elements, and CAX4 for quadrilateral element of bilinear shape function in ABAQUS models. Higher order elements with quadratic shape types, six-node CAX6 and eight-node CAX8 element types, can also be selected if desired. Model quality checks described in Section 4.4 apply to axisymmetric models as well. Element type selection, for example, "TYPE=CAX4" in ABAQUS, is the key determinant for analysis software to choose appropriate element formulations from equations (5.34) and (5.35). Nodal loads, corresponding to distributed pressure load on the axisymmetric inner boundary surface in Figure 5.3, are calculated from the effective circumferential area around the axis of symmetry.

5.4 STRESS ANALYSIS OF THREE-DIMENSIONAL BODIES – CONCENTRATED LOADS

5.4.1 STRESSES IN A SEMI-INFINITE SOLID SUBJECTED TO CONCENTRATED NORMAL FORCE ON THE BOUNDARY

Polar coordinate description of stress field, as presented in Section 5.1, can be used to analyze the stress response of three-dimensional bodies subjected to concentrated loads with an axis of symmetry. Figure 5.5 shows a concentrated load "P" acting on the vertex of a 3D wedge having a vertical axis of symmetry "x". Following analytical function, for elastic stress distribution inside the body, is applicable to points (r,θ) away from the load application point:

$$\varphi = cPr\theta \sin(\theta) \qquad (5.36)$$

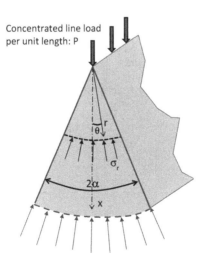

FIGURE 5.5 Concentrated load acting on the vertex of 3D wedge.

Compatibility condition of equation (5.18) is satisfied by the stress function of equation (5.36), where c is an unknown constant. Substituting the stress function from equation (5.36) into equations (5.6), we get the following definitions for the stress components in polar coordinate system:

$$\sigma_r = 2cP\frac{\cos(\theta)}{r}, \quad \sigma_\theta = 0, \quad \tau_{r\theta} = 0 \tag{5.37}$$

The equilibrium condition between applied force P and the resultant vertical force on a cylindrical surface at a distance r from the load application point can be written as follows:

$$2\int_0^\alpha (\sigma_r.\cos\theta)r.d\theta = -P \tag{5.38}$$

Substitution of the expression for σ_r from equation (5.37) into equation (5.38) leads to the definition of constant c as follows:

$$c = -\frac{1}{(2\alpha + \sin 2\alpha)} \tag{5.39}$$

Stresses inside the wedge, defined in equations (5.37), can be re-written as follows:

$$\sigma_r = -\frac{P\cos(\theta)}{r\left(\alpha + \dfrac{1}{2}\sin 2\alpha\right)}, \quad \sigma_\theta = 0, \quad \tau_{r\theta} = 0 \tag{5.40}$$

Stresses in a semi-infinite solid, subject to a normal load on its horizontal surface (Figure 5.6), can be obtained by substituting $\alpha = \pi/2$ in equations (5.40):

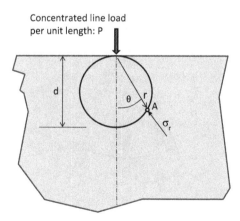

FIGURE 5.6 Radial stress inside a semi-infinite solid subject to concentrated normal load on the horizontal surface.

$$\sigma_r = -\frac{2P}{\pi} \cdot \frac{\cos(\theta)}{r}, \quad \sigma_\theta = 0, \quad \tau_{r\theta} = 0 \tag{5.41}$$

Considering a circle of diameter d, with center on the line of applied normal load on the horizontal surface of semi-infinite solid (Figure 5.6), location of point A on the perimeter is given by $r = d.\cos(\theta)$. Substituting it in equation (5.41), radial stress on the boundary of circle in the semi-infinite solid is thus given by the following well-known equation:

$$\sigma_r = -\frac{2P}{\pi d} \tag{5.42}$$

Equation (5.42) implies that radial stress is same at all points on the perimeter of circle in Figure 5.6, except at the load application point, and the magnitude of stress decreases inversely with increasing distance d.

5.4.2 STRESSES IN A SOLID BEAM SUBJECTED TO CONCENTRATED LATERAL FORCES

Stress derivations presented for semi-infinite solid can be effectively used to estimate stresses inside a solid beam under concentrated lateral loading condition. Figure 5.7 shows a simply supported beam of depth h, length L, and section width b (in normal direction of the planar view). Using the member reference coordinate system x–y, as shown in the figure, the bending stress distribution on beam section at the mid-span is given by the following equation:

$$\sigma'_x = \frac{M.y}{I} = \frac{(PL/4).y}{b.h^3/12} = \frac{3PL}{bh^3} y \tag{5.43}$$

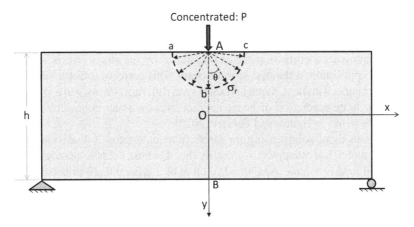

FIGURE 5.7 Stress field inside a simply supported beam at the vicinity of applied concentrated vertical load.

Using the definition of radial stress, σ_r, from equation (5.41), the resultant horizontal force acting on one quadrant of the cylindrical surface "abc" of Figure 5.7 can be defined as in the following:

$$\int_0^{\pi/2} \left(\sigma_r . \sin\theta \right) r . d\theta = \int_0^{\pi/2} \left(\frac{2P}{\pi} \cos\theta \sin\theta \right) d\theta = \frac{P}{\pi} \qquad (5.44)$$

Effects of this horizontal force on beam section at mid-span can be represented by an axial stress (equation 5.45) and a bending stress (equation 5.46) as defined in the following:

$$\sigma_x'' = \frac{P}{\pi bh} \qquad (5.45)$$

$$\sigma_x''' = \frac{\left(-\dfrac{P}{\pi} . \dfrac{h}{2} \right) y}{\left(bh^3 \Big/ 12 \right)} = -\frac{6P}{\pi bh^2} y \qquad (5.46)$$

Combining the stress components defined by equations (5.43), (5.45), and (5.46), resultant normal stress on beam section at mid-span is given by the following:

$$\sigma_x = \frac{3P}{bh^3} \left(L - \frac{2h}{\pi} \right) y + \frac{P}{\pi bh} \qquad (5.47)$$

Substituting $y = h/2$ in equation (5.47), normal stress at point B on beam section (Figure 5.7) is given by

$$\sigma_x = \frac{3PL}{2bh^2} \left(1 - \frac{4}{3\pi} . \frac{h}{L} \right) \qquad (5.48)$$

Term $(3PL/2bh^2)$ in equation (5.48) represents the stress value given by elastic bending stress of beam, and the term in parenthesis on the right-hand side of this equation represents a correction factor introduced by the stress effects of concentrated load application at the mid-span of beam. This correction factor value is significant for beams with large depth and short span (h/L ratio – not small). Discussions on more accurate prediction of beam internal stresses under concentrated applied loads can be found in Ugural and Fenster (2012).

Stress analysis of solids, using the stress function approach discussed so far in Chapters 3 and 5, has attempted to simplify the 3D stress fields to manageable two-dimensional field problems. Analytical solution of a general three-dimensional elasticity problem, involving equations (1.10, 2.5, 2.8–2.13, and 2.25), is usually not attempted. Stiffness-based finite element method provides a very attractive and effective analysis technique by simply expanding the two-dimensional formulations to three-dimensional space as discussed in the following Section 5.5.

5.5 FINITE ELEMENTS FOR STRESS ANALYSIS OF GENERAL 3D SOLIDS

Figure 5.8 shows a general solid subjected to arbitrary load and boundary conditions in three- dimensional space. Each node of hexahedral solid finite element possesses 3 translational degrees of freedom. Assuming simple element geometry aligned with coordinate directions x–y–z, shape functions for nodal deformations are given by the following equations:

$$H_1 = \frac{1}{8}\left(1+\frac{x}{a}\right)\left(1+\frac{y}{b}\right)\left(1+\frac{z}{c}\right)$$

$$H_2 = \frac{1}{8}\left(1-\frac{x}{a}\right)\left(1+\frac{y}{b}\right)\left(1+\frac{z}{c}\right)$$

$$\vdots$$

$$H_8 = \frac{1}{8}\left(1+\frac{x}{a}\right)\left(1-\frac{y}{b}\right)\left(1-\frac{z}{c}\right)$$

(5.49)

Internal deformation response of the element is interpolated from nodal displacements by using the shape functions of equations (5.49):

$$u(x,y) = H_1.u_1 + H_2.u_2 + ... + H_u.u_8 = \sum H_i.u_i$$
$$v(x,y) = H_1.v_1 + H_2.v_2 + ... + H_u.v_8 = \sum H_i.v_i$$
$$w(x,y) = H_1.w_1 + H_2.w_2 + ... + H_u.w_8 = \sum H_i.w_i$$

(5.50)

Strain–displacement relationship matrix [B] in equation (2.6) is a 6×24 matrix of shape function derivatives as defined in the following:

$$[B] = \frac{1}{8abc}\begin{bmatrix}(b+y)(c+z) & -(b+y)(c+z) & .. & .. & .. & .. & .. & .. \\ .. & .. & .. & .. & .. & .. & .. & .. \\ .. & .. & .. & .. & .. & .. & .. & .. \end{bmatrix}$$

(5.51)

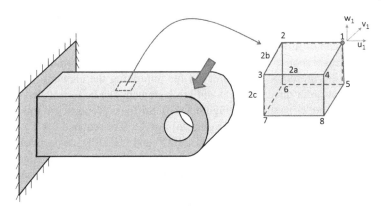

FIGURE 5.8 Displacement degrees of freedom in finite element model of three-dimensional solids.

For general orientation of solid element in 3D space, shape functions in equations (5.49) and the derivative terms in [B] matrix definition (equation 5.51) are re-written by using iso-parametric coordinates (r,s,t) – similar to the two-dimensional description presented in Section 3.6. Relationship between Cartesian model coordinate system (x, y, z), and the iso-parametric element definition in (r, s, t), is defined by expanding the Jacobian from equation (3.70) into 3D system:

$$[J] = \begin{bmatrix} \dfrac{\partial x}{\partial r} & \dfrac{\partial y}{\partial r} & \dfrac{\partial z}{\partial r} \\[2ex] \dfrac{\partial x}{\partial s} & \dfrac{\partial y}{\partial s} & \dfrac{\partial z}{\partial s} \\[2ex] \dfrac{\partial x}{\partial t} & \dfrac{\partial y}{\partial t} & \dfrac{\partial z}{\partial t} \end{bmatrix} \tag{5.52}$$

where coordinates (x, y, z) inside an element are defined in terms of the nodal coordinates by using the 3D interpolation functions written in terms of (r, s, t). Effects of element shape on accuracy of stiffness matrix calculations, discussed in Sections 3.6 and 3.7, are equally applicable to 3D solid elements as well. It is important to maintain regular cubic shape of solid elements in 3D finite element models.

Stress–strain relationship matrix [C] for 3D stress state is defined by equation (2.25). Stiffness matrix in equation (2.44) has 24 terms involving fourth-order coordinate terms arising from pre- and post-multiplication of the [B] matrix terms from equation (5.51). Numerical calculations of the stiffness terms in equation (3.84) can produce exact integration of the stiffness values by using $2 \times 2 \times 2$ Gauss integration rule for the eight-node hexahedral finite element in Figure 5.9(a). ABAQUS input data for this eight-node solid element example will be as follows:

*ELEMENT, ELSET=*setname1*, TYPE=C3D8
1, 7,8, 5, 6, 3, 4, 1, 2

Like the behavior of linear 2D solid elements discussed in Section 4.2, fully integrated 3D solid elements, using node-to-node linear shape functions, tend to provide artificially stiff response under bending and shear. Reduced one-point integration of the eight-node linear solid element can be specified by choosing element type "C3D8R" instead of fully integrated "C3D8", but with the risk of experiencing zero energy hourglass response mechanism with reduced integration formulation. Similar to the discussion presented in Section 3.8 for improving the element deformation behavior, higher order quadratic 3D solid brick element can be formulated by introducing mid-side nodes as shown in Figure 5.9(b). Stiffness properties of this 20-node higher element can be calculated by using full $3 \times 3 \times 3$ or reduced $2 \times 2 \times 2$ Gauss integration rule (ABAQUS element types C3D20 and C3D20R, respectively). Two-dimensional formulations for constant stress–strain triangular element, presented in Section 3.9, can also be expanded to define constant stress–strain 3D element of tetrahedral shape as shown in Figure 5.9(c) (ABAQUS element type "C3D4"). Stiffness properties of this element can be calculated by using one-point integration rule while

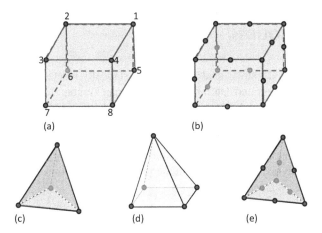

FIGURE 5.9   (a) Linear hexahedral solid element (C3D8), (b) Higher order 20-node brick element (C3D20), (c) linear tetrahedral element (C3D4), (d) linear pyramid (C3D5), and (e) higher order 10-node tetrahedron element (C3D10).

the higher order 10-node tetrahedron element of Figure 5.9(e) may use four points for full integration or one point for faster reduced integration (ABAQUS element types C3D10 and C3D10R, respectively).

5.6   THREE-DIMENSIONAL FEA MODEL PREPARATION AND ELEMENT QUALITY CHECKS

Finite element analysis model of a three-dimensional solid body can be developed from 3D CAD data by following the (a)–(g) model preparation steps described in Section 4.4. For error-free 3D model preparation, HyperMesh session at the beginning should be launched by selecting ABAQUS-standard 3D template if ABAQUS is expected to be used in subsequent FEA analysis. The "3D" menu panel of HyperMesh presents all necessary mesh generation task buttons, including the "elem types" for pre-selecting the type of 3D elements (Figure 5.9) to be used during geometry discretization. Interactive mesh generation technique, described in Section 4.4 for 2D solids, is also available to adapt the refinement of mesh grid in potential areas of high stress concentration in 3D analysis problems. Quality check of 3D meshes requires rigorous scrutiny of both external and internal sectional views of the solid bodies. Appropriate element type selection, through "TYPE" declaration in ABAQUS input file, will set the internal calculations in motion based on numerical process described in Section 5.5. Part property definition, through "*SOLID SECTION" key word in ABAQUS, will include associated material name. Loads can be applied directly as concentrated loads at nodes along the global coordinate directions. Surface pressure loads can also be applied on 3D faces of solid elements. ABAQUS will internally convert the surface pressure loads to equivalent nodal loads for solving the matrix equilibrium equation (1.5). Data structure for ABAQUS model of 3D solids looks very similar to the example shown for 2D example in Section 4.4.7.

5.7   PRACTICE PROBLEMS – STRESS ANALYSIS OF AXISYMMETRIC AND 3D SOLIDS

PROBLEM 1

Section 1.8 presented a stress analysis model of statically indeterminate solid beam, subjected to a concentrated lateral load at mid-span (Figure 1.20). Finite element analysis result, using the ABAQUS model file, "SOLIDBEAM_COARSE_MESH. inp" (data file available at a website noted in the preface of this book), was compared with elementary beam theory prediction in Practice Problem-1 of Section 1.8. Calculate the normal stress, σ_x, at the bottom most point of beam mid-span ($x = 100$) by using the analytical solution technique presented in Section 5.4. Compare this result with those obtained in Section 1.8.

PROBLEM 2

Figure 5.10 shows the cut segment of a long steel pipeline subjected to an internal fluid pressure of 0.01 GPa. Ignoring the external boundary conditions along the length of the pipe, predict stress distribution through the thickness of pipe by using a plane-strain finite element model on the cross-sectional plane perpendicular to the member axis direction. Assume $E = 210$ GPa, $\nu = 0.3$. Compare the finite element analysis results with analytical predictions obtained by using the axisymmetric stress analysis formulations presented in Section 5.2.

PROBLEM 3

Figure 5.11 shows a modified design of pipe segment from Figure 5.10, with an added circumferential steel reinforcement of 10 mm thickness – integrally joined to the main pipe body over a length of 100 mm. Using an axisymmetric finite element model on the cross-sectional plane along the member axis direction, predict the effect of added reinforcement on the stress distribution inside the pipe wall for internal pressure loading of 0.01 GPa. Compare the predicted stress results with those obtained in Problem-2.

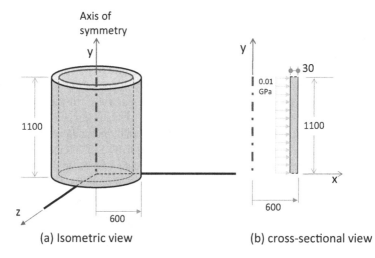

(a) Isometric view   (b) cross-sectional view

FIGURE 5.10   Cut segment of a steel pipeline (dimensions are in mm), subjected to uniform internal fluid pressure of 0.01 GPa.

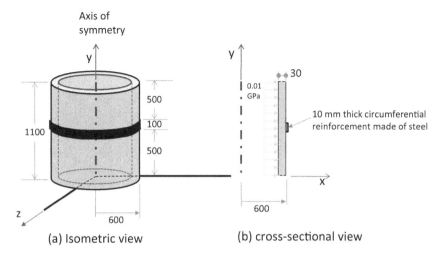

FIGURE 5.11 Locally reinforced pipe segment subjected to uniform internal fluid pressure of 0.01 GPa.

PROBLEM 4

Figure 5.12 shows simplified geometry of a storage tank made of concrete ($E = 40$ GPa, $\nu = 0.2$) sitting freely on frictionless base support. Assuming the tank is filled with water, calculate the stress distribution inside the tank wall using an axisymmetric finite element analysis model. What simplified hand calculations can be done to verify the finite element analysis results?

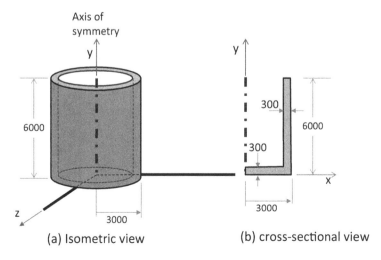

FIGURE 5.12 Simplified geometry of a free-standing open cylindrical tank of uniform wall thickness (dimensions in mm).

6 Deformation Analysis of Beams for Axial, Bending, Shear, and Torsional Loads

SUMMARY

Beams are long slender members that carry transverse loads by producing resistance to bending deflection of member axis. The standard solid mechanics description of bending stress distribution in beams is reviewed in Section 6.1. The shear stress distribution, associated with the bending response of beams, is discussed in Section 6.2. Section 6.3 analyzes the transverse normal stress response that is generally ignored in the analysis of long slender beams. Section 6.4 presents a detailed review of the torsional stress responses of prismatic members. The membrane analogy technique for calculating the torsional response properties of both open and closed section beams has been discussed in detail. The combined stress response of beams for axial, bending, shear, and torsional load effects is calculated from the simple superposition of the component values (discussed in Section 6.5). Section 6.6 presents the well-known Euler–Bernoulli beam theory that relates the lateral deflection of beam axis to bending deformation mode without considering the effect of transverse shear deformation of the material. Limitation of the beam bending deflection analysis, based on Euler–Bernoulli beam theory, is also discussed in this section. Stress analysis of curved beam profiles is presented in Section 6.7 for the completeness of the solid mechanics-based beam analysis technique.

A key component of this chapter, i.e. finite element stiffness formulation for beam resistance mechanisms related to axial, bending, and torsional deformation modes, is described in Section 6.8. The axial and torsional stiffness properties of prismatic beam elements are derived by using independent linear interpolations of the associated nodal DOF. The internal transverse deformation of beam is calculated with cubic interpolations of nodal DOF associated with transverse and rotation deformation modes. The relationship between internal strain and transverse deformation is defined directly based on the Euler–Bernoulli beam theory. The resulting beam element stiffness properties, thereby, turn out to be the same values available from direct stiffness analysis of straight-profile beam structures. The stiffness matrix of two-node prismatic beam element is, thus, directly calculated based on section properties and element length – without having to use the numerical integration technique presented in equation (3.84). This beam element formulation, often referred to as Euler–Bernoulli beam element, provides accurate results for slender beams (with length-to-depth ratio greater than 20). The contribution of shear deformation in the material becomes significant for smaller span-to-depth ratios. Beam stiffness formulations, including shear deformation in the material, are presented in Section 6.9 based on linear interpolation of nodal response variables. The use of linear shape function makes the structural response artificially stiff. Corrective actions to reduce shear locking have also been discussed in Section 6.9. Modeling options of different beam element types, with specific references to the ABAQUS element library, have been discussed in Section 6.10. Finally, Section 6.11 presents practice problems on stress analysis of beam-type structural members.

6.1 BENDING STRESSES IN A BEAM

Figure 6.1 shows the deformation profile of a beam under pure bending load. Following the Euler–Bernoulli theory of beam deflections (Popov 1978), normal stress on the beam cross-section, made of single homogeneous material, can be expressed as a linear function of distance y from beam's neutral axis while all other stress components are zero. The stress field inside homogeneous beam for pure bending load can be described by the following equation:

$$\sigma_x = -\mu y, \quad \sigma_y = \sigma_z = \tau_{xy} = \tau_{xz} = \tau_{yz} = 0 \tag{6.1}$$

where μ is a constant. In absence of body forces ($F_x = F_y = F_z = 0$), stress functions of equation (6.1) readily satisfy the stress equilibrium equations (1.10). These stress functions, with the stress–strain relationships of equations (2.25), also satisfy the deformation compatibility equations (2.8)–(2.13). Stress functions (equation 6.1), with the following boundary conditions (equations 6.2), thus represent the exact solution for pure bending problem of Figure 6.1:

$$\int (\sigma_x).dA = \mu.\int y.dA = 0, \quad \int y.(\sigma_x).dA = -\mu\int y^2.dA = M \tag{6.2}$$

Integral in the second expression of equation (6.2) represents the moment of inertia (I) of beam cross-section about the axis of bending rotation. Using the expression for constant μ from equation (6.2), bending stress on beam cross-section (equation 6.1) can be re-written in the following form (equation 6.3):

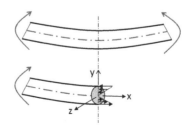

FIGURE 6.1   Bending stresses on a beam section.

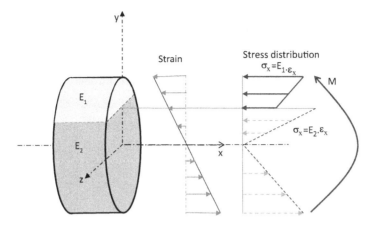

FIGURE 6.2 Bending stress distribution on a beam section made up of two dissimilar materials with elastic moduli of E_1 and E_2.

$$\sigma_x = -\frac{M.y}{I} \qquad (6.3)$$

This is the familiar bending stress formula for a straight beam subjected to pure bending load condition. Assumption of linear bending stress variation over the depth of beam section holds true for beams made of single homogeneous material. Composite beams, constructed by continuous bonding of dissimilar materials, do not experience linear stress variation through beam depth. In elastic bending response analysis of composite beams, plane sections of beams are assumed to remain plane during bending deformation resulting in linear variation of strain through beam depth (Figure 6.2). Using Hooke's law, bending stresses in the dissimilar materials are calculated by using the material-specific elastic modulus as shown in Figure 6.2. For analysis simplicity, composite material construction is replaced with a single material section by scaling the lateral section dimension of a substituted material area with factor (E_2/E_1), where E_2 is the elastic modulus of substituted material and E_1 is the modulus of substituting material. Moment of inertia I is calculated from the hypothetical cross-sectional dimensions of equivalent single material geometry, and the bending stress values are calculated by using the familiar equation (6.3).

6.2 STRESSES DUE TO TRANSVERSE SHEAR

When a beam is bent by transverse loads, usually both a bending moment (M) and a shear force (V) act on beam cross-section. Figure 6.3 shows a beam segment of length dx subjected to bending moments M and $M + dM$, respectively, on the left and right sides of the beam element. Normal stresses caused by these bending moments can be defined by using the bending stress formulation given in equation (6.3). Considering the bending stresses caused by the right-side bending moment $M + dM$, the resultant normal force acting on the area $\bar{A}$, representing part of the beam section above a distance y from the neutral axis, can be expressed as in the following equation (6.4):

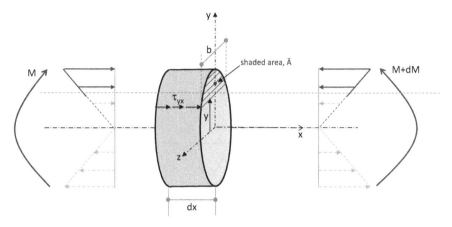

FIGURE 6.3   Beam subjected to bending moment variation over a segment of length dx.

$$\int_{\bar{A}} \sigma_x.dA = -\int_{\bar{A}} \left(\frac{M+dM}{I} \right) y.dA \qquad (6.4)$$

Similarly, considering the bending stresses caused by the left-side bending moment M, the resultant normal force on the area $\bar{A}$ on the left-side of the beam segment is as follows:

$$\int_{\bar{A}} \sigma_x.dA = \int_{\bar{A}} \left(\frac{M}{I} \right) y.dA \qquad (6.5)$$

Difference between the normal forces from two sides (equations 6.4 and 6.5) will cause internal shear stresses (τ_{yx}) over the area ($b.dx$), where b is the width of the beam section at distance y from the neutral axis. Assuming a uniform distribution of shear stresses over the area, the overall equilibrium of the part of beam segment above the position y can be expressed by the following equation:

$$\tau_{yx}.b.dx + \int_{\bar{A}} \left(\frac{M}{I} \right) y.dA - \int_{\bar{A}} \left(\frac{M+dM}{I} \right) y.dA = 0 \qquad (6.6)$$

Re-arranging the terms in equation (6.6) gives:

$$\tau_{yx} = \frac{dM}{dx} \cdot \frac{1}{Ib} \int_{\bar{A}} y.dA \qquad (6.7)$$

The variation of bending moment over beam segment length dx represents the shear force V acting on the beam section (Popov 1978). And the integral term on the right side of equation (6.7) represents the first moment of the beam section area $\bar{A}$ about the neutral axis. Writing Q for that integral term, equation (6.7) can be re-written in the following form:

$$\tau_{yx} = \tau_{xy} = \frac{VQ}{Ib} \tag{6.8}$$

where the equality condition, $\tau_{yx} = \tau_{xy}$, comes from the property of shear stress distribution in three-dimensional bodies as discussed in Section 1.5. Equation (6.8), thus, explicitly describes the relationship between shear force V acting on a beam section and the shear stress τ_{xy} measured at a distance y from the neutral axis. For a rectangular beam section of width b and depth $2h$, the shear stress at a distance y from the neutral axis can be obtained from equation (6.8) by substituting $I = b(2h)^3/12$, and $Q = b(h^2 - y^2)/2$:

$$\tau_{xy} = \frac{3}{4} \cdot \frac{V}{bh^3} \left(h^2 - y^2 \right) \tag{6.9}$$

Equation (6.9) indicates a parabolic distribution of shear stress over the beam depth, with maximum value occurring at the neutral axis ($y = 0$) given by:

$$\tau_{max} = \frac{3}{2} \cdot \frac{V}{2bh} \tag{6.10}$$

where $2bh$ is the cross-sectional area of the rectangular beam section. Maximum shear stress is, thus, 1.5 times the average shear stress for a rectangular beam section. Equation (6.8) can be used to predict shear stress distribution on beam cross-sections of general shape. Figure 6.4(a) shows an I-section beam. Shear stress at a distance y from the neutral axis, for an applied shear force V, can be expressed as follows:

$$\tau_{xy} = \frac{V}{It} \left[\frac{b}{2} \left(h^2 - h_1^2 \right) + \frac{t}{2} \left(h_1^2 - y^2 \right) \right] \tag{6.11}$$

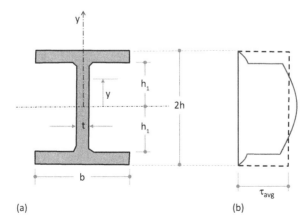

(a)   (b)

FIGURE 6.4   (a) Cross-section of an I-section beam; (b) shear stress distribution over beam cross-section.

where t is the width of the beam section where shear stress is measured. For the I-section beam, t will be replaced by b when stress is measured inside the flange – resulting in a very small shear stress value in the wide flange area. Figure 6.4(b) shows the distribution of shear stress over the beam section following the parabolic function of equation (6.11). In typical engineering practice, small shear stresses in the flanges are ignored, and the average shear stress is calculated by assuming a uniform distribution over the extended depth of beam web:

$$\tau_{avg} = \frac{V}{2th} \tag{6.12}$$

Average shear stress distribution over the beam web, shown by the dotted line in Figure 6.4(b), provides a reasonable estimate for actual engineering decisions instead of using the more rigorous estimate given by equation (6.11).

6.3 TRANSVERSE NORMAL STRESS IN A BEAM

Figure 6.5 shows a cantilever beam subjected to a uniformly distributed transverse load of p per unit length. Assuming the beam section to be rectangular shape with dimension b in the normal direction of view plane, the equilibrium state of a differential element in coordinate y-direction (Figure 6.5(c)) can be expressed as follows:

$$\sigma_y.b.dx = \int_y^h \left(\frac{\partial \tau_{xy}}{\partial x}.dx \right).b.dy \tag{6.13}$$

Writing the shear force in the section of cantilever beam at a distance x from the left end as, $V = p.x$, shear stress in the beam can be obtained from equation (6.9) as follows:

$$\tau_{xy} = \frac{3}{4}.\frac{p.x}{bh^3}\left(h^2 - y^2\right) \tag{6.14}$$

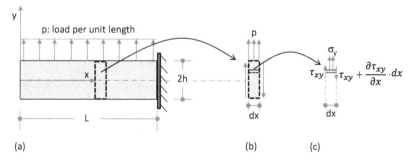

(a)   (b)   (c)

FIGURE 6.5   (a) A cantilever beam subjected to a uniform transverse load of p per unit length; (b) transverse loads on segment of length dx; and (c) stresses in y-direction on a differential element of length dx and height dy.

Substituting equation (6.14) into equation (6.13), and upon integration of the right-hand side, the following expression is obtained for the transverse normal stress (Ugural and Fenster 2012):

$$\sigma_y = \frac{p}{b}\left[\frac{1}{2} + \frac{3}{4}\left(\frac{y}{h}\right) - \frac{1}{4}\left(\frac{y}{h}\right)^3\right] \tag{6.15}$$

The highest value of σ_y is obtained at the upper surface $(y = h)$:

$$\sigma_y^{max} = \frac{p}{b} \tag{6.16}$$

Maximum value of bending stress on the beam cross-section, given by equation (6.3), is expressed as

$$\sigma_x^{max} = \pm\frac{(p.L.L/2).h}{b.(2h)^3/12} == \pm\frac{3}{4}.\frac{p}{b}.\left(\frac{L}{h}\right)^2 \tag{6.17}$$

Ratio between the maximum transverse normal stress (equation 6.16) and the maximum bending stress (equation 6.17) is given by the following:

$$\frac{\sigma_y^{max}}{\sigma_x^{max}} = \frac{4}{3}.\left(\frac{h}{L}\right)^2 \tag{6.18}$$

For typical slender beams with a proportion of $L > 20h$, equation (6.18) indicates a very small ratio between transverse normal stress and bending stress. It is, thus, customary to assume $\sigma_y \approx 0$ in slender beams subjected to transverse loading.

6.4   TORSIONAL RESPONSE OF A BEAM

Deformation response of a prismatic circular member under torsional load is generally assumed to possess three characteristics: (i) plane sections, perpendicular to the longitudinal axis of the member, remain plane after the application of torque; (ii) shearing strain varies linearly from zero at the center to maximum on the outer surface; and (iii) material is homogeneous and obeys Hooke's law ($\tau = G.\gamma$). Based on the second and third assumptions, the torsional stress, at a distance r from the center point of the circular section in Figure 6.6, can be expressed as follows:

$$\tau = \frac{r}{\rho}.\tau_{max} \tag{6.19}$$

where ρ is the radius of the circular section, and τ_{max} is the maximum torsional stress on the outer surface. Considering the equilibrium between applied torque T and the resultant of torsional stresses τ, we obtain:

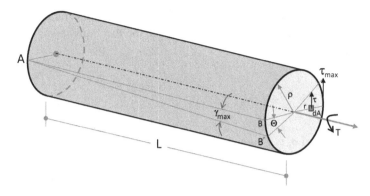

FIGURE 6.6   Torsional stress and angular rotation in a prismatic circular section member.

$$T = \int r.(\tau.dA) \qquad (6.20)$$

where dA is a small area at a distance r from the center of the circular cross-sectional area. Combining equations (6.19) and (6.20) gives:

$$T = \frac{\tau_{max}}{\rho}.\int r^2.dA \qquad (6.21)$$

The integral on the right-hand side represents a property of member cross-section – commonly known as the polar moment of inertia. Substituting J for the integral expression, and after rearranging the terms, the following well-known expression is obtained for torsional stress in a circular section prismatic member:

$$\tau_{max} = \frac{T\rho}{J} \qquad (6.22)$$

Using Hooke's law, the maximum shear strain can be obtained from equation (6.22):

$$\gamma_{max} = \frac{\tau_{max}}{G} = \frac{T\rho}{JG} \qquad (6.23)$$

Relating the shear strain with the angle of twist Θ (Figure 6.6):

$$\gamma_{max} = \Theta.\frac{\rho}{L} \qquad (6.24)$$

Combining equations (6.23) and (6.24), we obtain:

$$\frac{T}{\Theta} = \frac{JG}{L} \qquad (6.25)$$

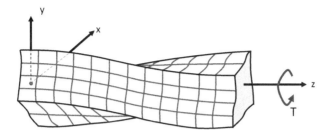

FIGURE 6.7 Torsional deformation of a rectangular beam section.

The term JG is commonly referred to as the torsional rigidity of a member. Relationships (6.19)–(6.25) are valid for circular section members (both solid and hollow). However, plane sections of noncircular members, e.g. the rectangular section in Figure 6.7, do not remain plane during torsional deformation (Timoshenko and Goodier 1982). Equation (6.25), defining the relationship between applied torque and angular twist response, can be generally used for noncircular section members, after taking account of warpage effects in the definition of section property J. Stress function-based semi-inverse method, used by Saint Venant (Timoshenko and Goodier 1982), provides an effective solution to this problem by assuming that the general torsional deformation is a superposition of section rotation and section warpage effects. Considering the coordinate origin at the left end of beam in Figure 6.7, rotation (θ) of cross-section at a distance z from coordinate origin can be related to the deformations in x–y plane as follows:

$$u = -y.\theta \qquad\qquad v = x.\theta \qquad\qquad (6.26)$$

Warpage-induced deformation of the member section is expressed by a function ψ:

$$w = \theta.\psi\left(x,y\right) \qquad\qquad (6.27)$$

Displacement field equations (6.26 and 6.27) lead to the following expressions for strains in the beam:

$$\varepsilon_x = \varepsilon_y = \varepsilon_z = \gamma_{xy} = 0$$

$$\gamma_{xz} = \frac{\partial w}{\partial x} + \frac{\partial u}{\partial z} = \theta.\left[\frac{\partial \psi}{\partial x} - y\right] \qquad\qquad (6.28)$$

$$\gamma_{yz} = \frac{\partial w}{\partial y} + \frac{\partial v}{\partial z} = \theta.\left[\frac{\partial \psi}{\partial y} + x\right]$$

Assumptions for warpage-induced displacement field, thus, lead to no distortion in x–y plane representing member cross-section. Using Hooke's law, stress components corresponding to the strain values of equations (6.28) are obtained as follows:

$$\sigma_x = \sigma_y = \sigma_z = \tau_{xy} = 0$$

$$\tau_{xz} = G\theta.\left[\frac{\partial \psi}{\partial x} - y\right] \qquad (6.29)$$

$$\tau_{yz} = G\theta.\left[\frac{\partial \psi}{\partial y} + x\right]$$

Equations (6.29) represent a pure shear problem defined by τ_{xz} and τ_{yz}. Substituting the expressions from equations (6.29) into the stress equilibrium equations (1.10), and ignoring the body force terms ($F_x = F_y = F_z = 0$):

$$\frac{\partial \tau_{xz}}{\partial z} = 0 \qquad \frac{\partial \tau_{yz}}{\partial z} = 0 \qquad \frac{\partial \tau_{xz}}{\partial x} + \frac{\partial \tau_{yz}}{\partial y} = 0\,s \qquad (6.30)$$

The first two conditions in equation (6.30) are satisfied since τ_{xz} and τ_{yz} are independent of z (equations 6.29). The third condition of equation (6.30) can be satisfied by defining the stress components τ_{xz} and τ_{yz} in terms of a yet-to-be-determined stress function ϕ:

$$\tau_{xz} = \frac{\partial \varphi}{\partial y} \qquad \tau_{yz} = -\frac{\partial \varphi}{\partial x} \qquad (6.31)$$

Substituting the expressions from equations (6.31) into (6.29), we obtain:

$$\frac{\partial \varphi}{\partial y} = G\theta.\left[\frac{\partial \psi}{\partial x} - y\right]$$

$$\frac{\partial \varphi}{\partial x} = -G\theta.\left[\frac{\partial \psi}{\partial y} + x\right] \qquad (6.32)$$

Differentiating the first expression of equation (6.32) with respect to y, the second with respect to x, and adding the 2nd to the first, the following differential equation is obtained (where $F = -2G\theta$):

$$\frac{\partial^2 \varphi}{\partial x^2} + \frac{\partial^2 \varphi}{\partial y^2} = F \qquad (6.33)$$

Stress function ϕ assumed for a given torsional problem must satisfy the compatibility condition (equation 6.33) as well as the stress boundary conditions (equations 1.14). The first and second expressions in equations (1.14) are readily satisfied by the stress components (equations 6.29). For zero applied boundary stress in z-direction ($p_z = 0$), the third expression of equations (1.14) reduces to:

$$\tau_{xz}.l + \tau_{yz}.m = 0 \qquad (6.34)$$

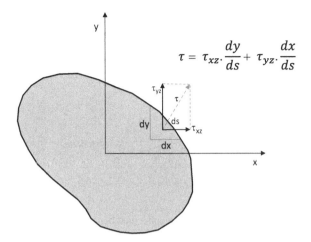

FIGURE 6.8   Boundary condition for torsional stress distribution on a member section.

Using the stress component definitions from equations (6.31), and the direction cosine definitions at a point on the boundary, $\ell = dy/ds$ and $m = -dx/ds$ (Figure 6.8), equation (6.34) is re-written as

$$\frac{\partial \varphi}{\partial y} \cdot \frac{dy}{ds} + \frac{\partial \varphi}{\partial x} \cdot \frac{dx}{ds} = \frac{d\varphi}{ds} = 0 \tag{6.35}$$

Equation (6.35) shows that the derivative of stress function ϕ on the boundary is zero – implying that the torsional stress must follow the tangential direction on the boundary. Now considering the overall equilibrium between applied torque T and the torsional stresses on a member cross-section, we have:

$$T = \iint \left(x.\tau_{yz} - y.\tau_{xz} \right) dx.dy = -\iint \left(x.\frac{\partial \varphi}{\partial x} + y.\frac{\partial \varphi}{\partial y} \right) dx.dy \tag{6.36}$$

Integrating equation (6.36) by parts, and imposing the condition $\phi = 0$ at the boundary (Timoshenko and Goodier 1982), gives:

$$T = 2\iint \varphi.dx.dy \tag{6.37}$$

Equation (6.37) implies that the magnitude of torque T is equal to twice the volume under stress function ϕ. Analytical solution of equations (6.33), (6.35), and (6.36) is tedious. An alternative technique is to rely on the similarity between the torsion problem and membrane deflection problem (Figure 6.9), and to use the membrane deflection solution as a surrogate for the torsional response. Under the normal internal pressure p, the responses of membrane in Figure 6.9 are represented by deflection z and a boundary traction of S per unit length. Considering the membrane deflection in the x–z plane, slopes at the two ends of a differential element are described by equation (6.38):

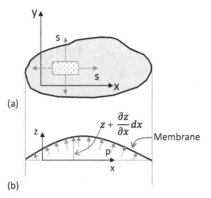

FIGURE 6.9 (a) Cross-sectional area of a solid member in the x–y plane with an inflated membrane over the same area and (b) x–z section view of the membrane under internal pressure p.

$$\beta = \frac{\partial z}{\partial x}, \qquad \beta + d\beta = \frac{\partial z}{\partial x} + \frac{\partial^2 z}{\partial x^2}.dx \qquad (6.38)$$

Similarly, considering a section of inflated membrane in the y–z plane gives:

$$\alpha = \frac{\partial z}{\partial y}, \qquad \alpha + d\alpha = \frac{\partial z}{\partial y} + \frac{\partial^2 z}{\partial y^2}.dy \qquad (6.39)$$

Using the equations (6.38 and 6.39) for membrane slope, the equilibrium state in normal direction of differential membrane element can be written as

$$-\left(S.dy\right).\frac{\partial z}{\partial x} + S.dy\left(\frac{\partial z}{\partial x} + \frac{\partial^2 z}{\partial x^2}.dx\right) - \left(S.dx\right).\frac{\partial z}{\partial y}$$
$$+\left(S.dx\right)\left(\frac{\partial z}{\partial y} + \frac{\partial^2 z}{\partial y^2}.dy\right) + p.dx.dy = 0 \qquad (6.40)$$

Upon simplification of equation (6.40), membrane deflection equation is obtained as follows:

$$\frac{\partial^2 z}{\partial x^2} + \frac{\partial^2 z}{\partial y^2} = -\frac{p}{S} \qquad (6.41)$$

Membrane deflection equation (6.41) is similar to the compatibility equation for torsional problem (equation 6.33), with z analogous to ϕ, $1/S$ to G, and p to 2θ. Solution of membrane deflection problem (experimental or analytical) is commonly used to determine the solution for torsional problem. A very effective use of the membrane analogy is in the torsion problem solution of thin-walled members

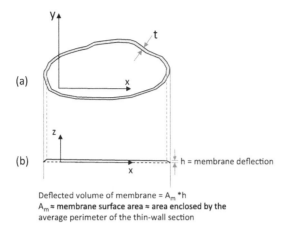

(a)

(b)   h = membrane deflection

Deflected volume of membrane = A_m *h
A_m ≈ membrane surface area ≈ area enclosed by the
average perimeter of the thin-wall section

FIGURE 6.10 (a) Thin-wall closed section of a member subjected to torsion and (b) membrane analogy model of thin-wall section.

(Figure 6.10). Assuming uniform wall thickness for the arbitrary section geometry in the figure, average torsional stress flowing through the wall is given by the average slope of membrane over the thickness of thin-walled member:

$$\tau = membrane\ slope, \left(\frac{\partial z}{\partial x}\right) = \frac{h}{t} \qquad (6.42)$$

Membrane deflection, h, in equation (6.42) is defined by

$$h = \frac{deflected\ volume\ of\ membrane}{membrane\ surface\ area} = \frac{T/2}{A_m} \qquad (6.43)$$

where T is the magnitude of applied torsion equal to twice the volume of deflected membrane. Combining equations (6.42) and (6.43), average torsional stress flowing through the thickness of thin-walled closed section member in Figure 6.10 is given by the following equation (6.44):

$$\tau = \frac{T}{2A_m t} \qquad (6.44)$$

Equation (6.44) presents an elegant solution for the torsional stress in a member of arbitrary cross-section. The powerfulness of membrane analogy becomes more evident when a thin-walled open section, shown in Figure 6.11, is considered. Ignoring the curvature in the y-direction, the membrane deflection equation (6.41) can be re-written in the following form for narrow member section:

$$\frac{\partial^2 z}{\partial x^2} = -\frac{p}{S} \qquad (6.45)$$

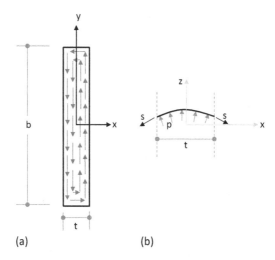

FIGURE 6.11 (a) Torsional stresses in a thin-wall member and (b) membrane analogy for the torsional response.

Integrating equation (6.45) twice, and inserting the boundary conditions of $dz/dx = 0$ at $x = 0$ and $z = 0$ at $x = t/2$, membrane deflection equation is obtained as follows:

$$z = \frac{1}{2} \cdot \frac{P}{S} \cdot \left[\left(\frac{t}{2} \right)^2 - x^2 \right] \tag{6.46}$$

Volume under the membrane surface is given by

$$V = \iint z.dx.dy = \frac{1}{12} \cdot \frac{P}{S} . bt^3 \tag{6.47}$$

Replacing the membrane analogous terms in equation (6.47), torsional resistance of member (equation 6.37) can be obtained as

$$T = 2.V = \frac{1}{3} . bt^3 . G\theta \tag{6.48}$$

Writing the total angle of twist over the member length as, $\Theta = \theta * L$, equation (6.48) can be re-written in the following-form:

$$\frac{T}{\Theta} = \left[\frac{1}{3} . bt^3 \right] \frac{G}{L} = \frac{J_e . G}{L} \tag{6.49}$$

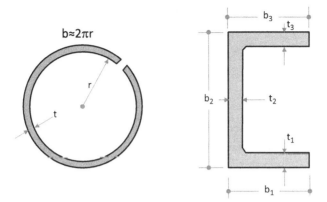

FIGURE 6.12 Thin-walled open section members.

Comparing equation (6.49) with (6.25), the term J_e represents the effective polar moment of inertia of a narrow rectangular section that is analogous to the polar moment inertia of a circular cross-section. The value of J_e for thin-walled general open section members (Figure 6.12) can be obtained by extending the definition from equation (6.49) to the following general form:

$$J_e = \sum \frac{1}{3} bt^3 \tag{6.50}$$

Angular twist of member under torsion can be obtained from equation (6.49) by using the effective polar moment of inertia definition from equation (6.50). Average torsional stress flowing through thin-wall section is given by equation (6.44) where effective area is given by: Σbt. For the open circular arc section of Figure 6.12(a), effective polar moment of inertia is $J_e = (2/3)\pi r t^3$. Using the integral definition in equation (6.21), J for a closed circular section is: $2\pi r^3 t$. Ratio between these two expressions is $3.(r/t)^2$. For a thin-wall tube section, with $r/t = 20$, torsional rigidity of the closed section turns out to be 1200 times the value of an open section. The expression for J_e in equation (6.50), derived based on the membrane analogy of Figure 6.11, is applicable to thin narrow sections. Analysis method can be extended to consider bi-directional curvature of a membrane deflection simulating the torsional response of a general rectangular section (Timoshenko and Goodier 1982). Figure 6.13 shows that the expression for polar moment of inertia of a general rectangular section approaches that of equation (6.50) as the proportion of rectangular section approaches that of a thin narrow section. Finite element calculation of stiffness properties requires the evaluation of member section property J_e. Membrane analogy-based definition of J_e, explained in the above, is directly used to define the torsional stiffness of a member of arbitrary cross-section by using equation (6.49).

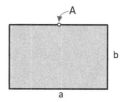

Polar moment of inertia of solid rectangular section: $J = \alpha \cdot ab^3$

b/a	1.0	0.667	0.5	0.1	→0
α	0.141	0.196	0.229	0.312	0.333

FIGURE 6.13   Polar moment of inertia of a solid rectangular section.

6.5   BEAM RESPONSE TO COMBINED LOAD EFFECTS

Beam response to individual actions of bending, transverse shear, and torsion has been described in Sections 6.1–6.4. Stress response of a beam under axial load is similar to that of a bar or truss member – as described by equation (2.36). Resistance mechanisms against axial and bending modes of deformation are assumed to be uncoupled (independent of each other). Normal stresses acting on a beam cross-section, calculated separately for axial and bending load effects, are superposed to get the resultant stress values. Figure 6.14 shows an eccentric normal load P_1 acting on a beam cross-section. Effects of this force on beam axis can be expressed by the following three force components:

$$F_x = P_1 \qquad M_y = -P_1.z_1 \qquad M_z = -P_1.y_1 \qquad (6.51)$$

Normal stresses caused by bending moments M_y and M_z can be calculated separately by using equation (6.3), and be combined with that caused by axial force F_x to get the

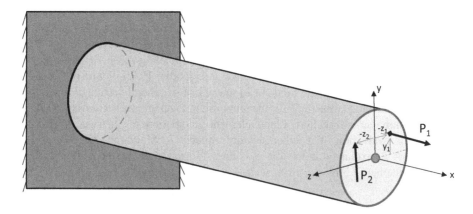

FIGURE 6.14   Eccentric normal load acting on a beam section.

resultant normal stress at any point on the beam cross-section. Effects of transverse load P_2 acting on beam cross-section of Figure 6.14 can be expressed by the following load components:

$$V_y = P_2 \qquad\qquad T_x = -P_2.z_2 \qquad\qquad M_z = -P_2.x \qquad (6.52)$$

Bending stress acting on a beam cross-section, at a distance x from the loading plane of cantilever beam, can be obtained from equation (6.3). Shear stresses caused by transverse shear force V_y can be calculated by using equation (6.8), and those caused by torsion T_x can be calculated by using equations (6.19) and (6.22). Torsion value T_x in equation (6.52) is calculated by multiplying the transverse load P_2 with z_2 – the offset distance of load action from the vertical axis of symmetry y. Effective torsion calculation, for member section without a vertical axis of symmetry, requires the calculation of "shear center" of a beam section (Popov 1978). Figure 6.15 shows a transverse load P acting through the centroid of a non-symmetric beam section, and the associated bending moment M_z acting on the section about z-axis. Bending stress (σ_x) and the internal shear stress (τ_{zx}) on a segment of the beam flange are shown on a separate free-body diagram. The flow of complementary shear stress (τ_{xz}) through beam flange, as well as the flow of stress (τ_{xy}) through beam web, is shown on the beam cross-section. As evident from the directions of stress flow through the C-section beam, shear stresses produce a twisting effect about the member axis passing through the centroid. In order to prevent twisting effect on the beam, the transverse force P should be applied through the point S so that net torsional effect on the beam section will be "zero". Point S is the "shear center" of beam section – a point away from the centroid of non-symmetric section such that transverse force applied through that point will produce zero twist about the beam axis. Considering the equilibrium between external and internal effects, the distance to shear center is given by the following equation (6.53) where parameters refer to the dimensions of C-section beam in Figure 6.15:

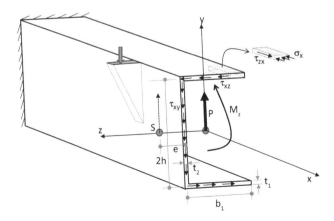

FIGURE 6.15 Transverse load and bending moment on a non-symmetric beam section and the resulting shear flow through the beam section.

$$e = \frac{3}{2} \cdot \frac{b_1^2 t_1}{\left(h t_2 + 3 b_1 t_1 \right)} \tag{6.53}$$

Shear center can, thus, be located by using the dimensions of beam cross-section. In finite element models, beam element properties generally refer to the centroid of section. Effects of surface pressure loads are represented by equivalent loads, moment and torsion referring to the centroidal axis of member. The concept of shear center is not directly used in finite element formulations. However, the concept is reviewed here because of its importance in beam design to minimize the potential torsional effects produced by transverse load applications.

6.6 ELASTIC BENDING DEFLECTION OF BEAMS

Combining equations (6.1) and (6.3) with the generalized Hooke's law for stress–strain relations (equations 2.18 and 2.24), the following expressions are obtained for the strain responses of the beam under pure bending load:

$$\varepsilon_x = -\frac{M.y}{EI_z}, \quad \varepsilon_y = \varepsilon_z = \vartheta \cdot \frac{M.y}{EI_z}, \quad \gamma_{xy} = \gamma_{xz} = \gamma_{yz} = 0 \tag{6.54}$$

where ϑ is Poisson's ratio and EI_z is the flexural rigidity. Overall axial deformation of the beam under bending load is assumed zero. From the beam deflection profile of Figure 6.16, relative rotation of beam section can be related to normal strain as follows:

$$d\Lambda = -\frac{\varepsilon_x dx}{y} \tag{6.55}$$

Combining equation (6.55) with the first part of equation (6.54), the following equation is obtained between beam axis rotation and applied moment M:

$$\frac{d\Lambda}{dx} = \frac{M}{EI_z} \tag{6.56}$$

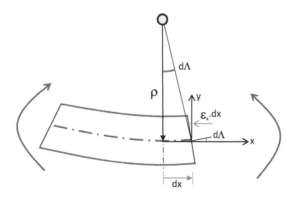

FIGURE 6.16   Pure bending deformation of a beam element.

Slope of the deflected beam axis at any point x can be expressed as derivative of the transverse deflection v as follows:

$$\Lambda = \frac{dv}{dx} \tag{6.57}$$

Combining equations (6.56) and (6.57), the following well-known expression is obtained between the beam deflection v and applied bending moment M, where symbol I is used instead of I_z for the sake of simplicity:

$$\frac{M}{EI} = \frac{d^2v}{dx^2} \tag{6.58}$$

Double integration of equation (6.58) with respect to coordinate variable x gives the following general expression for beam deflection under pure bending load effect:

$$EI.v(x) = \int \left[\int M.dx \right].dx + c_1.x + c_2 \tag{6.59}$$

Here, c_1 and c_2 are integration constants. Equation (6.59) can be used to determine deflection response from known moment distribution function and related boundary conditions. For the cantilever beam example of Figure 3.5, moment M in Equation (6.59) can be replaced by $P.x$, where P is the applied load at the left end of cantilever beam. Equation (6.59) for that specific example, thus, takes the following form:

$$EI.v(x) = \frac{P.x^3}{6} + c_1.x + c_2 \tag{6.60}$$

Boundary conditions of zero displacement and zero rotation at the right-end support ($x = L$) of beam in Figure 3.5 lead to the following expressions for the constants c_1 and c_2:

$$EI.\frac{\partial v}{\partial x} = \left[\frac{P.L^2}{2} + c_1 \right] = 0 \rightarrow c_1 = -\frac{P.L^2}{2} \tag{6.61}$$

$$EI.v = \left[\frac{P.L^3}{6} + c_1.L + c_2 \right] = 0 \rightarrow c_2 = \frac{P.L^3}{3} \tag{6.62}$$

Substitution of the expressions for constants c_1 and c_2, from equations (6.61) and (6.62) into equation (6.60), gives the following expression for deflection of axis ($y = 0$) of the cantilever beam in Figure 3.5:

$$v(x) = \frac{P.x^3}{6EI} - \frac{P.L^2.x}{2EI} + \frac{P.L^3}{3EI} \tag{6.63}$$

Deflection equation (6.63) for beam axis, derived from bending deformation only, matches exactly with first three terms of equation (3.56) that has been derived from plane-stress functions for cantilever beam problem. The last term in equation (3.56), representing the beam axis deflection due to shear deformation of the material, is obviously not captured by the bending deflection equation (6.63). In the cantilever beam example of Figure 3.5, shear deformation in the material will accumulate over the length of cantilever, thus, resulting into a higher deflection value compared to that given by equation (6.63). Deflection at the left end ($x = 0$) of cantilever beam axis, considering both bending and shear deformation in the material, is obtained from equation (3.56):

$$v_{(x=0, y=0)} = \frac{PL^3}{3EI} + \frac{Ph^2L}{2GI} \tag{6.64}$$

Ratio between the two terms on the right-hand side of equation (6.64), second term representing the shear deformation and the first one representing the bending deformation in material, leads to the following metric for the measurement of a beam's slenderness:

$$\frac{Ph^2L/2GI}{PL^3/3EI} = \frac{3}{2} \cdot \frac{E}{G} \left(\frac{h}{L}\right)^2 = \frac{3}{4}(1+\vartheta) \cdot \left(\frac{2h}{L}\right)^2 \approx \left(\frac{2h}{L}\right)^2 \tag{6.65}$$

where $2h$ is the depth of rectangular beam section and L is the span of beam in Figure 3.5. Evidently, for long slender beams, $L > 10*(2h)$, equation (6.65) indicates a very low contribution of shear deformation (<1%) to the overall beam deflection. Euler–Bernoulli beam theory, considering bending deformation only, provides an accurate estimate for the overall deflection for long slender beams. However, for short-span deep beams, ratio in equation (6.65) is not negligible – meaning that material shear deformation needs to be included in beam deflection calculations. Idea of shear correction factor application to the Euler–Bernoulli's beam deflection estimate is attributed to early 20th-century work by Timoshenko and Ehrenfest – commonly known as "Timoshenko Beam Theory" (Wikipedia.org 2020). Stiffness properties of beam elements, with and without material shear deformation effects, are discussed in Sections 6.8 and 6.9.

6.7   STRESS ANALYSIS OF CURVED BEAMS

Analysis techniques described thus far assumed straight profiles of beam axis. However, the basic assumptions of beam response mechanisms can be extended to consider initially curved profile of beam with the following strict assumptions:

1. Beam axis follows single-curvature profile both before and after bending
2. Beam cross-section at any point possesses an axis of symmetry pointing towards the center of curvature of the beam axis
3. Beam cross-sections originally normal to the beam axis remain so after bending. Deformation due to transverse shear is not considered in curved beam analysis.

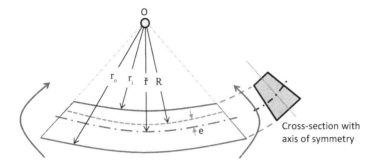

FIGURE 6.17   Pure bending deformation of a curved beam element.

Figure 6.17 shows an initially curved beam profile with radius of centroidal axis $\breve{r}$, radius of outer fiber r_o, and that of inner fiber r_i. Different initial lengths of inner and outer fibers of curved beam profile will produce different strain amplitudes at the extreme fibers – implying that neutral axis will not pass through the centroid of beam section. Distance of the neutral axis from center of curvature is given by the following equation (Ugural and Fenster 2012):

$$R = \frac{A}{\int \frac{dA}{r}} \tag{6.66}$$

where dA is an infinitesimal area on beam cross-section at a distance r from the center of curvature O, and A is the area of beam cross-section. The distance between centroidal axis and neutral axis is equal to:

$$e = \breve{r} - R \tag{6.67}$$

The tangential stress at distance r from the center of curvature is given by the following equation:

$$\sigma_\theta = -\frac{M(R-r)}{Aer} \tag{6.68}$$

Curved beam formula (equation 6.68) for the calculation of bending stress is commonly known as Winker's formula (developed by E Winkler 1835–1888). Radius of neutral axis R (equation 6.66) for standard beam section shapes is often available in reference literature (Ugural and Fenster 2012). Equation (6.68) provides a simple but reasonably accurate method for the calculation of bending stress in a curved beam. Application of the method is, however, limited to simple example cases meeting the restrictive assumptions listed earlier in this section. Review of curved beam analysis is included here for the sake of completeness of theory of elasticity approaches for beam response analysis. General structural geometries subjected to combined axial, transverse and bending load effects can be accurately modelled and analyzed by using 3D finite elements described in Section 5.5.

6.8   STIFFNESS PROPERTIES OF PRISMATIC EULER–BERNOULLI BEAM ELEMENTS

A "prismatic" member is one in which cross-section properties (shape, dimensions, material, etc.) do not change along the member length. For analysis efficiency, a prismatic beam can be represented by a skeletal (line) element (Figure 6.18). It is assumed that the cross-sectional dimensions are smaller than the dimension along member length. A general beam element in 3D space will have 12 degrees of freedom (6 at each end). Figure 6.18(b) shows 6 DOF at each end – aligned with global coordinate directions x–y–z. For member resistance calculations, deformations in beam local coordinate directions (Figure 6.18(c)) need to be derived from the global displacement DOF. Coordinate transformation relationships between the global and local DOF can be expressed as follows where (l,m,n) are direction cosines defined in Figure 6.18:

$$
\begin{Bmatrix} u_1 \\ u_2 \\ u_3 \\ u_4 \\ u_5 \\ u_6 \\ u_7 \\ u_8 \\ u_9 \\ u_{10} \\ u_{11} \\ u_{12} \end{Bmatrix} =
\begin{bmatrix}
l_1 & m_1 & n_1 & 0 & 0 & 0 & 0 & 0 & 0 & 0 & 0 & 0 \\
l_2 & m_2 & n_2 & 0 & 0 & 0 & 0 & 0 & 0 & 0 & 0 & 0 \\
l_3 & m_3 & n_3 & 0 & 0 & 0 & 0 & 0 & 0 & 0 & 0 & 0 \\
0 & 0 & 0 & l_1 & m_1 & n_1 & 0 & 0 & 0 & 0 & 0 & 0 \\
0 & 0 & 0 & l_2 & m_2 & n_2 & 0 & 0 & 0 & 0 & 0 & 0 \\
0 & 0 & 0 & l_3 & m_3 & n_3 & 0 & 0 & 0 & 0 & 0 & 0 \\
0 & 0 & 0 & 0 & 0 & 0 & l_1 & m_1 & n_1 & 0 & 0 & 0 \\
0 & 0 & 0 & 0 & 0 & 0 & l_2 & m_2 & n_2 & 0 & 0 & 0 \\
0 & 0 & 0 & 0 & 0 & 0 & l_3 & m_3 & n_3 & 0 & 0 & 0 \\
0 & 0 & 0 & 0 & 0 & 0 & 0 & 0 & 0 & l_1 & m_1 & n_1 \\
0 & 0 & 0 & 0 & 0 & 0 & 0 & 0 & 0 & l_2 & m_2 & n_2 \\
0 & 0 & 0 & 0 & 0 & 0 & 0 & 0 & 0 & l_3 & m_3 & n_3
\end{bmatrix} *
\begin{Bmatrix} U_1 \\ U_2 \\ U_3 \\ U_4 \\ U_5 \\ U_6 \\ U_7 \\ U_8 \\ U_9 \\ U_{10} \\ U_{11} \\ U_{12} \end{Bmatrix}
\tag{6.69}
$$

Writing the vectors and matrix of equation (6.69) in compact form gives:

$$\{u_i\} = [T]\{U_i\} \tag{6.70}$$

Matrix [T] is orthogonal, i.e. its inverse is equal to its transpose (Cook et al. 1989). Upon calculation of the element stiffness matrix [k] in local coordinate system, matrix [T] can be used to transform it to global coordinate system:

$$[K] = [T]^T [k][T] \tag{6.71}$$

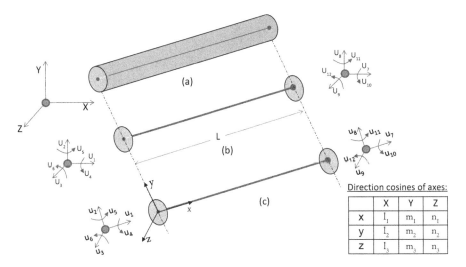

FIGURE 6.18 (a) A general beam element in *X–Y–Z* global reference system; (b) skeletal line element representation of the beam with DOF in global directions; and (c) DOF in element local reference system (*x, y, z*) and the direction cosines of axes.

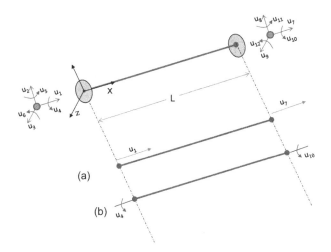

FIGURE 6.19 Local deformation of a beam element: (a) DOF associated with axial deformation mode and (b) DOF associated with torsional deformation mode.

Local stiffness matrix [*k*] is assembled from the superposition of beam resistances for axial, shear, bending, and torsional deformation modes that have been discussed earlier. Axial deformation of a beam element can be calculated from DOF u_1 and u_7 (Figure 6.19(a)) by using linear interpolation functions shown in Figure 2.11. Resistance to that axial deformation is defined by the member cross-sectional area (A) and elastic modulus of material (E). Stiffness matrix for beam resistance to axial deformation is identical to that of a truss or cable element defined by equation (2.57):

$$[k]_{axial} = \begin{bmatrix} \dfrac{AE}{L} & \dfrac{-AE}{L} \\[2mm] \dfrac{-AE}{L} & \dfrac{AE}{L} \end{bmatrix} \tag{6.72}$$

where L is the element length. Linear interpolation functions of Figure 2.11 can also be used to calculate the torsional deformation of beam caused by angular twists u_4 and u_{10} at the two ends (Figure 6.19(b)). Following the analytical steps, described in Section 2.8 for axial deformation mode, the stiffness matrix of beam element associated with torsional resistance mechanism can be defined by extending the definition from equation (6.49):

$$[k]_{torsional} = \begin{bmatrix} \dfrac{JG}{L} & \dfrac{-JG}{L} \\[2mm] \dfrac{-JG}{L} & \dfrac{JG}{L} \end{bmatrix} \tag{6.73}$$

where G is the shear modulus of material and J is the effective polar moment of inertia of beam cross-section (suffix e is dropped for the sake of simplicity). Shear modulus for homogeneous isotropic material is calculated from Young's modulus and Poisson's ratio by using equation (2.23). Polar moment of inertia of beam section is calculated by using the case-specific formulations described in Section 6.4.

Bending deformation of a 3D beam under transverse loading can occur in two coordinate planes – (x,y) and (x,z). Figure 6.20(a) shows the general deformation profile in (x,y) plane. The transverse displacement (v) at any point on the beam axis can be expressed in terms of the nodal DOF (u_2, u_6) and (u_8, u_{12}) as follows:

$$v = H_2.u_2 + H_6.u_6 + H_8.u_8 + H_{12}.u_{12} \tag{6.74}$$

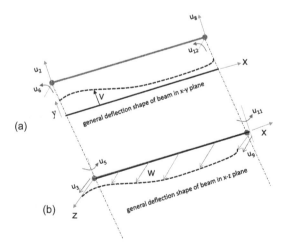

FIGURE 6.20   (a) DOF associated with transverse deformation of beam in x–y plane and (b) those associated with deformation in x–z plane.

where H_2, H_6, H_8, and H_{12} are the interpolation functions (also known as shape functions) that can be described with third-degree polynomials (for four nodal displacement DOF – u_2, u_6, u_8, and u_{12}). Beam deflection shape functions, with appropriate boundary conditions as shown in Figure 6.21, are given by the following cubic functions:

$$
\begin{aligned}
H_2 &= 1 - \frac{3x^2}{L^2} + \frac{2x^3}{L^3} \\[2mm]
H_6 &= x - \frac{2x^2}{L} + \frac{x^3}{L^2} \\[2mm]
H_8 &= \frac{3x^2}{L^2} - \frac{2x^3}{L^3} \\[2mm]
H_{12} &= -\frac{x^2}{L} + \frac{x^3}{L^2}
\end{aligned}
\tag{6.75}
$$

Using the Euler–Bernoulli's beam deformation theory (discussed in Section 6.6), normal deformation at any point on a beam cross-section can be defined by the following equation:

$$
u = -y.\frac{\partial v}{\partial x}
\tag{6.76}
$$

Corresponding normal strain at the same point is given by

$$
\varepsilon = \frac{\partial u}{\partial x} = -y.\frac{\partial^2 v}{\partial x^2}
\tag{6.77}
$$

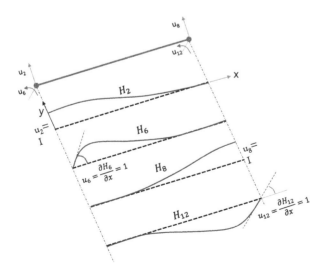

FIGURE 6.21 Shape functions for transverse deformation of beam in x–y plane.

Combining equations (6.74), (6.75), and (6.77), normal strain on beam cross-section can be expressed in terms of the nodal DOF (u_2, u_6, u_8, and u_{12}) as follows:

$$\varepsilon_x = \frac{-y}{L^3}\left[(12x-6L) \quad L(6x-4L) \quad -(12x-6L) \quad L(6x-2L)\right] * \begin{Bmatrix} u_2 \\ u_6 \\ u_8 \\ u_{12} \end{Bmatrix} \quad (6.78)$$

Re-writing equation (6.78) in the familiar finite element terms:

$$\varepsilon_x = \left[B\right] * \begin{Bmatrix} u_2 \\ u_6 \\ u_8 \\ u_{12} \end{Bmatrix} \quad (6.79)$$

where strain–displacement relationship matrix [B] is a function of x and y:

$$\left[B\right] = \frac{-y}{L^3}\left[(12x-6L) \quad L(6x-4L) \quad -(12x-6L) \quad L(6x-2L)\right] \quad (6.80)$$

Stress–strain relationship for bending deformation mode of beam is given by the Hooke's law:

$$\sigma_x = E.\varepsilon_x \quad (6.81)$$

The familiar stress–strain relationship matrix for bending deformation, thus, takes the following form:

$$\left[C\right] = \left[E\right] \quad (6.82)$$

Substituting expressions (6.80) and (6.82) into equation (2.44), and inserting differential volume $dV = A.dx$ (where A is the cross-sectional area of prismatic beam element), stiffness matrix of beam element for bending mode of deformation in (x,y) plane is obtained as follows:

$$\left[k\right]_{xy} = \int \int \left[B\right]^T . \left[C\right].\left[B\right].dV = \begin{bmatrix} \dfrac{12EI_z}{L^3} & \dfrac{6EI_z}{L^2} & \dfrac{-12EI_z}{L^3} & \dfrac{6EI_z}{L^2} \\[2ex] \dfrac{6EI_z}{L^2} & \dfrac{4EI_z}{L} & \dfrac{-6EI_z}{L^2} & \dfrac{2EI_z}{L} \\[2ex] \dfrac{-12EI_z}{L^3} & \dfrac{-6EI_z}{L^2} & \dfrac{12EI_z}{L^3} & \dfrac{-6EI_z}{L^2} \\[2ex] \dfrac{6EI_z}{L^2} & \dfrac{2EI_z}{L} & \dfrac{-6EI_z}{L^2} & \dfrac{4EI_z}{L} \end{bmatrix} \quad (6.83)$$

where I_z is the moment of inertia of beam cross-section defined as

$$I_z = \int y^2 . dA \qquad (6.84)$$

Following a similar approach, stiffness matrix of beam for transverse deformation mode in $(x–z)$ plane (Figure 6.20(b)) can be obtained as follows:

$$\left[k\right]_{xz} = \left[\int \left[B\right]^T.\left[C\right].\left[B\right].dV\right] = \begin{bmatrix} \dfrac{12EI_y}{L^3} & \dfrac{6EI_y}{L^2} & \dfrac{-12EI_y}{L^3} & \dfrac{6EI_y}{L^2} \\[2ex] \dfrac{6EI_y}{L^2} & \dfrac{4EI_y}{L} & \dfrac{-6EI_y}{L^2} & \dfrac{2EI_y}{L} \\[2ex] \dfrac{-12EI_y}{L^3} & \dfrac{-6EI_y}{L^2} & \dfrac{12EI_y}{L^3} & \dfrac{-6EI_y}{L^2} \\[2ex] \dfrac{6EI_y}{L^2} & \dfrac{2EI_y}{L} & \dfrac{-6EI_y}{L^2} & \dfrac{4EI_y}{L} \end{bmatrix} \qquad (6.85)$$

where I_y is the moment of inertia of beam cross-section defined as

$$I_y = \int z^2 . dA \qquad (6.86)$$

Combining the stiffness contributions from four resistance mechanisms, equation (6.72) for axial deformation, equation (6.73) for torsion, equation (6.83) for bending in (x, y) plane, and equation (6.85) for bending in (x,z) plane, overall stiffness matrix of the beam with reference to its 12 local DOF (Figure 6.18(c)) is defined by the following equation (6.87):

$$[k]= \begin{bmatrix}
\frac{AE}{L} & 0 & 0 & 0 & 0 & 0 & -\frac{AE}{L} & 0 & 0 & 0 & 0 & 0 \\
0 & \frac{12EI_z}{L^3} & 0 & 0 & 0 & \frac{6EI_z}{L^2} & 0 & -\frac{12EI_z}{L^3} & 0 & 0 & 0 & \frac{6EI_z}{L^2} \\
0 & 0 & \frac{12EI_y}{L^3} & 0 & \frac{6EI_y}{L^2} & 0 & 0 & 0 & -\frac{12EI_y}{L^3} & 0 & \frac{6EI_y}{L^2} & 0 \\
0 & 0 & 0 & \frac{JG}{L} & 0 & 0 & 0 & 0 & 0 & -\frac{JG}{L} & 0 & 0 \\
0 & 0 & \frac{6EI_y}{L^2} & 0 & \frac{4EI_y}{L} & 0 & 0 & 0 & -\frac{6EI_y}{L^2} & 0 & \frac{2EI_y}{L} & 0 \\
0 & \frac{6EI_z}{L^2} & 0 & 0 & 0 & \frac{4EI_z}{L} & 0 & -\frac{6EI_z}{L^2} & 0 & 0 & 0 & \frac{2EI_z}{L} \\
-\frac{AE}{L} & 0 & 0 & 0 & 0 & 0 & \frac{AE}{L} & 0 & 0 & 0 & 0 & 0 \\
0 & -\frac{12EI_z}{L^3} & 0 & 0 & 0 & -\frac{6EI_z}{L^2} & 0 & \frac{12EI_z}{L^3} & 0 & 0 & 0 & -\frac{6EI_z}{L^2} \\
0 & 0 & -\frac{12EI_y}{L^3} & 0 & -\frac{6EI_y}{L^2} & 0 & 0 & 0 & \frac{12EI_y}{L^3} & 0 & -\frac{6EI_y}{L^2} & 0 \\
0 & 0 & 0 & -\frac{JG}{L} & 0 & 0 & 0 & 0 & 0 & \frac{JG}{L} & 0 & 0 \\
0 & 0 & \frac{6EI_y}{L^2} & 0 & \frac{2EI_y}{L} & 0 & 0 & 0 & -\frac{6EI_y}{L^2} & 0 & \frac{4EI_y}{L} & n_2 \\
0 & \frac{6EI_z}{L^2} & 0 & 0 & 0 & \frac{2EI_z}{L} & 0 & -\frac{6EI_z}{L^2} & 0 & 0 & 0 & \frac{4EI_z}{L}
\end{bmatrix}$$

$$(6.87)$$

Stiffness matrix of prismatic Euler–Bernoulli beam element, with a straight geometric profile, is thus explicitly calculated by using the material and geometric property information in equation (6.87). Equation (6.71) is then used to transform the local stiffness $[k]$, calculated from equation (6.87), to global coordinate reference system (Figure 6.18(b)). Stiffness matrix of each element, calculated by following the above steps, is assembled to the global stiffness matrix of a multi-element model for load–deflection analysis. In the calculation of beam stiffness for transverse load effects, consideration has been given to lateral deformation caused by bending deformation only (equations 6.83 and 6.85). Element formulation based on this assumption is generally referred to as Euler–Bernoulli beam element. As discussed in Section 6.6, this assumption (of ignoring the contribution of shear deformation) provides accurate results for slender beams (equation 6.65).

6.9 STIFFNESS PROPERTIES OF BEAMS INCLUDING SHEAR DEFORMATION

Equation (6.74) has expressed the transverse deformation of internal points of a beam for coupled effects of transverse and rotational responses at the nodes. Using the linear interpolation functions from Figure 2.11, and considering the transverse deformation response in x–y plane for the beam in Figure 6.20(a), internal deformation responses can be calculated by separately interpolating the nodal responses for transverse and rotational DOF:

$$v = \begin{bmatrix} H_2 & H_8 & 0 & 0 \end{bmatrix} \begin{Bmatrix} u_2 \\ u_8 \\ u_6 \\ u_{12} \end{Bmatrix} \tag{6.88}$$

$$\beta = \begin{bmatrix} 0 & 0 & H_6 & H_{12} \end{bmatrix} \begin{Bmatrix} u_2 \\ u_8 \\ u_6 \\ u_{12} \end{Bmatrix} \tag{6.89}$$

where v is transverse deformation at any point inside the beam, and β is the rotation of beam section at that point. Defining the curvature of beam axis as the first derivative of β with respect to coordinate x:

$$\frac{\partial \beta}{\partial x} = \begin{bmatrix} 0 & 0 & \frac{\partial H_6}{\partial x} & \frac{\partial H_{12}}{\partial x} \end{bmatrix} \begin{Bmatrix} u_2 \\ u_8 \\ u_6 \\ u_{12} \end{Bmatrix} = \begin{bmatrix} B_\beta \end{bmatrix} \{u_i\} \tag{6.90}$$

where $\{u_i\}$ is the vector of nodal displacements and rotations, and $[B_\beta]$ is the relationship matrix defining curvature at selected beam internal point. Defining the flexural rigidity of prismatic beam by EI (from equation 6.56), and using the curvature-to-nodal displacement relationship from equation (6.90), stiffness of the beam for bending mode of deformation can be calculated by using the standard definition from equation (2.44):

$$\left[k\right]_{EI} = \int \left[B_\beta\right]^T * \left[EI\right] * \left[B_\beta\right]dx \tag{6.91}$$

A key assumption in Euler–Bernoulli beam theory is that beam section normal to neutral axis remains normal during bending deformation. However, shear deformation in the material, when taken into consideration (Figure 6.22), leads to deviation from that normality condition. Considering that shear deformation of the material contributes partially to the total rotation response of the beam section (Figure 6.22), shear strain can be related to v and β as:

$$\gamma = \frac{\partial v}{\partial x} - \beta \tag{6.92}$$

Using interpolation relationships from equations (6.88) and (6.89):

$$\frac{\partial v}{\partial x} = \begin{bmatrix} \dfrac{\partial H_2}{\partial x} & \dfrac{\partial H_8}{\partial x} & 0 & 0 \end{bmatrix} \begin{Bmatrix} u_2 \\ u_8 \\ u_6 \\ u_{12} \end{Bmatrix} = \begin{bmatrix} B_v \end{bmatrix}\{u_i\} \tag{6.93}$$

$$\beta = \begin{bmatrix} 0 & 0 & H_6 & H_{12} \end{bmatrix} \begin{Bmatrix} u_2 \\ u_8 \\ u_6 \\ u_{12} \end{Bmatrix} = \begin{bmatrix} H_\beta \end{bmatrix}\{u_i\} \tag{6.94}$$

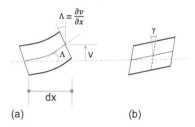

(a)   (b)

FIGURE 6.22   (a) General rotational response of a beam section; (b) rotation due to shear deformation.

Substituting the relationships from equations (6.93) and (6.94) into equation (6.92), relationship between internal shear strain and nodal response variables is defined by the following equation:

$$\gamma = \left[B_v - H_\beta \right] \{ u_i \}$$ (6.95)

Using the standard definition of stiffness matrix from equation (2.44), beam stiffness corresponding to the shear deformation of material (equation 6.95) can be defined by the following equation:

$$\left[k \right]_{GA} = \int \left[B_v - H_\beta \right]^T \left[\xi GA \right] * \left[B_v - H_\beta \right] dx$$ (6.96)

where G is the shear modulus of material, A is the cross-sectional area of beam, and ξ is a correction factor for assuming uniform shear stress distribution over beam cross-section (Figure 6.4(b)), in lieu of more rigorous stress distribution defined by equation (6.8). For a rectangular beam cross-section example, shear strain distribution is given by the parabolic function (equation 6.9). Equating the shear strain energy corresponding to the parabolic shear stress distribution, to that of average shear stress distribution ($\tau_a = V/A$), the shear stiffness correction factor is found to be $\xi = 5/6$ (Bathe 1996). Correction factors for different beam cross-sections were derived by Cowper (1966). Combining equations (6.91) and (6.96), stiffness matrix corresponding to bending and shear deformations in (x,y) plane (Figure 6.20(a)) is given by equation

$$\left[k \right]_{xy} = \left[k \right]_{EI} + \left[k \right]_{GA}$$ (6.97)

Following the approach described above, a similar stiffness matrix of the beam, considering both bending and shear deformations of beam in (x,z) plane, can also be derived. Similar to the description of Euler–Bernoulli beam element presented in Section 6.8, overall stiffness matrix of a beam element in its local reference axis system can be assembled by combing the contributions from bending and shear deformations with that of axial deformation (equation 6.72) and that of torsional deformation equation (6.73):

$$\left[k \right] = \left[k \right]_{AE} + \left[k \right]_{JG} + \left[k \right]_{EI} + \left[k \right]_{GA}$$ (6.98)

The standard coordinate transformation rule (equation 6.71) is then used to transform the local stiffness $[k]$ to global coordinate reference system (Figure 6.18(b)). Shear deformation in beam response was first introduced by Timoshenko and Ehrenfest (Wikipedia.org 2020); but beam formulations considering shear deformation in the material is commonly referred to as "Timoshenko" beam element.

Stiffness matrices in equations (6.91) and (6.96) have been derived based on linear interpolation of translational and rotational responses at nodes, while the stiffness matrix formulation presented in equation (6.87) has been derived based on cubic interpolation of both transverse translation and rotational responses at the nodes. Use

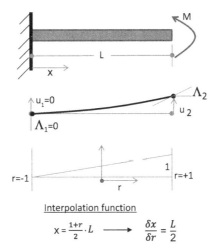

FIGURE 6.23   Linear interpolation of pure bending response of a cantilever beam.

of linear interpolation functions in stiffness calculations has its drawbacks. Considering the cantilever beam in Figure 6.23, subjected to pure bending moment M, the translational and rotational responses at the free end (u_2 and Λ_2) are calculated by using equation (6.58):

$$\Lambda_2 = M/EI * L; \quad u_2 = ML^2/2EI \tag{6.99}$$

Using the linear interpolation function shown in the figure, internal transverse displacement and rotational responses at any point x tun out to be:

$$v = \frac{1+r}{2} u_2 \qquad \qquad \beta = \frac{1+r}{2} \Lambda_2 \tag{6.100}$$

Inserting the expressions from equation (6.100) into equation (6.92), and using the coordination relationship shown in Figure 6.23, internal shear strain in the beam element is given by

$$\gamma = \frac{u_2}{L} - \frac{1+r}{2} \Lambda_2 \tag{6.101}$$

Combining equations (6.99) and (6.101), shear strain inside the beam element of Figure 6.23 is found to be:

$$\gamma = -\frac{ML}{2EI} r \tag{6.102}$$

Per the principle of solid mechanics, shear strain (and stress) in a beam under pure bending load should be "zero". Use of the linear interpolation function in the analysis of cantilever beam in Figure 6.23, however, produces a shear strain inside the beam except at $r = 0$. Non-zero shear strain, introduced by the linear interpolation function, is commonly referred to as parasitic shear or "shear locking" of beam. Element stiffness formulation based on linear interpolation functions (equation 6.96) produces artificially stiff structural response – similar to the phenomenon discussed in Section 4.2 for 2D solids elements. A potential remedy for the shear locking behavior of linear beam elements is to calculate shear strain at the element center ($r = 0$); and assume it to remain constant through the element length. This action enforces "zero" shear condition for a pure bending condition without sacrificing actual shear strain that may appear due to other general loading conditions. Alternative beam element formulations, with shear flexibility contribution, apply a scaling factor to the shear rigidity term (ξGA) in equation (6.96) to limit the shear stiffness value as the beam slenderness increases. Example scaling factor, used in ABAQUS beam element formulation, is given by the following equation (Dassault Systems 2020b):

$$scale\ fcator = \frac{1}{1 + 0.25 * \dfrac{A.L^2}{12I}} \tag{6.103}$$

where L is the beam span, A is the cross-sectional area, and I is the moment of inertia. Yet another approach for considering shear flexibility contribution in beam stiffness formulation is to apply a shear correction factor to the stiffness terms in equations (6.83 and 6.85) that have been derived by using cubic interpolation functions:

$$[k] = \frac{EI}{L^3(1+\psi)} \begin{bmatrix} 12 & 6L & -12 & 6L \\ 6L & (4+\psi)L^2 & -6L & (2-\psi)L^2 \\ -12 & -6L & 12 & -6L \\ 6L & (2-\psi)L^2 & -6L & (4+\psi)L^2 \end{bmatrix} \tag{6.104}$$

The shear correction factor in equation (6.104) is defined by equation (6.105) (Logan 2012):

$$\psi = \frac{12EI}{\xi GAL^2} \tag{6.105}$$

Comparing the terms in equation (6.104) with those in equations (6.83 and 6.85), it is evident that the shear deformation in beam reduces the magnitude of stiffness terms associated with transverse displacement and rotational response modes. Contribution of shear deformation can be neglected for slender beam: $L > 10*(2h)$ in equation (6.65).

6.10   ANALYSIS OF BEAMS AND FRAMES WITH FEA SOFTWARE PACKAGES

Dedicated software packages such SAP2000 (CSIAmerica.com) and STAAD Pro (Bentley.com) are widely used in design and analysis of structural frames.

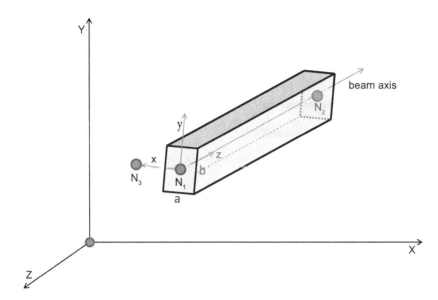

FIGURE 6.24   Local reference system of a beam element in ABAQUS software.

General-purpose finite element analysis packages (e.g. ABAQUS, ANSYS, etc.) also provide equally capable element library and modeling tools for analysis of skeletal structural frames. Figure 6.24 shows a straight beam element with ABAQUS-specific local reference coordinate definitions. Element geometry in three-dimensional space is described by the following syntax for ABAQUS analysis model description:

> *ELEMENT, TYPE=B31, ELSET=*setname*
> *Element_ID_no*, N_1, N_2, N_3

where nodes N_1, N_2, and N_3, defined in global coordinate system (X,Y,Z), also define the local orientation of the beam element – nodes N_1 and N_2 define the element axis direction (z), and node N_3 the local direction x as shown in Figure 6.24. The element type selection "B31" refers to the beam element formulation, described in Section 6.9, that uses linear interpolation functions with material shear deformation included. Shear locking behavior in the element is reduced by scaling the shear rigidity term with scaling factor defined in equation (6.103). Rectangular beam section example, shown in Figure 6.24, is defined in ABAQUS input file by using the following syntax:

> *BEAM SECTION, ELSET= *setname*, MATERIAL= *matname*,
>   SECTION=RECT
> a,b

where dimensions a and b refer to the specific local orientation of rectangular section identified in Figure 6.24. Section properties specific to axial, shear, bending, and torsional response modes are internally calculated by ABAQUS based on the section

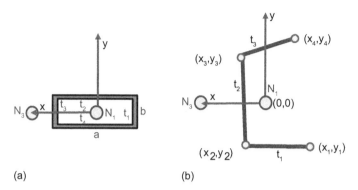

(a)   (b)

FIGURE 6.25   Examples of beam section dimensions in local reference system for ABAQUS beam elements: (a) thin-wall closed box section, and (b) thin-wall open section.

definition provided with the element input data. ABAQUS provides a wide variety of options to define beam section dimensions including thin-walled closed and open sections. Section dimensions for the closed box section example, shown in Figure 6.25(a), are defined by using the following ABAQUS input syntax:

> *BEAM SECTION, ELSET= *setname*, MATERIAL= *matname*, SECTION=BOX
> a,b,t_1,t_2,t_3,t_4

where a and b are side dimensions of rectangular box section and t_1,..t_4 are the thickness values of side walls as identified in Figure 6.25(a). Following data input syntax describes the dimensions of three-sided arbitrary thin-walled section example shown in Figure 6.25(b):

> *BEAM SECTION, ELSET= *setname*, MATERIAL= *matname*, SECTION=ARBITRARY
> 3, x_1, y_1, x_2, y_2, t_1
> x_3, y_3, t_2
> x_4, y_4, t_3

where the number of segments in the arbitrary section is identified at the beginning of the first data input line, followed by local coordinates (x_1,y_1) of the first corner point, coordinates (x_2,y_2) for the second point, and the thickness t_1 of first segment. Each additional segment of arbitrary section is defined in additional data lines listing the coordinates (x_i,y_i) followed by thickness t_i of that segment. Polar moment of inertia of thin-walled open section is calculated by using membrane analogy (equation 6.50).

Two-node Euler–Bernoulli beam element, ignoring the shear deformation in material, can be selected by choosing "TYPE=B33" in input data for element description. ABAQUS beam element library also includes a three-node element (B32) that uses quadratic shape functions (Figure 2.15), and it includes the material shear deformation in stiffness formulation following the same developments presented in Section 6.9. Geometry of three-node element is described as follows:

> *ELEMENT, TYPE=B32, ELSET=*setname*
> *Element_ID_no*, N_1, N_2, N_3, N_4

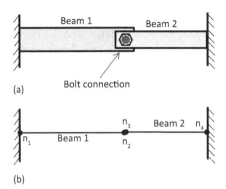

FIGURE 6.26 (a) Assembly of two-beam elements with moment-free bolted joint, and (b) elements defined with separate nodes n2 and n3 at the joint location.

where N_1 and N_3 are end nodes, N_2 is the mid-side node, and N_4 refers to the local coordinate direction x defining section orientation in 3D space. Extra node description, N_3 in two-node element and N_4 in three-node element, is not required in two-dimensional analysis of frames where local reference direction x is assumed normal to the model description plane. Element types for two-dimensional model simulation are "B21" in lieu of "B31" for linearly interpolated two-node beam element with shear deformation; "B23" in lieu of "B33" for Euler–Bernoulli beam element; and "B22" in lieu of "B32" for three-node element with shear deformation.

In frame assemblies, degrees of freedom at shared nodes experience resistances (stiffness contributions) from all connecting elements, thus, producing fully coupled motion among the joining members. However, not all joints in beam assemblies are designed to produce fully coupled motions among the elements. For example, single bolt connection between two beam parts in Figure 6.26 keeps translational motions coupled while allowing relative rotation between the two joining ends. Coupling of selected DOF between connecting members are modeled in ABAQUS by assigning coincident separate nodes to the joining elements, and then by constraining only the selected DOF. Input parameters, for example problem of Figure 6.26, can be described as follows:

*ELEMENT, TYPE=B21, ELSET=*set1*
Beam-1, n1, n2
Beam-2, n3, n4
*ELEMENT, TYPE=CONN2D2, ELSET=*set2*
El-id, n2, n3
*CONNECTOR SECTION, ELSET=*set2*
JOIN

where n2 and n3 are coincident nodes attached to beam elements Beam-1 and Beam-2, respectively. A special joint element is defined, of TYPE=CONN2D2, for connectivity between two nodes (n2 and n3) on a two-dimensional plane. Property of the connector element (*El-id*) is defined by selecting "JOIN" keyword in the connector element property description which implies to ABAQUS that the nodes connected to element (*El-id*) are kinematically coupled for translational DOF. Rotational DOF at nodes n2 and n3 remain un-coupled, meaning that the nodes can rotate independent of each other. ABAQUS joint element library includes a long list of various other options to define selective coupling between adjoining elements. More discussion on the modelling of joints in structural analysis is presented in Chapter 8.

6.11 PRACTICE PROBLEMS: LOAD–DEFLECTION ANALYSIS OF BEAMS

PROBLEM 1

Figure 6.27 shows a solid rectangular section cantilever beam subjected to an in-plane transverse load of $P = 10$ kN at the free end. Determine vertical deflection at the load application end by using the following analysis models:

    a. Analytical solution with plane stress idealization as presented in Section 3.4
    b. Beam element model with and without shear deformation consideration
    c. Plane-stress finite element model of the solid beam structure

Comment on the result differences, and on the reliability of analysis results (b) and (c) relative to the analytical solution given by (a).

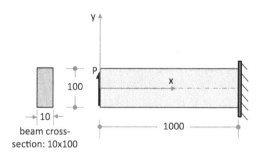

FIGURE 6.27   A rectangular steel beam ($E = 210$ GPa, $\nu = 0.3$) is subjected to an in-plane transverse load of $P = 10$ kN at the left end while the right end is fully constrained (all dimensions in mm).

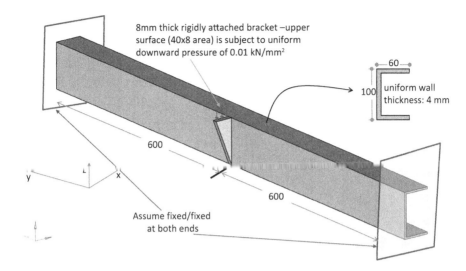

FIGURE 6.28   Thin-wall C-section beam loaded with a shear bracket at mid-span.

PROBLEM 2

Figure 6.28 shows a 1200 mm long C-section aluminum beam subjected to a uniform pressure load applied on a loading bracket seamlessly attached to the side of main beam. Geometry data of the beam (including loading bracket) is given in the file "3D-C-beam-with-loading-bracket.stp" (support data is available at a website mentioned in the preface of this book). Assume that both main beam and loading bracket are made of aluminum (E = 70 GPa, ν = 0.33). Determine the shear stress flow through the flanges of the beam at mid-span section using the following analysis models:

    a. Analytical solution for the combined effects of shear and torsion as discussed in Section 6.5
    b. Beam element model of the problem
    c. 3D solid element model of the beam and bracket assembly

Comment on the result differences caused by the different modeling assumptions.

PROBLEM 3

Figure 6.29 shows a solid steel hook of circular cross-section – subjected to a force of P = 0.5 kN. Geometry data (Solid_Hook_Geometry.stp) is available in the support data website. Determine the normal stress distribution on the section A-B using finite element simulation technique. Calculate the tangential stress at point A using hand calculations for curved beam theory; and compare the result with that of finite element simulation model.

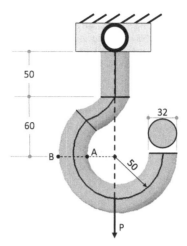

FIGURE 6.29   Stress analysis of a solid circular section steel hook (dimensions in mm). Assume elastic material properties: $E = 210$ GPa, $\nu = 0.3$

PROBLEM 4

Geometry of a long beam, including section dimensions, material property, loading and boundary conditions are described in Figure 6.30. Beam possesses a moment-free joint at point C. Predict the vertical deflection at joint C by analyzing the structure using beam and joint elements discussed in Section 6.10.

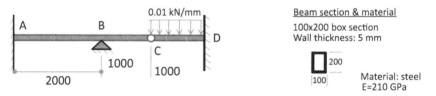

Ends A and D are fully constrained
Support @ B provides constraint against vertical deflection
Internal hinge @ C releases moment resistance at that point
Distributed vertical load of 0.01 kN/mm is applied over the segment CD
Assume constant box x-section and constant material properties.

FIGURE 6.30   Statically indeterminate beam with an internal moment-free joint at point C.

7 Analysis of 3D Thin-Wall Structures (Plates and Shells)

SUMMARY

Plates and shells are three-dimensional solid material components that have thickness values much smaller in comparison to the surface dimensions. Finite element simulation of such structures with 3D solid elements requires too many elements to be used. Moreover, bending response of lower order solid elements generally tends to be stiff, thus under-predicting the displacement response of the structures. The alternative to 3D solid element use is to discretize the mid-surface of thin-walled parts to surface-based elements that are capable of producing bending and shear resistance to out-of-plane loads. Fundamentals of bending stress–strain and deflection responses of flat plates are reviewed in Sections 7.1 and 7.2. Plate bending theory, however, explains only one part of the shell resistance mechanism. Traditional structural shells have been conceived and used over centuries to provide high resistance in the tangential mid-plane direction. Section 7.3 presents stress analysis of a general shell – highlighting the dominant role of membrane stresses in the shell response mechanism. Evidently, the in-plane membrane resistance mechanism can be simulated by using the plane-stress formulations presented in Chapter 3. Stiffness calculations for the plate bending response are developed in Section 7.4. The flat shell element formulation, developed from superposition of the in-plane membrane resistance and the out-of-plane plate bending resistance mechanisms, is presented in Section 7.5. Like the deep beam behavior, discussed in Chapter 6, material shear deformation in thick shells can accumulate to a significant part of the deflection response of structures. The stiffness formulation of flat shell elements, with added consideration of material shear deformation, is presented in Section 7.6. Many software implementations provide flat shell elements to be used for simulation of both planar and curved shell structures. It is generally expected that refined finite element mesh, with quadrilateral and triangular shaped flat elements, can adequately simulate the response of curved shell structures. However, huge volume of research work, partly driven by desire for theoretical purity, and partly for improved accuracy, has eventually led to the development and implementation of iso-parametric curved shell element formulations in some software packages. The salient features of curved shell elements are presented in Section 7.7. General topics of finite element mesh quality checks and integration rules of shell elements are discussed in Section 7.8. A review of ABAQUS-specific curved shell element options is presented in Section 7.9. Practice problems for shell structural analyses are presented in the final Section 7.10.

7.1 BENDING STRESSES AND STRAINS IN A PLATE

Figure 7.1 shows the roof panel of a vehicle subjected to a hypothetical distributed load in the normal direction of surface. Action of this applied load is resisted by material resistances to bending and shear – similar to the behavior of beams discussed in Chapter 6. However, the roof panel in this example cannot be modeled as a beam-like discrete line element because of its large planar dimensions. Neither can it be modeled as a three-dimensional solid body providing resistances to translational degrees of freedom only. Special analytical techniques are required to analyze the bending and shear deformation responses of plates and shells (Timoshenko and Woinowsky-Krieger 1970). Considering the plate element of uniform thickness in local z-direction, deformation degrees of freedom at a material point on the mid-surface of plate can be represented by translation in the normal direction of mid-surface and two rotations as shown in Figure 7.2. Rotational degree of freedom about local z-axis is ignored assuming rigid rotational response in the x–y plane of the plate. Extending the Euler–Bernoulli's bending deformation theory of beams,

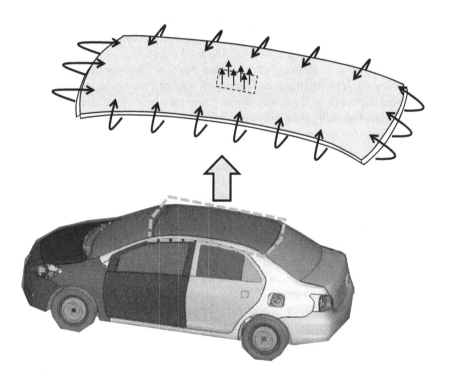

FIGURE 7.1   Thin sheet metal roof panel of a vehicle – capable of providing resistance to bending under normal load on the surface (vehicle FEA model: courtesy of NHTSA 2020).

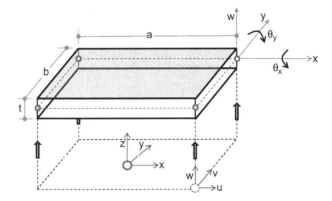

FIGURE 7.2   A thin plate ($t < b/20$) with mid-surface in the x–y plane – generated by collapsing the 3D solid in the z-direction.

out-of-plane pure bending response of a plate can be idealized with the following assumptions – commonly known as Kirchoff's plate bending theory:

- Deflection of mid-surface is small compared to thickness
- Lines normal to mid-surface before deformation remain normal after deformation (shear deformation in the material is ignored)
- No mid-surface straining occurs due to plate bending
- Stress component normal to mid-surface is zero

Based on these assumptions, the general strain–displacement relationships of equations (2.4) can be re-written for plate bending case as follows:

$$\varepsilon_x = \frac{\delta u}{\delta x}, \quad \varepsilon_y = \frac{\delta v}{\delta y}, \quad \gamma_{xy} = \frac{\delta u}{\delta y} + \frac{\delta v}{\delta x},$$

$$\varepsilon_z = \frac{\delta w}{\delta z} = 0, \quad \gamma_{yz} = \frac{\delta w}{\delta y} + \frac{\delta v}{\delta z} = 0, \quad \gamma_{xz} = \frac{\delta w}{\delta x} + \frac{\delta u}{\delta z} = 0 \tag{7.1}$$

Considering the simplified case of one-directional bending of a plate (Figure 7.3), rotation of plate section in y–z plane, x-axis being normal to that plane, is related to the transverse deflection of mid-surface w by equation (7.2):

$$\theta = \frac{\partial w}{\partial x} \tag{7.2}$$

Assumption of plate section normal to the mid-surface remaining normal through the deformation phase leads to a linear variation of deformation through thickness of plate:

$$u = -z.\frac{\partial w}{\partial x} \tag{7.3}$$

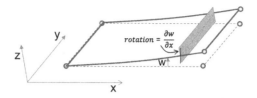

FIGURE 7.3   Bending rotation of plate-section in y–z plane.

Similarly, considering the simplified case of uni-modal bending of plate caused by rotation of section x–z (y-axis being normal to that plane), deformation inside the plate in the y-direction is defined by the following equation:

$$v = -z \cdot \frac{\partial w}{\partial y} \qquad (7.4)$$

Substituting expressions from equations (7.3) and (7.4) into equations (7.1), non-zero strain terms are given by the following equations:

$$
\begin{aligned}
\varepsilon_x &= \frac{\delta u}{\delta x} = -z \cdot \frac{\partial^2 w}{\partial x^2} \\
\varepsilon_y &= \frac{\delta v}{\delta y} = -z \cdot \frac{\partial^2 w}{\partial y^2} \\
\gamma_{xy} &= \frac{\delta u}{\delta y} + \frac{\delta v}{\delta x} = -2z \cdot \frac{\partial^2 w}{\partial x \, \partial y}
\end{aligned}
\qquad (7.5)
$$

Using the Hooke's law, stress–strain relationships for bi-directional plate bending response can be defined as follows (equations 7.6):

$$
\begin{aligned}
\sigma_x &= \frac{E}{1-v^2}\left(\varepsilon_x + v\varepsilon_y\right) \\
\sigma_y &= \frac{E}{1-v^2}\left(\varepsilon_y + v\varepsilon_x\right) \\
\tau_{xy} &= G\gamma_{xy}
\end{aligned}
\qquad (7.6)
$$

where strains ε_x, ε_y, and γ_{xy} vary in the z-direction through the plate thickness (equations 7.5). Similar to the beam response, bending stresses on a plate section (made of homogeneous isotropic material) vary linearly in the z-direction from the mid-surface, and the shear stress distribution takes a parabolic shape (Figure 7.4).

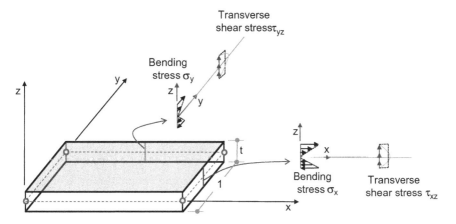

FIGURE 7.4   Bending and shear stresses through the thickness of plate.

7.2 ANALYTICAL SOLUTIONS FOR PLATE BENDING DEFLECTIONS

Integrating the bending the stresses (σ_x) over y–z plane of plate section, moment (M_x) per unit strip length (Figure 7.4) can be defined by

$$M_x = \int_{-t/2}^{t/2} z.\sigma_x.dz = \frac{E}{1-v^2} \int_{-t/2}^{t/2} z.\left(\varepsilon_x + v\varepsilon_y\right).dz \tag{7.7}$$

Substituting the expressions for ε_x and ε_y from equations (7.5), and after conducting partial integration with respect to z, equation (7.7) takes the following form:

$$M_x = \frac{-Et^3}{12\left(1-v^2\right)}.\left(\frac{\partial^2 w}{\partial x^2} + v\frac{\partial^2 w}{\partial y^2}\right) = -D.\left(\frac{\partial^2 w}{\partial x^2} + v\frac{\partial^2 w}{\partial y^2}\right) \tag{7.8}$$

where D [$= Et^3/12(1-v^2)$] is known as flexural rigidity of a plate made with homogeneous isotropic material. Inside the right-hand side expression of equation (7.8), the second-order differential terms represent the change in curvature of the plate in the x- and y-direction, respectively. Similarly, by integrating the contributions of other two stress components (σ_y and τ_{xy}), we get the following expressions for moments per unit strip length (equations 7.9 and 7.10):

$$M_y = -D.\left(\frac{\partial^2 w}{\partial y^2} + v\frac{\partial^2 w}{\partial x^2}\right) \tag{7.9}$$

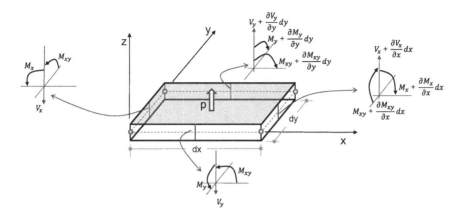

FIGURE 7.5 Shear forces and bending moments on the edges of a uniformly loaded plate element.

$$M_{xy} = -D.(1-v).\frac{\partial^2 w}{\partial x\, \partial y} \tag{7.10}$$

Considering the equilibrium of vertical forces acting on a differential plate element of dimensions (dx, dy) (Figure 7.5), we obtain:

$$\frac{\partial V_x}{\partial x}.dx.dy + \frac{\partial V_y}{\partial y}.dx.dy + p.dx.dy = 0 \tag{7.11}$$

Factoring out the non-zero multiplication term $(dx.dy)$, equation (7.11) takes the following reduced form:

$$\frac{\partial V_x}{\partial x} + \frac{\partial V_y}{\partial y} + p = 0 \tag{7.12}$$

Similarly, considering the equilibrium of moments about x and y axes separately, we arrive at the following equations:

$$\frac{\partial M_{xy}}{\partial x} + \frac{\partial M_y}{\partial y} - V_y = 0 \tag{7.13}$$

$$\frac{\partial M_{xy}}{\partial y} + \frac{\partial M_x}{\partial x} - V_x = 0 \tag{7.14}$$

Combining equations (7.8), (7.9), (7.10), (7.12), (7.13), and (7.14), bending deflection response of a plate, subjected to a uniform distributed load, is given by a single differential equation:

$$\frac{\partial^4 w}{\partial x^4} + 2\frac{\partial^4 w}{\partial x^2.\partial y^2} + \frac{\partial^4 w}{\partial y^4} = \frac{p}{D} \tag{7.15}$$

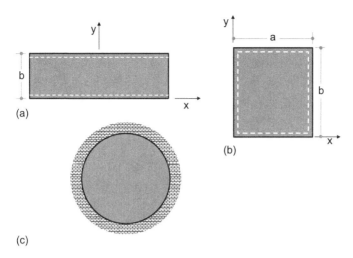

FIGURE 7.6 (a) Narrow rectangular plate simply supported on long sides; (b) rectangular plate simply supported on all sides; and (c) circular plate with rigid support on the perimeter.

Integration of the differential equation (7.15), with the use of appropriate boundary conditions, provides the plate deflection response $w(x,y)$. Once the deflection function $w(x,y)$ is known, strains and stresses inside the plate can be determined by using the relationships described earlier in this section. For example, the deflection of narrow rectangular plate of width b in Figure 7.6(a), with simply supported long edges and a distributed load of $p = p_o \sin(\pi y/b)$, is obtained by integrating equation (7.15) as follows:

$$w = \left(\frac{b}{\pi}\right)^4 . \frac{p_0}{D} . \sin\left(\frac{\pi y}{b}\right) \tag{7.16}$$

where p_o refers to the peak amplitude of non-uniform load along the plate centerline at $y = b/2$. Additional solutions for elementary plate deflection examples can be found in Timoshenko and Woinowsky-Krieger (1970) and Ugural and Fenster (2012). Deflection of the simply supported rectangular plate in Figure 7.6(b), subjected to a uniformly distributed pressure p, is given by the following equation (Ugural and Fenster 2012):

$$w = \frac{16.p_0}{\pi^6 D} \sum_{m}^{\infty} \sum_{n}^{\infty} \frac{\sin(m\pi x/a).\sin(n\pi y/b)}{mn.\left[(m/a)^2 + (n/b)^2\right]^2}, \qquad m,n = 1,3,5\ .... \tag{7.17}$$

For a uniformly loaded square plate ($a = b$), with simple supports at the edges, maximum deflection at mid-span is obtained from equation (7.17) as follows:

$$w_{max} \approx 0.0443.p_o.\left(\frac{a^4}{Et^3}\right) \tag{7.18}$$

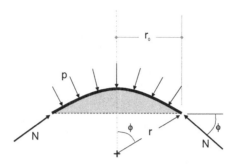

FIGURE 7.7   Segment of a spherical shell subjected to uniform external pressure (p).

And the maximum bending at the mid-span is given by

$$M_{x,max} = M_{y,max} \approx 0.0472.p_o.a^2 \tag{7.19}$$

Similarly, maximum bending responses of a circular plate (Figure 7.7(c)) of radius a, with moment–resistant rigid support on the perimeter, are given by

$$\text{Maximum deflection}\left(@\,\text{plate center}\right): w_{max} = \frac{p_o.a^4}{64D} \tag{7.20}$$

$$\text{Maximum bending moment}\left(@\,\text{the boundary}\right): M_r = -\frac{p_o.a^2}{8} \tag{7.21}$$

where p_o is the uniform pressure applied on the circular plate. Analytical solutions for elementary plate bending examples are useful in the early designs of structural members that can be simplified to be represented by one of the known examples. These solutions can also be used to approximately verify the finite element analysis results of more complex structural problems.

7.3 IN-PLANE MEMBRANE STRESS RESISTANCE OF A SHELL

Figure 7.7 shows a portion of a spherical shell of radius r and thickness t, subjected to a uniform pressure p. The uniform external pressure generates an internal "membrane" force per unit length (N), acting in the tangential direction of mid-surface (normal direction of radial cross-sectional plane). From the consideration of overall system equilibrium in vertical direction:

$$\left(2\pi r_o\right).N.Sin\phi = p.\left(\pi.r_o^2\right) \rightarrow N = \frac{pr_o}{2.Sin\phi} = \frac{pr}{2} \tag{7.22}$$

Membrane stress inside the spherical shell of thickness t, caused by the in-plane tangential force (N), is found to be:

$$\sigma_n = -\frac{N}{t} = -\frac{pr}{2t} \qquad (7.23)$$

where $-$ve sign implies compressive stress on the shell section. Ignoring the normal stress on the mid-surface of spherical shell, membrane strain caused by the bi-directional membrane stresses is given by

$$\varepsilon_n = \frac{\sigma_n}{E} - \vartheta . \frac{\sigma_n}{E} = -\frac{1-\vartheta}{E} . \frac{pr}{2t} \qquad (7.24)$$

Reduced circumference of spherical shell, because of the compressive normal strain (ε_n), is defined by

$$2\pi \left(r + r.\varepsilon_n\right) = 2\pi.r\left(1 + \varepsilon_n\right) = 2\pi.r' \qquad (7.25)$$

where r' is the new radius of the spherical shell. Change in curvature of the shell can be calculated from:

$$\left(\frac{1}{r'} - \frac{1}{r}\right) = \frac{1}{r}\left(\frac{1}{1+\varepsilon_n} - 1\right) = -\frac{\varepsilon_n}{r}\left(\frac{1}{1+\varepsilon_n}\right) = -\frac{\varepsilon_n}{r}\left(1 - \varepsilon_n + \varepsilon_n^2 - .....\right) \qquad (7.26)$$

Ignoring the higher order terms of ε_n in equation (7.26), and using the expression from equation (7.24), change in curvature of the spherical shell under uniform external pressure is defined by the following equation:

$$\left(\frac{1}{r'} - \frac{1}{r}\right) = -\frac{\varepsilon_n}{r} = \frac{1-\vartheta}{E} . \frac{p}{2t} \qquad (7.27)$$

Using the expression for the change in curvature from equation (7.27) into the equation for plate bending moment (equation 7.8), bending moment inside the spherical shell is defined as follows:

$$M_r = \frac{-Et^3}{12\left(1-v^2\right)}\left(\frac{\partial^2 w}{\partial x^2} + v\frac{\partial^2 w}{\partial y^2}\right) = \frac{-Et^3}{12\left(1-v^2\right)}\left(\frac{1-\vartheta}{E} + v.\frac{1-\vartheta}{E}\right) . \frac{p}{2t} = -\frac{pt^2}{24} \qquad (7.28)$$

Maximum bending stress caused by the bending moment M_r, acting on a rectangular segment of unit width and thickness t, is given by equation (7.29):

$$\sigma_b = \frac{6.M_r}{t^2} = -\frac{p}{4} \qquad (7.29)$$

Taking the ratio between membrane stress σ_n (equation 7.23) and bending stress σ_b (equation 7.29) for spherical shell under uniform external pressure, we obtain:

$$\frac{\sigma_n}{\sigma_b} = -\frac{2r}{t} \qquad (7.30)$$

For typical shell dimensions of $r \gg t$, equation (7.30) indicates that membrane stress dominates the internal resistance mechanism of the shell under external surface pressure. Stiffness properties of shell, thus, depend on combined resistance mechanisms of plate bending (Section 7.4) and plane-stress effects (Section 3.5).

7.4 BENDING STIFFNESS OF FLAT PLATE ELEMENT

As discussed in Section 7.1 and illustrated in Figure 7.2, deformation degrees of freedom for bending response of a plate involve transverse deflection (w) and two rotations θ_x and θ_y. Numerous formulations have been proposed in the literature for deformation response of plate elements (Logan 2012). In one of the most basic plate bending formulations, transverse displacement response of four-node plate element, having a total of 12 DOF, is defined by a polynomial function with 12 coefficients:

$$\begin{aligned}
w = {} & a_1 + a_2x + a_3y + a_4x^2 + a_5xy + a_6y^2 \\
& + a_7x^3 + a_8x^2y + a_9xy^2 + a_{10}y^3 + a_{11}x^3y + a_{12}xy^3
\end{aligned} \qquad (7.31)$$

This function satisfies the first basic requirement of meeting the compatibility condition defined by equation (7.15) for zero external normal load ($p = 0$). Equation (7.31), however, is complete up to third order (with first 10 terms). The fourth-order terms x^3y and xy^3 are chosen to ensure displacement continuity along inter-element boundaries. This function does not ensure slope continuity along inter-element boundaries. Full compatibility is impossible to obtain with simple polynomial expressions involving only 3 DOF at the corner nodes (Zienkiewicz and Taylor 1991). Despite this limitation, convergence properties of many such non-conforming formulations have been successfully proven in the literature.

Taking partial derivatives of expression in equation (7.31), with respect to coordinate variables x and y, deformation responses of the reference mid-surface of a plate can be expressed by following relations:

$$\begin{Bmatrix} w \\ \dfrac{\partial w}{\partial y} \\ \dfrac{\partial w}{\partial x} \end{Bmatrix} = \begin{bmatrix} 1 & x & y & x^2 & xy & y^2 & x^3 & x^2y & xy^2 & y^3 & x^3y & xy^3 \\ 0 & 0 & 1 & 0 & x & 2y & 0 & x^2 & 2xy & 3y^2 & x^3 & 3xy^2 \\ 0 & 1 & 0 & 2x & y & 0 & 3x^2 & 2xy & y^2 & 0 & 3x^2y & y^3 \end{bmatrix} \times \begin{Bmatrix} a_1 \\ a_2 \\ .. \\ .. \\ a_{12} \end{Bmatrix} \qquad (7.32)$$

Writing the matrix equations (7.32) at four node locations, we get the following expressions for deformation responses at the nodes:

$$
\begin{Bmatrix} w_i \\ \theta_{xi} \\ \theta_{yi} \\ w_j \\ .. \\ .. \end{Bmatrix} =
\begin{bmatrix}
1 & x_i & y_i & x_i^2 & x_i y_i & y_i^2 & x_i^3 & x_i^2 y_i & x_i y_i^3 & y_i^3 & x_i^3 y_i & x_i y_i^3 \\
0 & 0 & 1 & 0 & x_i & 2y_i & 0 & x_i^2 & 2x_i y_i & 3y_i^2 & x_i^3 & 3x_i y_i^2 \\
.. & & & & & & & & & & & \\
.. & & & & & (12 \times 12 \ matix) & & & & & & \\
.. & & & & & & & & & & & \\
.. & .. & .. & .. & & .. & & .. & .. & .. & .. & ..
\end{bmatrix}
\times
\begin{Bmatrix} a_1 \\ a_2 \\ a_3 \\ .. \\ .. \\ a_{12} \end{Bmatrix}
$$

(7.33)

where (x_i, y_i) are the coordinates of node i. Denoting the (12×12) multiplication matrix on the right-hand side of equation (7.33) as $[P]$, equations (7.33) for the nodal displacement responses are re-written as follows:

$$
\begin{Bmatrix} w_i \\ \theta_{xi} \\ \theta_{yi} \\ w_j \\ .. \\ .. \end{Bmatrix} = \begin{bmatrix} P \end{bmatrix} \times \begin{Bmatrix} a_1 \\ a_2 \\ a_3 \\ .. \\ .. \\ a_{12} \end{Bmatrix}
$$

(7.34)

Polynomial coefficients, $a_1 \ldots a_{12}$, can be calculated by inverting the relationship matrix [P]:

$$
\{a\} = \begin{bmatrix} P \end{bmatrix}^{-1} \times \{u_i\}
$$

(7.35)

where $\{u_i\}$ is vector of nodal displacements and rotations (w_i and θ_i). Using the displacement function (equation 7.31), curvatures of plate bending can be defined as follows:

$$
\begin{Bmatrix} -\dfrac{\partial^2 w}{\partial x^2} \\ -\dfrac{\partial^2 w}{\partial y^2} \\ -2.\dfrac{\partial^2 w}{\partial x \, \partial y} \end{Bmatrix} =
\begin{bmatrix}
0 & 0 & 0 & -2 & 0 & 0 & -6x & -2y & 0 & 0 & -6xy & 0 \\
0 & 0 & 0 & 0 & 0 & -2 & 0 & 0 & -2x & -6y & 0 & -6xy \\
0 & 0 & 0 & 0 & -2 & 0 & 0 & -4x & -4y & 0 & -12x^2 & -12y^2
\end{bmatrix}
\times
\begin{Bmatrix} a_1 \\ a_2 \\ .. \\ .. \\ a_{12} \end{Bmatrix}
$$

(7.36)

Writing the multiplication matrix on the right-hand side of equation (7.36) in condensed form as $[Q]$, curvatures of plate bending can be defined by

$$
\begin{Bmatrix} -\dfrac{\partial^2 w}{\partial x^2} \\ -\dfrac{\partial^2 w}{\partial y^2} \\ -2.\dfrac{\partial^2 w}{\partial x \, \partial y} \end{Bmatrix} = \begin{bmatrix} Q \end{bmatrix} \times \{a\}
$$

(7.37)

Substituting the relationship from equation (7.35) into equation (7.37), we get the following relationship between plate curvature and nodal displacement vector:

$$
\left\{
\begin{array}{c}
-\dfrac{\partial^2 w}{\partial x^2} \\[2mm]
-\dfrac{\partial^2 w}{\partial y^2} \\[2mm]
-2.\dfrac{\partial^2 w}{\partial x\,\partial y}
\end{array}
\right\} = \left[Q\right] \times \left[P\right]^{-1} \times \{u_i\} = \left[B\right] \times \{u_i\}
\tag{7.38}
$$

where $[B] = [Q].[P]^{-1}$ is the familiar finite element matrix relating nodal displacement DOF to internal deformation responses. Moment curvature relationships by combining equations (7.8), (7.9), and (7.10) can be written in the following matrix form:

$$
\left\{
\begin{array}{c}
M_x \\ M_y \\ M_{xy}
\end{array}
\right\} = \dfrac{Et^3}{12\left(1-v^2\right)}
\begin{bmatrix}
1 & v & 0 \\
v & 1 & 0 \\
0 & 0 & \dfrac{1-v}{2}
\end{bmatrix}
\cdot
\left\{
\begin{array}{c}
-\dfrac{\partial^2 w}{\partial x^2} \\[2mm]
-\dfrac{\partial^2 w}{\partial y^2} \\[2mm]
-2.\dfrac{\partial^2 w}{\partial x\,\partial y}
\end{array}
\right\} = \left[C\right].
\left\{
\begin{array}{c}
-\dfrac{\partial^2 w}{\partial x^2} \\[2mm]
-\dfrac{\partial^2 w}{\partial y^2} \\[2mm]
-2.\dfrac{\partial^2 w}{\partial x\,\partial y}
\end{array}
\right\}
\tag{7.39}
$$

where $[C]$ is the moment–curvature relationship matrix defined by elastic material properties (E and v) and plate thickness (t). Using the standard stiffness matrix formulation from equation (2.44), bending stiffness of a plate is calculated by integrating the properties over x–y plane of element:

$$
\left[K\right] = \iint \left[B\right]^T \left[C\right]\left[B\right] dx.dy
\tag{7.40}
$$

where $[B]$ and $[C]$ are element property matrices defined in equations (7.38) and (7.39). Numerical integration with iso-parametric element definition, discussed in Section (3.7), is used to calculate the stiffness in finite element analysis. Eventually, standard matrix method analysis of structure, with consideration of applied external loads and boundary conditions, provides the nodal responses of a finite element model of plates. Stiffness formulation details for triangular plate bending elements are not discussed in this book; those are available in Cook et al. (1989) and Zienkiewicz and Taylor (1991).

7.5   FLAT SHELL ELEMENT AS A COMBINATION OF PLATE BENDING AND MEMBRANE ELEMENTS

Curved shell surfaces are often represented by assembly of piece-wise linear flat elements (Figure 7.8). For the ideal case of 4 in-plane nodes, local directions x_1 and x_2

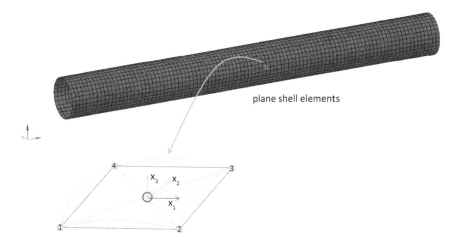

FIGURE 7.8 Curved shell surface is modeled by using four-node planar shell elements.

are tangential to the shell's mid-surface, and x_3 is normal to that mid-surface. Approximate assumptions are often made to define the planar surface of a shell. For example, in LS-DYNA software implementation (LSTC.COM 2020), shell mid-surface is assumed to be on the plane formed by two diagonals (connecting the opposite corner nodes), and the shell normal is assumed in the direction normal to that plane. Warped initial geometry, when idealized with a planar surface, leads to significant calculation errors. Stiffness properties of flat shell elements are generated by superposing the independent actions of in-plane stress resistance and plate bending resistance (Figure 7.9). Stiffness matrix of in-plane membrane resistance mechanism is formulated by using the iso-parametric element formulation procedure described in Sections (3.6) and (3.7). For four-node element configuration (Figure 7.9(a)),

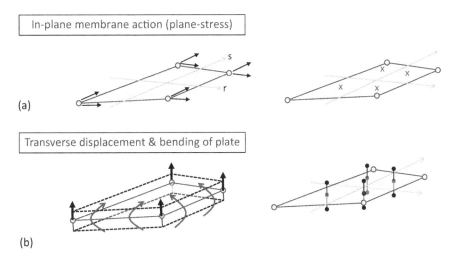

FIGURE 7.9 Resistance mechanisms in a shell: (a) membrane action and (b) plate bending.

membrane stiffness matrix is calculated by numerically adding the contributions of 4 Gauss integration points (equation 3.84):

$$\left[k_{membrane} \right] = \iint \left[B \right]^T \left[C \right] \left[B \right] dx.dy \qquad (7.41)$$

Effects of element geometry on result quality, discussed for plane-stress elements in Chapters 3 and 4, equally apply to shell analyses. Like the four-node plane-stress solids, full 2×2 integration of membrane stiffness matrix also leads to high in-plane stiffness for shell elements. Reduced integration in the element plane can circumvent that artificial stiffness effect, but it comes with the risk of producing zero-energy hourglass deformation mode in the element (Figure 3.13). Same planar integration rule (full- or reduced), used for the calculation of membrane stiffness matrix, can be used to calculate the bending stiffness matrix of constant-thickness plate element by using the integral formulation shown in equation (7.40). This bending stiffness equation, developed from moment–curvature relationships of equation (7.39), assumes that shell is made of single-layer homogeneous isotropic material. Stiffness matrix of composite shell, constructed with multiple layers of different materials, can be formulated by going back to the basic mechanics of plate bending response discussed in Section (7.1). Combining equations (7.5) and (7.38), bending-induced strains at a distance z from shell mid-surface are defined by

$$\begin{Bmatrix} \varepsilon_x \\ \varepsilon_y \\ \gamma_{xy} \end{Bmatrix} = \begin{Bmatrix} -z.\dfrac{\partial^2 w}{\partial x^2} \\ -z.\dfrac{\partial^2 w}{\partial y^2} \\ -2z.\dfrac{\partial^2 w}{\partial x \, \partial y} \end{Bmatrix} = z.\left[Q \right] \times \left[P \right]^{-1} \times \{u_i\} = \left[B_z \right] \times \{u_i\} \qquad (7.42)$$

where $[Q]$ and $[P]$ are element geometric property matrices, defined in equations (7.33) and (7.36); and $[B_z]$ is the familiar matrix notation relating the internal material strains to the nodal displacement and rotational response variables. Stress–strain relationships for plate bending response, given in equations (7.6), can be re-written in the following matrix form:

$$\begin{Bmatrix} \sigma_x \\ \sigma_y \\ \tau_{xy} \end{Bmatrix} = \begin{bmatrix} \dfrac{E}{1-v^2} & \dfrac{vE}{1-v^2} & 0 \\ \dfrac{vE}{1-v^2} & \dfrac{E}{1-v^2} & 0 \\ 0 & 0 & G \end{bmatrix} \times \begin{Bmatrix} \varepsilon_x \\ \varepsilon_y \\ \gamma_{xy} \end{Bmatrix} = \left[C_z \right] \times \begin{Bmatrix} \varepsilon_x \\ \varepsilon_y \\ \gamma_{xy} \end{Bmatrix} \qquad (7.43)$$

where $[C_z]$ represents the stress–strain relationship matrix corresponding to the specific material layer at distance z from the shell mid-surface. Taking the strain–displacement relationship matrix $[B_z]$ from equation (7.42), and stress–strain relationship $[C_z]$ from equation (7.43), bending stiffness matrix of composite shell is calculated by

$$\left[k_{plate\ bending} \right] = \iiint \left[B_z \right]^T \left[C_z \right] \left[B_z \right] dx.dy.dz \tag{7.44}$$

In addition to the numerical calculation points in the x–y plane (mid-surface) of element, multiple calculation points are added in thickness direction to take account of material property variation in composite shell construction (Figure 7.9(b)). Overall stiffness of shell element, for combined actions of membrane and plate bending resistance, is obtained by adding the contributions $[k_{membrane}]$ and $[k_{plate\ bending}]$, respectively, from equations (7.41) and (7.44). Shell element formulation, presented above, is basically following the Kirchoff's plate bending theory without considering shear deformation in thickness direction. Similar to the Euler–Bernoulli's beam deflection theory (discussed in Chapter 6), Kirchoff's plate bending formulation provides accurate estimation of deflection response for thin shells ($t < b/20$ in Figure 7.2).

7.6   SHEAR DEFORMATION IN PLATES

Plate response theory with material shear deformation in the thickness direction, commonly known as Mindlin's plate deformation theory, was introduced by Mindlin (1951) with a reference to earlier Russian publication by Uflyand (1948). A similar theory was proposed earlier by Reissner (1945). Both theories are intended for thick plates in which normal to the mid-surface remains straight but not necessarily perpendicular to the mid-surface. Mindlin's theory ignores normal stress through the thickness, thus, implying a plane-stress state in the plane of plate, while Reissner's theory does not invoke the plane-stress condition. It assumes the shear stress is quadratic through the thickness of plate, and the plate thickness may change during deformation. Stiffness formulation for Mindlin's plate element can be derived by following the same procedure presented in Section (6.9) for the derivation of beam stiffness properties including material shear deformation. A line that is straight and normal to the mid-surface before deformation is assumed o remain straight but not normal to the surface after deformation (Cook et al. 1989). Strains at a point not on the mid-surface are defined by rewriting equation (7.42) in the following form:

$$\left\{ \begin{array}{c} \varepsilon_x \\ \varepsilon_y \\ \gamma_{xy} \end{array} \right\} = \left\{ \begin{array}{c} -z.\dfrac{\partial \beta_x}{\partial x} \\[2mm] -z.\dfrac{\partial \beta_y}{\partial y} \\[2mm] -z.\left(\dfrac{\partial \beta_x}{\partial y} + \dfrac{\partial \beta_y}{\partial x} \right) \end{array} \right\} = \left[B_\beta \right] \times \left\{ u_i \right\} \tag{7.45}$$

where β_x and β_y are rotations of the reference line obtained from nodal rotational variables by using element-specific interpolation functions (similar to equation 6.94 presented for shear deformation analysis of beams). Stiffness matrix for plate curvature is calculated by inserting $[B_\beta]$ in equation (7.44):

$$\left[k_{plate\ bending} \right] = \int \int \int \left[B_\beta \right]^T \left[C_z \right] \left[B_\beta \right] dx.dy.dz \tag{7.46}$$

Shear strains through thickness of plate are defined extending the definition from beam problem (equation 6.92):

$$\gamma_{yz} = \frac{\partial w}{\partial y} - \beta_y; \quad \gamma_{zx} = \frac{\partial w}{\partial x} - \beta_x s \tag{7.47}$$

where w is the transverse displacement of point (x, y) on the initial mid-surface. Expressing the transverse displacement w, and rotations β_x and β_y, in terms of nodal response variables $\{u_i\}$, equation (7.47) can be re-written in the following form:

$$\begin{Bmatrix} \gamma_{yz} \\ \gamma_{zx} \end{Bmatrix} = \begin{bmatrix} \dfrac{\partial}{\partial y} & -1 & 0 \\ \dfrac{\partial}{\partial x} & 0 & -1 \end{bmatrix} \begin{Bmatrix} w \\ \beta_y \\ \beta_x \end{Bmatrix} = \begin{bmatrix} \dfrac{\partial}{\partial y} & -1 & 0 \\ \dfrac{\partial}{\partial x} & 0 & -1 \end{bmatrix} . \left[H_i \right] . \{u_i\} = \left[B_\gamma \right] . \{u_i\} \tag{7.48}$$

where $[H_i]$ are element interpolation functions. Stress–strain relationships for through-thickness shear deformation are defined by

$$\begin{Bmatrix} \tau_{yz} \\ \tau_{zx} \end{Bmatrix} = \begin{bmatrix} G_{yz} & 0 \\ 0 & G_{zx} \end{bmatrix} \begin{Bmatrix} \gamma_{yz} \\ \gamma_{zx} \end{Bmatrix} = \left[C_\gamma \right] . \begin{Bmatrix} \gamma_{yz} \\ \gamma_{zx} \end{Bmatrix} \tag{7.49}$$

For homogeneous isotropic elastic material, shear modulus, $G_{yz} = G_{zx} = G$, is defined by equation (2.23) in terms of Young's modulus (E) and Poisson's ratio (ν). Stiffness matrix for material resistance to through-thickness shear deformation is obtained by using $[B\gamma]$ and $[C\gamma]$ in equation (2.44):

$$\left[k_{shear} \right] = \int \int \int \left[B_\gamma \right]^T \left[C_\gamma \right] \left[B_\gamma \right] dx.dy.dz \tag{7.50}$$

Overall stiffness matrix of a flat shell, is finally, obtained by adding the contributions of all three resistance mechanisms:

$$\left[k_{shell} \right] = \left[k \right]_{Equation\ (7.41)} + \left[k \right]_{Equation\ (7.46)} + \left[k \right]_{Equation\ (7.50)} \tag{7.51}$$

For thin shells ($t < b/20$ in Figure 7.2), third term in equation (7.51) can be ignored, thus, leaving only the membrane and plate bending resistance terms in the element stiffness formulation. The derivation of shell stiffness from superposition of membrane and plate bending actions, with or without through thickness shear deformation, assumes that the behavior of a continuously curved shell surface can be adequately represented by an assembly of piece-wise flat quadrilateral or triangular elements. Solutions obtained from such approximations tend to converge to theoretical analysis results as the size of elements in shell model decreases (Zienkiewicz and Taylor 1991). Many commercially available finite element simulation packages offer the option of flat shell elements only. Formulation for curved shell elements, although relatively more tedious, has also been implemented in several software packages.

7.7   CURVED SHELL ELEMENTS

Displacement-based iso-parametric formulation developed by Ahmad et al. (1970) is the standard choice in finite element literature for curved shell elements (Cook et al. 1989, Zienkiewicz and Taylor 1991, Bathe 1996). Three-dimensional geometry of a shell is represented by its mid-surface following the 3D curvature (Figure 7.10), and the straight lines in thickness direction are not constrained to remain normal to mid-surface during deformation, thus, allowing for shear deformation in the material. Coordinates (x, y, z) of any point inside the element at a distance t from mid-surface (Figure 7.10) are defined by

$$
\begin{Bmatrix} x \\ y \\ z \end{Bmatrix} = \sum H_i \begin{Bmatrix} x_i \\ y_i \\ z_i \end{Bmatrix} + \sum H_i.t.\frac{t_i}{2}.v_{3i} \tag{7.52}
$$

where, x_i, y_i, and z_i are Cartesian coordinates of nodes at mid-surface, t_i are shell thickness values at the nodes, v_{3i} is the direction cosine of normal to the mid-surface at the point under consideration, and H_i are coordinate interpolation functions specific to the nodal composition of an element geometry. For four-node quadrilateral

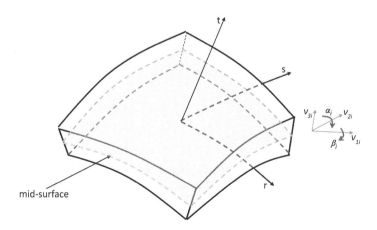

FIGURE 7.10   Curved shell element.

element shape, H_i are defined by equations (3.65), and for eight-node elements by equations (3.85). Similar formulations are also available in finite element literature for 3 and 6 node triangular elements. Using the same interpolation functions, displacements at a point inside the curved shell (u,v,w) are defined by

$$\begin{Bmatrix} u \\ v \\ w \end{Bmatrix} = \Sigma H_i \begin{Bmatrix} u_i \\ v_i \\ w_i \end{Bmatrix} + \Sigma H_i.t.\frac{t_i}{2}.v_{3i}.\begin{bmatrix} v_{1i} & -v_{2i} \end{bmatrix}\begin{Bmatrix} \alpha_i \\ \beta_i \end{Bmatrix} \qquad (7.53)$$

where $\{u_i, v_i, w_i\}$ are translational deformations of nodes on mid-surface, α_i and β_i are rotations of reference straight lines in thickness direction; and the direction cosines of orthogonal tangents to the mid-surface are denoted by v_{1i} and v_{2i}. Ignoring the strain in normal direction of mid-surface, strains in (x, y, z) coordinate system are defined from equations (2.4):

$$\begin{Bmatrix} \varepsilon_x \\ \varepsilon_y \\ \gamma_{xy} \\ \gamma_{xz} \\ \gamma_{yz} \end{Bmatrix} = \begin{Bmatrix} \dfrac{\partial u}{\partial x} \\ \dfrac{\partial v}{\partial y} \\ \dfrac{\partial u}{\partial y} + \dfrac{\partial v}{\partial x} \\ \dfrac{\partial u}{\partial z} + \dfrac{\partial w}{\partial x} \\ \dfrac{\partial v}{\partial z} + \dfrac{\partial w}{\partial y} \end{Bmatrix} \qquad (7.54)$$

Substituting the displacement expressions from equations (7.53) into (7.54), and using the coordinate transformation matrix J from equation (5.52), strain-to-nodal displacement relationships can be written in the following standard form for finite element calculations:

$$\begin{Bmatrix} \varepsilon_x \\ \varepsilon_y \\ \gamma_{xy} \\ \gamma_{xz} \\ \gamma_{yz} \end{Bmatrix} = \{\varepsilon\} = [B]\{u_i\} \qquad (7.55)$$

where $\{u_i\}$ is the vector of nodal displacement responses $(u_i, v_i, w_i, \alpha_i, \beta_i)$. Stress–strain relationships for curved shell elements are defined by Hookes' law:

$$
\begin{Bmatrix} \sigma_x \\ \sigma_y \\ \tau_{xy} \\ \tau_{xz} \\ \tau_{yz} \end{Bmatrix} = \{\sigma\} = \frac{E}{1-v^2} \begin{bmatrix} 1 & v & 0 & 0 & 0 \\ v & 1 & 0 & 0 & 0 \\ 0 & 0 & \dfrac{1-v}{2} & 0 & 0 \\ 0 & 0 & 0 & \dfrac{1-v}{2\xi} & 0 \\ 0 & 0 & 0 & 0 & \dfrac{1-v}{2\xi} \end{bmatrix} \times \begin{Bmatrix} \varepsilon_x \\ \varepsilon_y \\ \gamma_{xy} \\ \gamma_{xz} \\ \gamma_{yz} \end{Bmatrix} = [C]\{\epsilon\} \quad (7.56)
$$

where ξ is a shear correction factor for shear strain energy equivalence between actual parabolic distribution through thickness and the uniform distribution assumed in finite element calculations (analogous to the shear stress distribution in beams represented in equation 6.96). Relationship matrices [B] and [C], from equations (7.55) and (7.56), are used in equation (2.44) to determine the stiffness matrix of curved shell element. Numerical integration of equation (2.44) involves calculations on the element mid-surface as well as through the thickness direction – similar to the scheme shown in Figure 7.9. Single point integration in the thickness direction does not capture the bending resistance of a shell, thus, reducing it to a purely membrane element that can be used to analyze fabric roof systems such as the example shown in Figure 7.11.

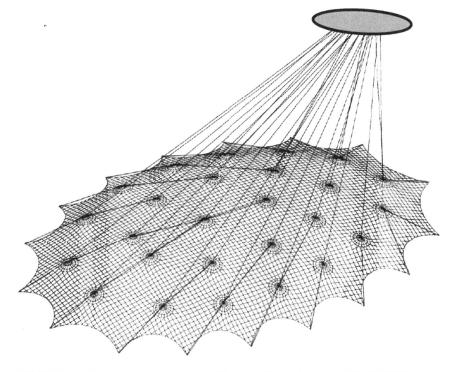

FIGURE 7.11   Membrane roof system in 3D space (Bhattacharjee and Chebl 1997).

7.8   SHELL ELEMENT MESH QUALITY AND INTEGRATION RULES

Finite element model preparation steps (a)-to-(g), described in Section 4.4 for 2D solids, are applicable to shell element model preparation as well. As evident from the above discussions in this chapter, a shell element is formulated from the superposition of numerical formulations related to in-plane membrane stress and out-of-plane bending resistance mechanisms. Element shape quality and integration rule effects, discussed in Chapters 3 and 4 for 2D solid elements, equally apply to the membrane stress resistance mechanism of shell element formulation. Quadrilateral shell elements, flat or curved, should be close to square shape in plan view. Excessive distortion of in-plane geometry significantly affects the quality of membrane stress response (similar to the effect highlighted for 2D plane-stress elements in Figure 3.12). Element shape quality metrics (Jacobian, aspect ratio, etc.) should be used to check the quality of shell element mesh before embarking on the journey to actual analysis task. An additional shape quality metric for shell elements is the warpage of shell plane as shown in Figure 7.12. When using flat shell elements in modeling curved surfaces, element warpage should be limited to no more than 10 degrees. Finer mesh grid with selective use of triangular elements tends to keep the geometric warpage under control while modeling complex 3D surfaces. The use of triangular shell element should be kept at minimum as it will make the in-plane bending response very stiff like its 2D counterpart (discussed in Section 3.9).

Fully integrated four-node quadrilateral shell element (Figure 7.9(b)) will also generate artificial high resistance to in-plane bending deformation – an inherent side effect of the linear assumption for in-plane deformation modes (Section 4.2, Figure 4.6). Reduced integration shell elements, one-point in-plane integration with multi-point through-thickness calculations (Figure 7.13(a)), have been successfully used in simulations of complex structural responses (Belytschko et al. 1984). Once again, the draw-back of in-plane reduced integration calculation is the hourglass zero energy

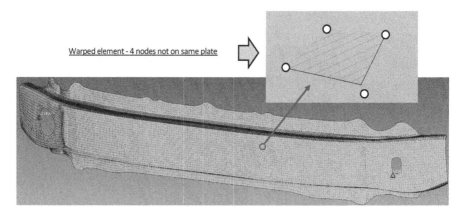

FIGURE 7.12   Warpage of shell element geometry.

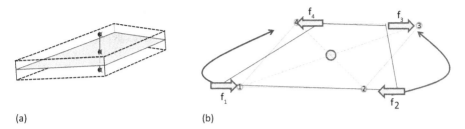

(a)   (b)

FIGURE 7.13   (a) Flat shell element with reduced in-plane integration rule; and (b) artificial numerical resistance addition to counteract zero energy deformation mode.

mechanism developing for in-plane bending mode (Figure 3.13). General-purpose finite element software packages usually contain internal numerical controls to minimize the risk of developing hourglass response mechanism. For example, the zero-energy in-plane bending response mechanism can be partially circumvented by adding artificial hourglass resistance forces (Figure 7.13(b)):

$$\begin{Bmatrix} f1 \\ f2 \\ f3 \\ f4 \end{Bmatrix} = -h_c * \mathrm{E} * \left(\Delta x \right) * \begin{Bmatrix} +1 \\ -1 \\ +1 \\ -1 \end{Bmatrix} \tag{7.57}$$

where $\{f_i\}$ is the vector of artificially added hourglass resistance forces depending on the hourglass deformation response at the nodes (Δx), h_c is a dimensionless penalty number (usually between 0 and 0.1), and E is the elasticity of material. Different variations of the hourglass formulation (equation 7.57) have been implemented in software packages. For example, LS-DYNA implementation (LSTC.COM 2021) derives hourglass forces based on nodal velocity response and material viscosity parameter, in lieu of relative displacement and elastic modulus, in dynamic response analysis of structures. Irrespective of the formulation details, hourglass resistance forces, acting at the nodal displacement DOF, are non-physical forces. These forces do absorb energy during element deformation. Reliable simulation software packages generally track the amount of artificial energy absorbed by the hourglass resistance mechanism. The reliability of a simulation model result, reporting more than 5% energy loss through hourglass mechanism, should be always questioned. A combination of multi-point integration rule for in-plane and out-of-plane bending mechanisms, with reduced integration rule for in-plane shear response, tends to minimize the negative effects of full and reduced integration techniques. Theory manuals of specific software packages generally contain the detailed information on what internal formulation techniques have been implemented to produce accurate simulation results.

7.9   ANALYSIS OF SHELLS WITH FEA SOFTWARE

Standard finite element simulation models for shell structures generally refer to the mid-surface geometry that can be easily extracted from 3D CAD data by using general-purpose model pre-processing software packages (such as HyperMesh and ANSA). The extracted mid-surface can be discretized into nodes and elements by using the surface meshing techniques of the pre-processing software. ABAQUS input model, for example, requires the standard 3D Cartesian coordinate data for nodes, with optional inputs for direction cosines of normal to the mid-surface at the node locations:

*NODES, NSET=*setname*
$N_1, x_1, y_1, z_1, l_1, m_1, n_1$
$N_2, x_2, y_2, z_2, l_2, m_2, n_2$
...

where x_i,y_i,z_i are Cartesian coordinates of node i, and l_i,m_i,n_i are direction cosines of the normal to mid-surface at that node. If the normals are not explicitly defined as part of the node definition, ABAQUS internally calculates the normal direction on element surfaces; and it defines the normal at a node as the average of the direction cosines of normal to adjoining element surfaces. ABAQUS element library includes curved shell elements of different nodal configurations: S4 for four-node shells, S3-for three-node shells, S8 for eight-node shells, etc. Example shell element in Figure 7.14 will be described in ABAQUS model file by following the standard input syntax:

*ELEMENT, TYPE=S4, ELSET=*setname*
Element_ID_no, N_1, N_2, N_3, N_4

where N1, N2, etc., are the element corner nodes. Number of nodes to describe element geometry depends on the shell element types (for example, three nodes for type

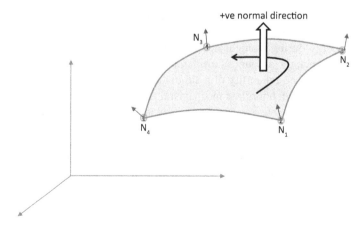

FIGURE 7.14   Four-node curved shell element with mid-surface normals at the corner nodes.

"S3" and eight nodes for element type "S8"). Reduced integration formulations for ABAQUS shell elements are selected by using the relevant element type name (for example, S4R for reduced integration of four-node elements; and S8R for eight-node elements). Uniform shell thickness and number of integration points through the thickness direction are defined by the following description:

*SHELL SECTION, MATERIAL=*matname*, ELSET=*setname*
thickness value, number of integration points through thickness
Variable thickness in the shell elements can be defined by using the "NODAL THICKNESS" option in shell section description:
*SHELL SECTION, MATERIAL=*matname*, NODAL THICKNESS, ELSET=*setname*
thickness value, number of through-thickness integration

Thickness values at the nodes are described by the *NODAL THICKNESS data block:

*NODAL THICKNESS
Node ID or Set ID, thickness value
Node ID or Set ID, thickness value
....

The sequence of node numbering in element connectivity description defines the direction of normal to an element surface (Figure 7.14). A surface load acting in the direction of normal is given a positive sign in distributed load description with *DLOAD in ABAQUS models:

*DLOAD
element ID or Set ID, P, surface pressure value
element ID or Set ID, P, surface pressure value
....

7.10   PRACTICE PROBLEMS: LOAD–DEFLECTION ANALYSIS OF SHELLS

PROBLEM 1

3D geometry of an extruded aluminum C-section component is shown in Figure 7.15. Exterior dimensions of the C-section are 60 mm × 100 mm, with uniform wall thickness of 4 mm; and the end-to-end length of the member is 1200 mm. Geometry data of the component is available in the file 3D-C-Beam.stp. External loading, boundary conditions and material properties have been described in the Figure 7.15. Conduct finite element analysis of the component using shell elements; and plot the contour of normal stresses on the x–z section at the mid-span of the member. Estimate, with hand calculations, the maximum bending stress at mid-span of member, and comment on the reason for differences between FEA and hand calculation results

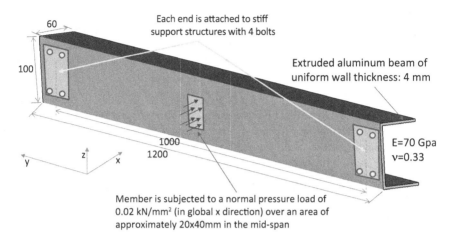

FIGURE 7.15   Practice Problem-1: stress analysis of a thin-wall C-section member.

PROBLEM 2
Re-analyze thin-walled C-section beam structure in Figure 6.28 with shell elements. Compare the shell model results with those obtained in Practice problem-2 in Section 6.11.

PROBLEM 3
Dent resistance of an automotive body panel is measured by a force resistance of >0.15 kN per 0.1 mm deformation when a normal pressure load is applied uniformly over a circular area of 51 mm diameter. Conduct dent resistance analysis at the mid-span of vehicle roof panel shown in Figure 7.1. Assume that 0.9 mm thick aluminum panel (E = 70 GPa, ν = 0.33) is riveted along the edges to the perimeter frames at every 60 mm. Curved shell surface of roof panel, with a rise of about 70 mm above the two far ends, covers an opening area of approximately 1460 × 1050 mm. 3D geometric profile data of roof panel's mid-surface is available in support data site (Roof_Panel.step).

PROBLEM 4
Figure 7.16 shows an automotive bumper beam with its material properties, boundary constraints and an arbitrary external loading condition. Geometric shape data of the component is available in data file (Bumper_Beam.step).
Develop a finite element analysis model of the structural component using triangular and quadratic shell elements. Verify the element qualities for shape distortion and warpage. Conduct analysis of the structure for applied loads and boundary

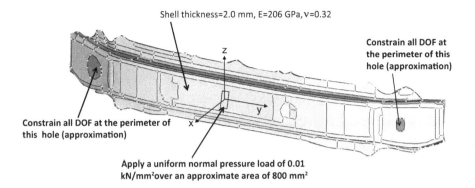

FIGURE 7.16   Practice Problem-4: Automotive bumper beam (Component design extracted from vehicle FEA model: courtesy of NHTSA 2020).

conditions; and plot a contour of global directional stresses σ_{yy} on the outermost layer (front surface) of the bumper beam shell. Compare the maximum stress value, obtained from contour plot, with hand calculated maximum bending stress value at mid-span assuming a straight beam profile of uniform cross-section.

8 Multi-Component Model Assembly

SUMMARY

Essential attributes of single-component finite element models have been discussed in Chapters 1–7. This chapter focuses on specific topics relevant for simulation models of multi-component assemblies. Element-to-element deformation compatibility, a special topic of concern during finite element modeling of parts having different behavioral characteristics, is discussed in Section 8.1. For interface deformation compatibility, the use of higher order elements is recommended in beam-solid and shell-solid assemblies. The important topic of modeling discrete inter-body connections is discussed in Section 8.2. Two commonly used inter-part rigid connection techniques, master-slave option using the elimination of slave DOF and the kinematic formulation based on Lagrangian multiplier approach, are reviewed in that Section. Discrete finite element definition of connector entity, based on deformable beam stiffness formulations, is introduced in Section 8.3. Section 8.4 extends this discrete connection modeling technique to mesh-independent implementation technique – a highly desirable feature in the preparation of large multi-part complex models. General part-to-part contact formulations, based on kinematic and penalty formulations, are introduced in Section 8.5. The contact formulation method is also a topic of high importance in nonlinear analysis of structures – discussed in Chapter 12. The alternative technique of modeling part-to-part thin-layer interfaces, with specially formulated solid elements, is presented in Section 8.6. Modular organization method for database management of large multi-part assemblies is discussed in Section 8.7. Result quality checks for multi-component finite element simulation models are briefly discussed in Section 8.8. Finally, a set of practice problems for the topics discussed in this chapter are presented in Section 8.9.

8.1 ELEMENT COMPATIBILITY AND CONVERGENCE OF SIMULATION RESULTS

Finite element formulations capable of providing resistance to various deformation modes have been presented in Chapters 2–7. Iso-parametric formulations inherently satisfy completeness requirement – that the displacement functions of an element must represent the rigid body displacements and constant strain states (Bathe 1996). Another fundamental requirement for monotonic convergence of finite element solutions is that the elements must be compatible. Compatibility is satisfied if the elements have the same nodes on the common interface, and same interpolation functions are used in adjoining elements. Structural component analysis examples in Chapters 1–7 have been modeled with single element types. Single element type usage guarantees that the elements in the model undergo compatible deformation responses. Compatible element selection, with error-free description of nodes, properties, and boundary conditions, virtually guarantees solution convergence towards theoretical solution provided that all active DOF in a model experience resistance from connecting elements and external boundary constraints. Number of active degrees of freedom in a simulation model depends on the resistance mechanism of selected element type. Models constructed with elements for 2D response simulation will have active DOF in the analysis plane only. For example, structural cable in Figure 2.18, providing resistance to axial deformation only, has been modeled with two truss elements in two-dimensional space. ABAQUS simulation, for example, keeps only translational DOF active at the nodes of a truss element model. Two-dimensional model using beam elements, e.g. the beam analysis model in Figure 6.30, has translational as well as rotational DOF active at the nodes. Automotive bumper beam problem in Figure 7.16, modeled with general 3D curved shell elements, has all 6 DOF active at each node. It is imperative that no active nodal DOF can undergo motion without resistance from connected elements. Some software packages by default keep all 6 DOF active at the nodes; and a user is required to explicitly define constraints ("zero" displacement boundary condition) on unused DOF. Model assembly as a whole unit is also required to have enough constraints against rigid body motion (zero energy mode). Figure 2.17 illustrates an example of rigid motion mechanism in an unstable analysis model.

Element-to-element deformation compatibility is not guaranteed when elements with different deformation response mechanisms are used to define an analysis model. Figure 8.1 shows an example where a structural assembly is defined by a combination of beam and cable components. Common joint (c) between beam and cable will have both translational and rotational motion. Cable element (modeled with single truss element in 2D space) will not provide any resistance to the rotational DOF at joint (C). However, bending resistance of the beam will provide resistance to that DOF. Input data for models comprising of more than one element type can be constructed, by including as many different element sets and properties as needed to describe the complete model. Partial ABAQUS input data for the problem in Figure 8.1 are listed in the following:

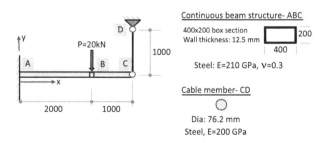

FIGURE 8.1   Mixed element-type assembly of beam and cable

```
*NODE, NSET=All-nodes
101,0.0,0.0,0.0
102,500.0,0.0,0.0
.....
108,3000.0,1000.0,0.0
*ELEMENT,TYPE=B21,ELSET=beams
1001,101,102
......
1006,106,107
*BEAM SECTION, MATERIAL=mat-beam, ELSET=beams, SECTION=BOX
400,200,12.5,12.5,12.5,12.5
*MATERIAL, NAME=mat-beam
*ELASTIC
210,0.3
*ELEMENT,TYPE=T2D2,ELSET=cable
9001,107,108
*SOLID SECTION, MATERIAL=mat-cable, ELSET=cable
4560.4
*MATERIAL, NAME=mat-cable
*ELASTIC
200,0.3
*BOUNDARY
101,1,6,0.0
108,1,2,0.0
*STEP, PERTURBATION
20 kN vertical downward load on beam
*STATIC
*CLOAD
105,2,-20.0
*END STEP
```

Beam element in the above analysis model provides resistance to rotational DOF at node #107 (corresponding to point C in Figure 8.1), while the truss element does not. This in-compatibility between discrete beam and truss elements does not pose any theoretical or computational challenge. It implies that element formulations do not need to be compatible in discrete structural element assembly provided that all DOF experience resistance to deformation. However, inter-element compatibility is a critical issue in modeling of continuum solids. It is tempting in engineering analysis to use beam and solid elements together (e.g. to model streel reinforced concrete structure), and to use shell and solid elements to model adhesively bonded thin sheet metal assemblies. Such incompatible element assemblies do not simulate continuous bonding at the element interfaces as intended when beam or shell undergoes bending deformation independent of the surrounding solid elements (Figure 8.2(a)). Increasing refinement of finite element mesh can reduce the effect of element-to-element incompatibility along interface boundaries. An effective alternative is to use higher order elements with mid-side nodes for maintaining interface deformation compatibility between beam and solid, or between shell and solid (Figure 8.2(b)). Use of specially

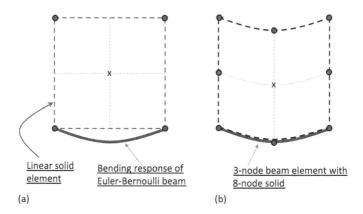

Linear solid   Bending response of   3-node beam element with
element   Euler-Bernoulli beam   8-node solid
(a)   (b)

FIGURE 8.2 (a) deformation incompatibility between Euler–Bernoulli beam and four-node solid sharing common nodes; (b) improved compatibility with higher order elements.

formulated transition elements has been proposed in the literature to connect elements of in-compatible deformation modes (Bathe 1996).

8.2 MODELING OF KINEMATIC JOINTS IN STRUCTURAL ASSEMBLIES

Figure 2.10(b) has shown a structural assembly where discrete truss elements are inter-connected at common nodes. Such structural joints, with fully coupled motion among the connecting members, can be simply modeled with common nodes without the need for any other special consideration. However, connecting member ends cannot be assigned to same node when some of the DOF at member ends are coupled while the others remain uncoupled. Beam assembly in Figure 6.26 represents an example where the joint between two beams keeps the translational motions coupled; and the rotational motions at the merging ends of two beams remain uncoupled. Some structural frame analysis software packages allow the definition of moment "release" corresponding to local rotational degree of freedom at end connections of beam elements. General-purpose finite element simulation software packages often allow the elements, merging at common structural joints, be modeled as independent entities – meaning that component elements will have their independent nodes; and the nodes merging to a joint will be coupled by using a kinematic joint entity. Coincident nodes n_2 and n_3 in Figure 6.26 can have same coordinates; and be part of independent elements merging to a common joint entity. Motions of these two nodes can be partially or fully coupled by defining kinematic constraint. A kinematic constraint basically enforces same motion on two or more DOF without allowing relative motion. Equation (8.1) in the following shows an example constraint formulation defining zero relative motion between two arbitrary DOF (u_j and u_k) in a model:

$$u_j - u_k = 0 \quad \rightarrow \quad \begin{bmatrix} 1 & -1 \end{bmatrix} \begin{Bmatrix} u_j \\ u_k \end{Bmatrix} = 0 \tag{8.1}$$

Multiple kinematic constraints in a large model can be defined by using matrix and vectors:

$$[e]_{p \times n} \cdot \{u\}_n = \{0\} \tag{8.2}$$

where $\{u\}$ is the vector of n number of active DOF in a model and $[e]$ is $p \times n$ matrix defining p number of constraint relationships (similar to that of equation 8.1) involving n number of DOF in a model. Finite element system equilibrium equation (1.5) and kinematic constraint relationships (8.2) can be combined to the following system of equations by using p number of Lagrangian multipliers (λ):

$$\begin{bmatrix} K_{n \times n} & e^T_{n \times p} \\ e_{p \times n} & 0 \end{bmatrix} \begin{Bmatrix} u_n \\ \lambda_p \end{Bmatrix} = \begin{Bmatrix} P_n \\ 0 \end{Bmatrix} \tag{8.3}$$

As long as constraint equations are linearly independent and $p < n$, standard matrix equation solvers can be used to solve equations (8.3) for the unknown displacements and Lagrangian multipliers (Bathe 1996). Lagrangian multipliers are in fact the internal force values required to enforce kinematic constraint conditions described by equation (8.2).

As discussed in Section 6.10, with ABAQUS input description for the example beam joint in Figure 6.26, nodes n_2 and n_3 are inter-connected with a kinematic connector element CONN2D2 with property type "JOIN".

*ELEMENT, TYPE=CONN2D2, ELSET=*set2*
El-id, n$_2$, n$_3$
*CONNECTOR SECTION, ELSET=*set2*
JOIN

This kinematic connector element ("JOIN") enforces same translational motion among the two joining nodes, while the rotational degrees of freedom at those nodes remain independent. Several other connector element types exist in ABAQUS library for defining node-to-node kinematic relationships. Kinematic connector elements can be considered as virtual elements that apply internal forces (represented by Lagrangian multipliers in equation 8.3) to eliminate relative motion among selected pair of DOF. These internal forces in connector elements can be post-processed for joint integrity assessment. Not all finite element analysis packages implement constraint formulation based on the Lagrangian multiplier approach.

Kinematic constraint definition, without involving connector elements, has a different implementation mechanism. In lieu of using the Lagrangian multiplier technique, general kinematic constraint formulations eliminate the slave degrees of freedom in most finite element implementations. Once a displacement degree of freedom is eliminated because of its "slave" relationship to a master reference DOF, additional displacement constraints (such as boundary conditions or other kinematic coupling definitions) cannot be applied to that again. Constraint definition based on elimination process of slave DOF permits the definition of constraints involving multiple slave DOF – commonly known as multi-point-constraint (MPC) in finite element literature. An MPC limits the motion of a group of nodes (often referred to as "slave" nodes) to that of a "master" or "reference" node. The reference (master) node has both translational and rotational degrees of freedom, and it can be subjected to boundary constraints if needed. ABAQUS input example for easy-to-define MPC type is presented in the following:

*KINEMATIC COUPLING, REF NODE=*ref_node_id*
n_i, n_j,.... etc. or *name of slave node set*

In the above example, all available DOF in the slave nodes are constrained to the DOF of the reference node. A reduced subset of the DOF can also be constrained by specifying the first and last DOF of the desired range:

*KINEMATIC COUPLING, REF NODE=*ref_node_id*
slave nodes, first dof, last DOF

Kinematic constraint can also be applied between a reference node and a set of elements on a surface:

*COUPLING, CONSTRAINT NAME=*a-name*, REF NODE=*ref_node*,
    SURFACE=*s-name*
*KINEMATIC
first dof, last DOF
*SURFACE, NAME=*s-name*, TYPE=ELEMENT
element ID or element set name

A group of nodes located on a surface can also be constrained to a reference node. Two surfaces can also be tied together. Each node on the first surface (the slave surface) will have the same values for its degrees of freedom as the point on the second surface (the master surface) to which it is closest. Discrete bonding of two or more surfaces, such as spot welding between sheet metals, can also be modeled using kinematic constraints; but it is more useful to model spot welds by using deformable connector elements (as discussed in Section 8.3).

Multi-point constraint definition in some software packages is identified as "rigid body" that binds a set of elements or a set of nodes to a reference node whose motion governs the motion of the entire rigid body. Rigid bodies can be used to model very stiff components, either fixed or undergoing large motions. The principal advantage of representing portions of a model with rigid bodies, rather than with deformable finite elements, is computational efficiency. Although some computational effort is required to update the motion of the nodes of the rigid body, and to assemble the concentrated and distributed loads, the motion of the rigid body is determined completely by a maximum of six degrees of freedom at the reference node. Element-level calculations are not performed for elements that are part of a rigid body, thus making the rigid-body definition very appealing for achieving computational efficiency in simulations of multi-body dynamic problems. For example, in complex models, elements far away from the particular region of interest could be included as part of a rigid body, resulting in faster run times. Although the motion of the rigid body is governed by the six degrees of freedom at the reference node, rigid bodies allow accurate representation of the geometry, mass, and rotary inertia of the parts. ABAQUS input syntax for a rigid body definition is shown in the following:

*RIGID BODY, REF NODE=*ref_node_id*
slave element set or node set

A reader can review users' guide of specific finite element software package, such as that of ABAQUS (Dassault Systems 2020b), to identify input syntax for different types of MPC definition. A general-purpose MPC definition enforces rigid body motion by eliminating the DOF at the slave nodes. As described earlier, connector elements in ABAQUS achieve the same objective without eliminating the slave DOF. Those elements provide the added advantage of calculating internal connector forces that can be monitored to assess the integrity of a joint. General-purpose MPC or RIGID BODY definitions, relying on the process of slave DOF elimination, do not provide the capability to monitor forces transmitted through the joint. The preference of joint element use, either kinematic formulation based on Lagrangian multiplier approach or the rigid body MPC approach, depends on the analysis objective. Large assemblies of relatively flexible structural components, such as the vehicle body structure shown in Figure 8.3, possess several thousand discrete spot welds that can

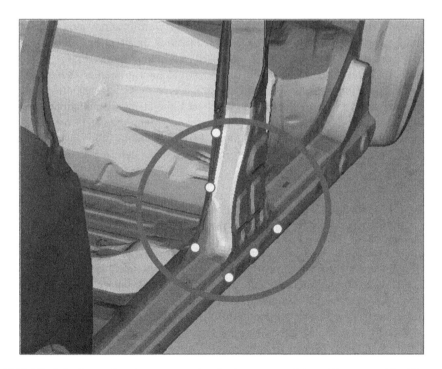

FIGURE 8.3 Example of an automotive body structure joint with spot-welds (Figure prepared from FEA model of vehicles – courtesy of NHTSA 2020).

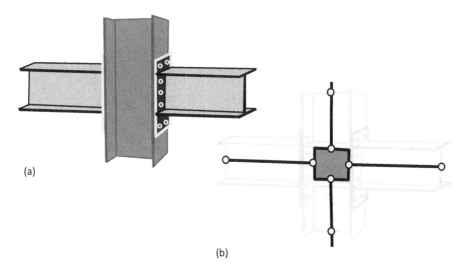

FIGURE 8.4   (a) Schematic view of joint details in a structural frame; and (b) idealized representation of a deformable joint element connected to primary frame members.

be modeled with rigid kinematic constraints to assess the overall system stiffness. However, structural integrity assessment at critical joint locations requires a more detailed stress–strain analysis of the joints themselves. Inherent flexibility of structural construction joints, such as the ones used in building frames (Figure 8.4), also requires an explicit representation of the non-rigid joint properties. Simulation of these structural joints, using kinematic constraint-type rigid elements, often leads to over-estimation of system stiffness. In practical finite element simulations, flexible joints are modeled by using special joint elements to approximately represent the deformable characteristics of physical joints. Section 8.3 in the following presents an enhanced form of connector element formulations using deformable finite element properties.

8.3   DEFORMABLE JOINT ELEMENTS FOR PART-TO-PART CONNECTIONS

Figure 8.5 shows a weld joint element "ab" connecting two plates "ABCD" and "PQRS". Local coordinate directions at end "a" are defined by the unit vectors e_1, e_2, e_3 shown in the figure, where direction e_3 is normal to the plane ABCD. Similarly, local directions at end "b" can be defined by another independent set of unit vectors with local direction 3 being normal to the plane PQRS. End "b" can be assumed to have same local coordinate directions of end "a" for the special case when two planes (ABCD and PQRS) happen to be parallel. Each end point of connector element (a and b) can have up to six degrees of freedom – three translations and three

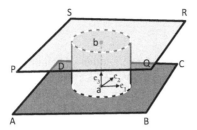

FIGURE 8.5 A weld element (*ab*) joining two plates ABCD and PQRS.

rotations. Displacements of points *a* and *b* are generally calculated by interpolating the responses of master surface nodes ABCD and PQRS, respectively. Stiffness properties of joint element ab can be formulated by considering 1–6 relative deformation modes between ends "*a*" and "*b*". Measuring the relative position of point "*b*", with respect to point "*a*", in local projection directions e_1, e_2, e_3, translational deformation inside the joint element is defined by the following equation:

$$u_1 = x - x_0; \quad u_2 = y - y_0; \quad u_3 = z - z_0 \tag{8.4}$$

where x_0, y_0, z_0 are the initial coordinates of node "*b*" relative to node "*a*" along the initial directions (e_1, e_2, e_3), and x, y, z are the relative position of point "*b*" after deformation. Similarly, relative rotational deformation of point "*b*" with respect to point "*a*" can be defined as

$$u_4 = \alpha^1 - \alpha_0^1; \quad u_5 = \alpha^2 - \alpha_0^2; \quad u_6 = \beta - \beta_0 \tag{8.5}$$

where α^1, α^2, and β are flexural and torsional deformation angles; and α_0^1, α_0^2, β_0 are the corresponding initial values. Input data syntax for describing 3D general joint element behavior in ABAQUS (ABAQUS Connector Elements, Dassault 2020b), considering both translational and rotational deformation modes, is given in the following:

*CONNECTOR SECTION, BEHAVIOR=*ab_elastic*, ELSET= *setname*
PROJECTION CARTESIAN, PROJECTION FLEXION-TORSION
orientation_a
orientation_b

where the option "PROJECTION CARTESIAN" specifies translational deformation calculations in the projected local vector directions following equation (8.4); and option "PROJECTION FLEXION-TORSION" specifies rotational deformation calculations following equation (8.5). Vector orientations at points "*a*", named "*orientation_a*", can be defined by listing 3 nodes as in the following:

*ORIENTATION, DEFINITION=NODES, NAME=*orientation_a*
$N_1,N_2,Node\_ID\_a$

where *Node_ID_a* specifies the node number of joint end point "*a*" as the origin of local coordinate system, node ID N_1 defines the local vector direction e_1, and N_2 defines the vector direction e_2 (Figure 8.5). Direction vectors for end connection point "*b*" can be defined independently, or it can follow the same definition of point "*a*". The orientation directions co-rotate with the rotation of the node to which they are attached. In element response calculations, the local directions are "centered" between the two local definitions at points "*a*" and "*b*".

In commonly used two-node joint element formulations, stiffness properties for the 6 internal joint deformation modes, described by equations (8.4) and (8.5), are defined by uncoupled stiffness terms as shown in the following equation:

$$[k]_{ab} = \begin{bmatrix} k_{11} & 0 & 0 & 0 & 0 & 0 \\ 0 & k_{22} & 0 & 0 & 0 & 0 \\ 0 & 0 & k_{33} & 0 & 0 & 0 \\ 0 & 0 & 0 & k_{44} & 0 & 0 \\ 0 & 0 & 0 & 0 & k_{55} & 0 \\ 0 & 0 & 0 & 0 & 0 & k_{66} \end{bmatrix} \tag{8.6}$$

where the first three diagonal terms represent stiffness values against three transla-tional deformation modes of the joint element, and the remaining three correspond to rotational deformation modes. Spring-like uncoupled stiffness property definition (equation 8.6) makes the joint element quite adaptable to special kinematic situations. For example, a moment-free deformable hinge connection can be simulated by assigning zero value to stiffness term corresponding to the relative twist between two end nodes. In equation (8.5), sixth deformation mode (u_6) represents that relative twist, meaning that k_{66} in equation (8.6) needs to be assigned zero value to create a torsion-free connection. In some software implementations, first local DOF is defined in the direction defined by end nodes "a" and "b", and the fourth diagonal stiffness term in equation (8.6) is associated with the relative twist mode of connector element. Joint element details are generally documented in software-specific users' manuals. ABAQUS input syntax for describing the stiffness terms of equation (8.6) is given in the following:

*CONNECTOR BEHAVIOR, NAME= *ab_elastic*
*CONNECTOR ELASTICITY, COMPONENT=1
k_{11}
*CONNECTOR ELASTICITY, COMPONENT=2
k_{22}
*CONNECTOR ELASTICITY, COMPONENT=3
K_{33}
*CONNECTOR ELASTICITY, COMPONENT=4

K_{44}
*CONNECTOR ELASTICITY, COMPONENT=5
K_{55}
*CONNECTOR ELASTICITY, COMPONENT=6
K_{66}

Stiffness properties in the above example have been assigned to an element set named "setname" which can contain one or more connector elements. Number of stiffness terms to be used depends on the formulation of specific joint elements. Connector behavior, described earlier with "PROJECTION CARTESIAN" and "PROJECTION FLEXION-TORSION" options, represents the most general connector element requiring all 6 stiffness terms. The description of connector elements in ABAQUS follows the standard element data input syntax:

*ELEMENT, TYPE=CONN3D2, ELSET=*setname*
Joint_Element_id_no, node_id_a, node_id_b
....

When no connector section data card (*CONNECTOR SECTION) is defined for an element set, ABAQUS uses the Lagrangian multiplier-based kinematic constraint formulation for the connector elements. DOF at end point "a" of connector element can be coupled to the nodes of base part ABCD by using the kinematic coupling formulation (*KINEMATIC COUPLING) described in Section 8.2:

*KINEMATIC COUPLING, REF NODE= *node_id_a*
N_A, N_B,.... etc. or *name of node set on part ABCD*

Similarly, DOF of node "b" can be coupled to the nodes of other base part PQRS. Stiffness terms defined in equation (8.6), multiplied by the relative deformation measurements from equations (8.4) and (8.5), provide the values of forces transferred through the connector element.

8.4   MESH-INDEPENDENT FASTENERS FOR PART-TO-PART CONNECTIONS

Many applications require the modeling of point-to-point connections between parts. These connections may be in the form of spot welds, rivets, screws, bolts, or other types of fastening mechanisms. There may be hundreds or even thousands of these connections in a large system model such as an automotive or airplane body structure. The fastener can be located anywhere between the parts that are to be connected regardless of the mesh. In other words, the location of the fastener can be independent of the location of the nodes on the surfaces to be connected. For example, connector element end points "a" and "b" in Figure 8.5 may need to be defined independent of the nodal points in base parts ABCD and PQRS. Meticulous manipulation of nodal positions in connected parts, to define a properly aligned pair of nodes on base parts, is not practical in complex structural simulation models involving several thousand inter-part connector elements. Use of kinematic coupling option to couple the DOF of end points of connector element "ab" (Figure 8.5) with the base part nodes, described in Section 8.3, provides a basic form of mesh-independent connector element description – meaning that connector element "ab" is defined at desired location independent of the mesh patterns of connected parts ABCD and PQRS. Mesh-independent description of connections is a basic feature required in finite element software packages for large-scale complex structural simulations. Many software packages provide very similar features for automatic generation of connection elements at user-specified locations. The number of individual connection elements is equal to the number of reference points specified by the user. Built-in functions in software packages identify the local nearest element surfaces to be connected including the connector orientation direction and kinematic coupling of DOF between connector element nodes and connected part nodes. In general, a reference point should be located as close to the surfaces being connected as possible. The reference node specifying the reference point can be one of the nodes on the connected surfaces or can be defined separately. ABAQUS determines the actual points where the fastener layers attach to the surfaces that are being connected by first projecting the reference point onto the closest surface. By default, ABAQUS projects each fastener reference point onto the closest surface along a directed line segment normal to the surface. Alternatively, a user can specify the projection direction if desired. Once the first connection point on the closest surface has been identified, the points on the other connection surface are determined by projecting the first connection point onto the other surface along the normal direction from the first connection point. The surfaces to be fastened can be specified using two different approaches – either by directly specifying the surfaces to be connected, or by letting the software package automatically identify the surfaces within a pre-defined search distance from the reference nodes. For surface-to-edge or edge-to-edge connections between parts, special optional attachment methods may need to be invoked.

Software packages allow either rigid or deformable connection element types to be included in automatic generation process. For example, deformable connector element "ab" in Figure 8.5 can be automatically generated in ABAQUS by specifying the following set of commands:

*NSET, NSET=*rfn*
Node_ID_a
*CONNECTOR SECTION, BEHAVIOR=*ab_elastic*, ELSET= *ss1*
PROJECTION CARTESIAN, PROJECTION FLEXION-TORSION
orientation_a
** BEHAVIOR= *ab_elastic* refers to the connector stiffness values defined earlier
     in section 8.3
*FASTENER, INTERACTION NAME=*nm1*, PROPERTY=*pp1*, ELSET=*ss1*,
     REFERENCE NODE SET=*rfn*
 ***blank line for default software choice of the projection direction*
surface_1_name, surface_2_name
*FASTENER PROPERTY, NAME= *pp1*

radius of connector element (a numerical value)
**blank line to indicate all DOF of connector node to be tied to base part

NSET=*rfn* refers to a set of nodes (one or more) to be used as reference nodes for automatic generation of connector elements. ELSET=*ss1* refers to an empty set of connector elements to be generated by ABAQUS at the reference node locations. Element surfaces to be connected are optionally identified by "*surface_1_name*" and "*surface_2_name*" in the above example, where these surfaces are predefined by *SURFACE command as discussed in Section 8.2. Alternatively, a user can specify a search radius in the *FASTERNER definition (SEARCH RADIUS=*R*) to let ABAQUS search for element surfaces that fall within a sphere of user-specified radius *R* with its center at the reference point. Projection of reference point onto the closest element surface is taken as the first point of connector element, and the local orientation direction e_3 (Figure 8.5) of connector is defined in the local normal direction of that closest surface. A user can override the default local system by specifying a local coordinate system name with "ORIENTATION=*orientation_name*" in the *FASTERNER definition card. Generally, the user-defined orientation should be such that the local e_3 direction of the orientation is approximately normal to the surfaces that are being connected; and the local e_1 and e_2 directions are approximately tangent to the surfaces that are being connected. By default, ABAQUS adjusts the user-defined orientation such that the local e_3 direction for each fastener is normal to the surface that is closest to the reference node for the fastener.

In automatically generated connector elements, the DOF of connection points are coupled to the average translational DOF of the connecting part nodes within a radius of influence. Rotational DOF of the connecting part nodes can also be included in coupling definition by using optional parameter "COUPLING=STRUCTURAL" in *FASTENER definition. The radius of influence, for the connected surface nodes to contribute to the motion of connection point, can be defined by specifying a numerical value with optional parameter "RADIUS OF INFLUENCE=" in *FASTENER definition. In absence of this parameter, ABAQUS computes a default value of the radius of influence internally, based on the fastener diameter (defined in *FASTENER PROPERTY) and the characteristic lengths of elements on connected part. The connector elements are given internally generated element numbers and assigned to the named user-specified element set (ELSET=*ss1* in the above example). Multiple connector element sets may need to be defined to specify unique stiffness values (equation 8.6) depending on the mechanical properties of the actual fasteners (bolts, rivets, spot welds, clinched joints, etc.).

8.5 SIMULATION OF PART-TO-PART INTERFACE CONTACTS

Initially separate bodies, rigid or deformable, may come into contact during analysis of load–deformation response. Changes to the interface contact condition in finite element simulation models can be simulated by tracking the relative motion between bodies. It is customary, in simulation software packages, to designate body parts as slaves and master (Figure 8.6), where the motions of slave part nodes are tracked with respect to the positions of surrounding master element surfaces. Lagrange

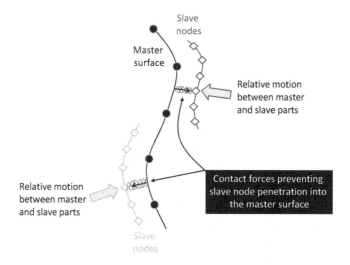

FIGURE 8.6   Master-slave concept in contact simulation of solid bodies.

multiplier method, discussed in Section 8.2, can be used to constrain the penetration of slave nodes into the master surface in the normal direction. Potential contacting bodies are defined upfront in ABAQUS simulation models by using the following data input syntax:

*CONTACT PAIR, INTERACTION=*interaction_property_name*
TYPE=SURFACE TO SURFACE
slave_surface_name, master_surface_name

In contact pair formulation, the Lagrange multiplier method is used for contact enforcement. In alternative contact formulations, commonly known as the penalty method, the relative penetration of slave node to master surface (deformation DOF u_3 in equation 8.4) is minimized by using a large contact stiffness value (k_{33} in equation 8.6). With this method, the contact force is proportional to the penetration distance; so some degree of penetration will occur. Penalty force on slave node is calculated as:

$$F_{penalty} = -k_{33}.u_{33} \tag{8.7}$$

An equal and opposite force is distributed on the nodes of master element surface. The penalty stiffness value k_{33} for a shell contact surface can be defined in terms of the bulk modulus K_i, and the face area A_i of the contact element (LS-DYNA Theory Manual, LSTC.COM 2021):

$$k_{33} = f_i * \frac{K_i.A_i}{\max\left(shell\ diagonal\right)} \tag{8.8}$$

where f_i is a scale factor for the interface penalty stiffness. ABAQUS/Standard, by default, sets the penalty stiffness to 10 times the representative underlying element stiffness. A low penalty stiffness typically results in better convergence of the solution, while the higher stiffness keeps the overclosure at an acceptable level as the contact pressure builds up. In LS-DYNA, the default option is to use the lower bulk modulus value of master and slave parts. This formulation may lead to a very low penalty stiffness value with undesirable inter-part penetrations during simulation of contacts between hard and very soft parts. Adjustment of penalty stiffness values to simulate contact between hard and soft bodies varies from software to software. Nonlinear contact stiffness formulations, as a function of the gap closure, can be used for improved stability of simulation models. In the nonlinear penalty method, the penalty stiffness increases linearly between regions of constant low initial stiffness and constant high final stiffness, resulting in a nonlinear pressure-overclosure relationship. Advantages of the penalty method, over Lagrange multiplier method, include solver efficiency and mitigation of over-constraint. ABAQUS input data syntax for selecting penalty method with linear contact stiffness formulation is described in the following:

*SURFACE INTERACTION, NAME=*any_name*
*SURFACE BEHAVIOR, PENALTY=LINEAR

In the linear penalty method, the so-called penalty stiffness is constant, so the pressure-overclosure relationship is linear. Relative sliding between contacting bodies can be considered, including frictional force transfer between the contacting bodies, by defining a friction coefficient in contact property definition:

*SURFACE INTERACTION, NAME= *any_name*
*FRICTION
0.4

where 0.4 is an example friction coefficient value assigned to "*any_name*".

In addition to the paired surface contact definition, described earlier, ABAQUS software also offers a general contact option to model surface-to-surface contact, edge-to-surface contact, edge-to-edge contact, and vertex-to-surface. General contact can also deal with self-contact as shown graphically in Figure 8.7. General contact uses the penalty method to enforce the contact constraints by default. The simplest way to define an all-inclusive general contact in ABAQUS is described in the following:

*CONTACT
*CONTACT INCLUSIONS, ALL EXTERIOR

where "ALL EXTERIOR" option includes all exterior element faces and all analytical rigid surfaces in the general contact definition. Commercially available software packages also offer an alternative convenient method of specifying the contact entities by using a pre-defined initial domain. Such domains, defined with a box, identify

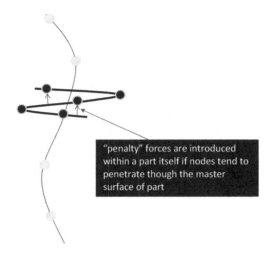

FIGURE 8.7 Self-contact in a body undergoing large deformation.

all parts that can potentially come into contact with one another during deformation. Master-slave relations are automatically defined by ABAQUS in general contact formulations. In ABAQUS/Standard, traditional pair-wise specifications of contact interactions generally result in more efficient analyses as compared to an all-inclusive self-contact approach of defining all potential contacts. Therefore, there is often a trade-off between ease of defining contact and analysis performance.

In model preparations for contact simulation, initial penetration between slave and master parts should be minimized, and physical intersection between contacting parts must be avoided with no exceptions (Figure 8.8). Initial positions of part surfaces must be placed with adequate offset, taking account of the part thickness values, to minimize penetration problems during model preparation. Software packages generally offer options to consider the reduced thickness of parts in contact interference calculations, thus, preventing numerical issues arising from artificial penetration and intersection issues. Refined mesh density is generally helpful for modeling parts with high curvature.

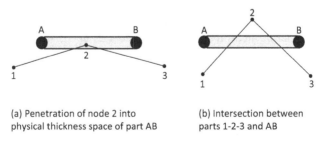

FIGURE 8.8 Initial penetration and intersection in finite element models.

8.6   THIN-LAYER INTERFACE ELEMENTS

Simulation of body contacts based on gap closure criterion, described in Section 8.5, involves numerical treatments internal to the software package in use. However, engineering analysis of bodies with embedded pre-existing joints requires an explicit definition of joint elements in the finite element simulation models. Interface joint regions (Figure 8.9) can be modeled with standard solid elements with unique material properties specific to the known behavior of joints. Simulation of joints with standard finite element formulations produces reasonably good results for mostly elastic systems with very small deformations occurring at the interface joints. However, standard finite element formulations, with coupled deformation response involving all nodal DOF, produce artificially stiff response when separation and sliding deformations inside joints dominate the system response. Special finite element formulations have been proposed in the literature for realistic simulation of joint separation and sliding that are encountered in soil-structure interfaces, rock joints, and construction joints in concrete dams (Desai et al. 1984, Tinawi et al. 1994).

General-purpose finite element software packages often contain special element formulations to simulate joint behavior. ABAQUS, for example, offers a library of cohesive joint elements to model the behavior of adhesive joints, interfaces in composites, and other situations where the integrity of interface joints may be of interest. The connectivity of cohesive elements is like that of continuum elements, but it is useful to think of cohesive elements as being composed of two faces separated by a thickness (Figure 8.10). The relative motion of the bottom and top faces measured along the thickness direction (local 3-direction for three-dimensional elements) represents opening or closing of the interface. The relative change in position of the bottom and top faces measured in the plane orthogonal to the thickness direction quantifies the transverse shear behavior of the cohesive element. In-plane membrane stresses are generally not considered in interface joint element formulations. In

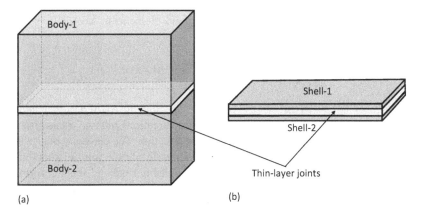

FIGURE 8.9   Pre-existing thin-layer joints: (a) between solid bodies, and (b) between thin shells.

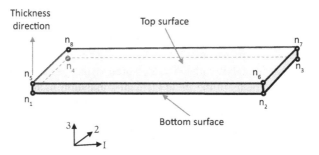

FIGURE 8.10   Finite element definition of a thin-layer interface element.

three-dimensional problems, the continuum-based constitutive model assumes one normal (through-thickness) strain, and two transverse shear strains at a material point. Element example, shown in Figure 8.10, can be described in ABAQUS as follows:

*ELEMENT, TYPE=COH3D8, ELSET=*set_1*
Eelement_ID, $n_1, n_2, n_3, n_4, n_5, n_6, n_7, n_8$

where nodes n_1, n_2, n_3, n_4 define the bottom surface and nodes n_5, n_6, n_7, n_8 define the top surface of interface element. The thickness direction is automatically determined by ABAQUS in normal direction from bottom surface to the top. A user can also define the local direction by using the optional ABAQUS command *ORIENTATION if desired. The interface zone is discretized with a single layer of cohesive elements through the thickness. Both top and bottom surface nodes of an interface element layer can be tied to the neighboring base material parts by using common node definitions. More generally, when the mesh in the interface zone does not match with the mesh of adjacent components, nodes of interface elements can be tied to adjacent base material nodes by using kinematic constraint formations described in Section 8.2. The constitutive behavior of interface cohesive elements in ABAQUS is defined by using the following property definition:

*COHESIVE SECTION, ELSET=*set_1*, MATERIAL=*mat_1*,
    RESPONSE=CONTINUUM

The choice of "RESPONSE=CONTINUUM" specifies to ABAQUS that interface element will have one normal strain (in thickness direction) and two transverse shear strains in a plane perpendicular to the thickness direction. The elasticity matrix defining the stress–strain relationship at material points inside the joint element can be defined by diagonal terms only – meaning uncoupled normal and shear deformation modes. By default, cohesive elements retain their resistance to compression even if their resistance to other deformation modes is completely degraded. As a result, the cohesive elements resist inter-penetration of the surrounding components even after the cohesive element has completely degraded in tension and/or shear. This approach works best when the top and the bottom faces of the cohesive element experience small relative sliding.

8.7 MODULAR ORGANIZATION OF DATA IN MULTI-COMPONENT MODEL ASSEMBLY

Single-component analysis models in Chapters 1–7 have been presented by describing all relevant model data in single files. A complex structural system (e.g. the vehicle model shown in Figure 7.1) may involve may different components, with many different element types and inter-connections. Description of several component model entities in one single file makes the management of model database very challenging, particularly when frequent model updates are required because of changes happening in a handful of parts. This challenge is mostly circumvented by the general-purpose finite element simulation software packages by allowing the assembly of large complex models from much smaller modular construction units. Basic geometric characteristics of components and subsystems may reside in separate model files while a master file will assemble all those into a single analysis model with relevant specifications for loads and boundary conditions. Component model files must adhere to certain modeling rules and restrictions to avoid conflicts; and also to facilitate easy execution of analysis exercises for multiple attributes. For example, finite element model of the vehicle roof panel in Figure 7.1 can be described in a data file (named Roof_Panel.inp) following ABAQUS input syntax:

*NODE, NSET=*Roof_panel_nodes*
100001,0.0,0.0,1000.0
100002,5.0,5.0,1000.0
.....
*NSET, NSET=*Roof_panel_boundary_nodes*
800001, 800002, ...
......
*ELEMENT, TYPE=S4, ELSET=*Roof_panel_shell_elements*
100001,100051,100099, 110026, 120345
......
*ELSET, ELSET=*Roof_panel_mid_span_elements*
731001, 731049, ...
....
*SHELL SECTION, MATERIAL=*Mat_roof_panel*, ELSET=
    Roof_panel_shell_elements
1.0, 5
*MATERIAL, NAME= *Mat_roof_panel*
*ELASTIC
210,0.3
*DENSITY
7.8e-06

Above data descriptions include nodes, elements, and material properties of the roof panel as well as entity sets of *Roof_panel_boundary_nodes* (list of all nodes corresponding to the attachment points of roof panel to the vehicle body structure) and *Roof_panel_mid_span_ elements* (list of a subset of shell elements at the mid-span of roof). Entity sets can be used to define loads, boundary conditions, connector elements, and contacts as needed in subsequent analysis models. Additional entity sets of nodes and elements can be defined that may or may not be used in subsequent analysis models. A simple load–deflection analysis model of the vehicle roof panel can be constructed in a separate ABAQUS input file by using the finite element model file *Roof_Panel.inp*:

```
** Main analysis file for load–deflection analysis of vehicle roof panel
*INCLUDE, INPUT= Roof_Panel.inp
*BOUNDARY
Roof_panel_boundary_nodes, 1, 6,
*STEP, PERTURBATION
*STATIC
*DLOAD
Roof_panel_mid_span_elements, P, -0.01
*END STEP
```

Above analysis model includes the component model file (*Roof_Panel.inp*), and an analysis model is constructed by using the node and element entity sets defined in the component model file. Any future changes to the design and finite element model of the component model will be contained within the component input data file only, thus keeping the main analysis file un-affected. Model assembly procedure, using *INCLUDE option, can be expanded to include multiple component models in a more complex model assembly:

```
** Main analysis file for load–deflection analysis of a multi-part assembly
*INCLUDE, INPUT= Part_1.inp
*INCLUDE, INPUT= Part_2.inp
. . . . . . .
*BOUNDARY
Node_set_1, 1, 6,
Node_set_2, 1, 6,
. . . .
*STEP, PERTURBATION
*STATIC
*DLOAD
Element_set_1, P, -0.01
Element_set_2, P, +0.03
. . . .
*END STEP
```

Model entities (nodes, elements, etc.) are defined by using pre-defined unique ID ranges in each component model to preclude conflicting IDs in the main assembly model. When a component model is changed during design iterations, changes are made to the relevant component input data file only, thus, keeping other components as well as the main analysis model practically unchanged.

8.8   RESULT QUALITY CHECKS

Result quality checks, discussed in Sections 3.10 and 4.5 for single-component finite element analysis models, are equally applicable to multi-component model assemblies as well. Multi-component models, however, present additional challenges due to discontinuous nature of stress–strain distributions at the interfaces and joints. Patch-test-like standard numerical simulation tests are not readily useful for quality checks of finite element models involving contacts and joints. Engineers can, however, make use of the powerful graphical visualization features of post-processing software tools that are often bundled with actual FEA solver products (such as LS-Pre-Post with LS_DYNA solver and ABAQUS-CAE with ABAQUS solver). Result database files, produced by the commonly used FEA solvers, can also be visualized by using general-purpose post-processing software packages, such as HyperView (Altair.com), META (Beta-CAE.com), and FEMAP (Siemens.com). For example, Figure 8.11 shows a snapshot of part-to-part penetrations occurring during simulation of load–deformation response – indicating potentially missing contact definition during the model preparation.

As discussed in Sections 8.2 and 8.3, simulation of inter-part connectivity by using deformable joint elements, or by using Lagrange-multiplier-based connector formulations, can directly produce output results of interaction forces flowing through these elements. However, commonly used alternative MPC formulations enforce slave-master constraint relationships by eliminating the slave DOF, thus missing the direct measurement of interaction forces. This shortcoming is overcome by defining virtual cut sections through structural assembly models for monitoring the internal forces on cut-sections. Ignoring the body force effects in the internal stress equilibrium condition ($F_i = 0$ in equations 1.10), resultant force effects on a plane cutting through the internal stress field of multi-element assembly (Figure 8.12) can be calculated by equation (8.9):

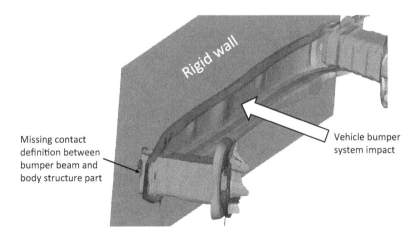

Missing contact
definition between
bumper beam and
body structure part

Vehicle bumper
system impact

FIGURE 8.11   Part-to-part penetration during simulation of load–deformation response.

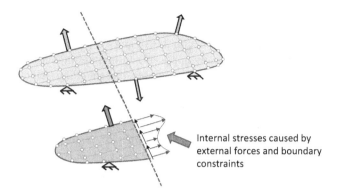

Internal stresses caused by external forces and boundary constraints

FIGURE 8.12   Internal stresses on the cut-plane of a finite element model.

$$\{P\} = \int [B]^T .\{\sigma\}.dV \qquad (8.9)$$

where $\{\sigma\}$ is the vector of internal stresses at element integration points, [B] the strain-displacement relationship matrix (equation 2.6), and $\{P\}$ the vector of internals forces and moments on the cut-plane – calculated by summing the contributions of all finite elements traversing the user-defined virtual cut-plane. Graphical postprocessors often include features to define virtual cut-planes for calculating internal forces from the finite element stress results saved in database files of FEA solvers. Commonly used FEA solvers also usually allow pre-defined virtual cut-planes in the model input files. Internal forces calculated on those pre-defined virtual planes can be optionally saved on the result database files, and those can be subsequently postprocessed by the graphical visualization software packages.

ABAQUS provides a simple method to create such an interior surface over the element facets, edges, or ends by cutting through a region of the model with a plane. The region can be identified using one or more element sets. The virtual cutting plane is defined by first specifying a point on the plane and a vector normal to the plane. ABAQUS then automatically forms a surface close to the specified cutting plane by selecting the element facets, edges, or ends of the continuum solid, shell, membrane, surface, beam, truss, or rigid elements in the selected region. The surface generated in this manner is an approximation for the cutting plane. ABAQUS input data syntax for defining a cut plane is described in the following:

*SURFACE, NAME=*surface_name*, TYPE=CUTTING SURFACE
x1, y1, z1, l,m, n
elset-1, elset-2, ...

where coordinates (*x1, y1, z1*) define the position of cut-plane and (*l, m, n*) define the direction cosines of normal to the cut-plane. A virtual internal surface is generated by cutting through the element sets named *elset-1* and *elset-2*, etc. A blank data line (without any element set name) generates a surface by cutting the whole model. Only the element nodal forces that lie on the positive side as defined by the normal to the cutting plane are included in the calculations. Integrated force and moments values on the virtual cut-plane can be requested in ABQUS by using the following output request command:

*INTEGRATED OUTPUT, NAME=*surface_name*

The integrated surface can also be defined at the interface between parts, and the output forces can be used to assess the force transmitted through contact between the parts.

The integrated output section definition does not impose any constraint on the component nodes. The average motion of nodes on the defined surface can be monitored by including a reference node definition.

8.9 PRACTICE PROBLEMS: ANALYSIS WITH MULTI-COMPONENT MODEL ASSEMBLIES

PROBLEM 1
Re-analyze Problem-2 (Figure 6.28), described in Section 6.11, using thin-layer interface elements (Section 8.6) between the main C-section beam and loading bracket. Graphically show the distribution of shear stress flowing through the interface zone. Determine the total shear flow through the interface by using a virtual section at the interface between parts.

PROBLEM 2
Re-analyze Problem-3 in Section 6.11 with a cut-section defined at location AB of Figure 6.29. Verify that resultant force and moment on cut-section match with the applied external loading on the hook.

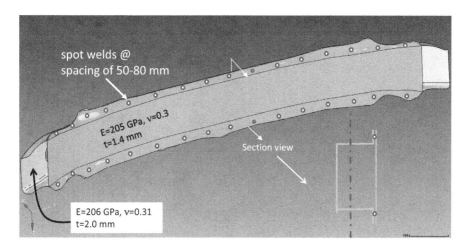

FIGURE 8.13 Spot-welded assembly of a reinforced automotive bumper beam.

PROBLEM 3

Automotive bumper beam in Problem-4 of Section 7.10 (Figure 7.16) is reinforced by adding a 1.4 mm steel reinforcement plate as shown in Figure 8.13. The back plate (Bumper_Back_Plate.step) is connected to the upper and lower flanges of the main beam (Bumper_Beam.STEP) with spot welds spaced approximately at 50–80 mm interval. Build a finite element simulation model of the spot-welded two-part assembly; and conduct load–deflection analysis of the assembly for the same loading and boundary conditions described in Problem-4 of Section 7.10. Show the distribution of normal stresses on a section through the mid-span of bumper beam. Compare the simulation results with hand calculations based on beam bending theory.

PROBLEM 4

Figure 8.14 shows an automotive connecting rod made of steel. Mesh model of the part is provided in Connecting_rod.inp. Model the compressive force distribution on the surface A–A–A as a uniform normal pressure of 0.01 kN/mm². The support condition on the rod can be described by normal displacement constraints on surface B–B–B. Conduct stress analysis of the part assuming linear elastic material response. Define a cross-section C–C–C through mid-height of the connecting and plot the distribution of normal stresses on the section.

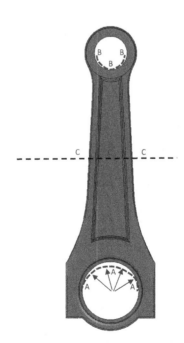

FIGURE 8.14   Automotive connecting rod made of steel ($E = 200$ GPa, $\nu = 0.3$) (FEA model created from the 3D CAD data available @ GRABCAD.COM 2020).

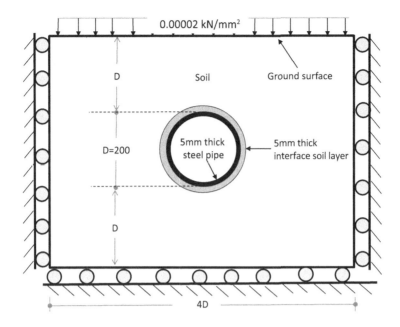

FIGURE 8.15 Cross-sectional view of a buried steel pipe (Assume: steel properties – $E =$ 210 GPa, $\nu = 0.3$; soil properties – $E = 0.02$ GPa, $\nu = 0.4$).

PROBLEM 5
Figure 8.15 shows a cross-sectional view of an underground steel pipe subjected to a uniform distributed surface load of 0.00002 kN/mm². Assuming plane-strain loading condition, develop a finite element analysis model of the soil-pipe system with thin-layer interface elements at the interface between soil and pipe. Conduct stress analysis of the system assuming linear elastic material response, assuming elastic properties of interface material to be same as that of surrounding soil. Produce a plot of shear stress distribution along the interface between soil and steel pipe.

9 Interpretation of Stress Analysis Results for Strength and Durability Assessment

SUMMARY

The basic purpose of finite element simulation is, obviously, to use the results for engineering decisions on functionality, strength and durability of products and structures. Books on finite element methods, as well as the software manuals and documents, primarily focus on element formulations and/or model preparation techniques. The correct interpretation of analysis results, based on advanced solid mechanics principles, is as important as the preparation of a good quality simulation model. This chapter presents few selected topics relevant for interpretation of the finite element stress analysis results. Section 9.1 starts with a brief review of the elastic material properties that are determined from standardized uniaxial material stress tests and are used directly in simulation models and in subsequent interpretation of results. Understandably, finite element simulation models deal with more than simple uniaxial stress field cases. Section 9.2 summarizes the stress component measurements that are predicted by different finite element formulations presented in earlier chapters. Material failure theories that form the analytical foundation for transformation of multiaxial stress field measurements to equivalent uniaxial predictor values are reviewed in Section 9.3. Commonly used stress field measurements, relevant for ductile and brittle materials, are specifically summarized in that section. Section 9.4 discusses the graphical post-processing technique for visualization of complex stress field responses. Fatigue life assessment of solids, based on finite element stress analysis results, is discussed in Section 9.5. It needs no mention that the reliability of finite element stress prediction, when used in conjunction with material strength criteria, is highly relevant for making good engineering conclusion. The sensitivity of finite element stress results to geometric discontinuities has been discussed in Chapter 3; and is revisited in Section 9.6 with additional case study results. Section 9.7 demonstrates the capability of standard finite element analysis model for predicting the stability of cracked solids with the use of linear elastic fracture mechanics metric, i.e. with the use of derived stress intensity factor from local stress analysis results. Finally, Section 9.8 presents practice problems to apply the special techniques covered in earlier sections of this chapter.

9.1   ENGINEERING PROPERTIES OF MATERIALS

Material specimens of standard size and shape are generally tested under uniaxial tensile loading condition (Figure 9.1) to determine the mechanical properties such as elastic modulus, yield strength, ultimate strength, failure strain, etc. ASTM Standard E8/E8M, for example, describes standards for specimen preparation and testing of metallic materials. Information generated from such tensile test is primarily used in material quality controls and in comparative study of different material grades. Material mechanical properties used in finite element models are also derived from the material specimen test data. Tensile force applied during testing is divided by the initial cross-sectional area of testing specimen (A_0) to determine the applied engineering stress "s" (Figure 9.2). Similarly, change in length measured over initial reference gage length (L_0) is used to calculate the engineering strain ("e") inside material (Figure 9.2). Stress–strain response, thus measured, with respect to initial geometric dimensions of material specimen, is plotted to produce engineering stress–strain diagram (Figure 9.2). Assuming no net volume change during load–deformation

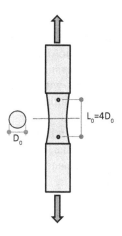

FIGURE 9.1   Uniaxial tensile testing of a material specimen.

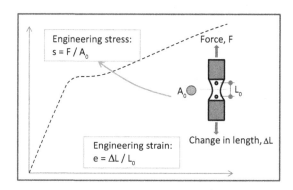

FIGURE 9.2   Engineering stress–strain response of a material test specimen.

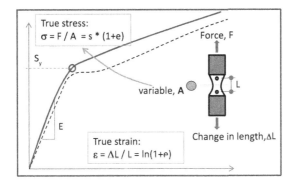

FIGURE 9.3   True stress–strain response calculated from engineering stress–strain response of a ductile material specimen.

response of isotropic ductile material specimen, true instantaneous stress (σ) and true strain (ε) in the material can be calculated from engineering stress–strain response by using the following expressions (Figure 9.3):

$$\varepsilon = \frac{\Delta L}{L} = \ln\left(1+e\right) \tag{9.1}$$

$$\sigma = \frac{F}{A} = s.\left(1+e\right) \tag{9.2}$$

Initial slope of the true stress–strain diagram defines the elastic modulus (E) of the material (Figure 9.3). Yield stress point S_y defines the departure from linear elastic stress–strain behavior to nonlinear material deformation (indicating accumulation of permanent plastic deformation in ductile metals). Uniaxial deformation response of material, for stress values below yield strength ($\sigma < S_y$), is defined by the elastic modulus (E). When uniaxial force is applied to a solid, it deforms in the direction of the applied force; and it also expands or contracts laterally depending on whether the force is tensile or compressive. If the solid is homogeneous and isotropic, and the material remains elastic under the action of the applied force, the lateral strain bears a constant relationship to the axial strain. This constant, called Poisson's ratio, is an intrinsic material property just like Young's modulus and shear modulus. ASTM Standard E132-17 is generally used to determine the Poisson's ratio from tensile load testing of material specimens having rectangular cross-section. As discussed in Chapters 2–7, elastic stress–strain responses of two- and three-dimensional stress fields can be defined by using E and ν with Hooke's law (discussed in Section 2.4).

9.2 STRESS–STRAIN RESULTS FROM LINEAR ELASTIC FINITE ELEMENT ANALYSIS OF SOLIDS

Stiffness properties of a finite element, defined by equation (2.44), involve the material stress–strain relationship matrix [C] and the strain–displacement relationship matrix $[B]$. Same relationship matrices, $[C]$ and $[B]$, are used to calculate the stress and strain values from the nodal displacement response values. Actual number of the stress and strain values, calculated at material points, depends on the type of finite element used in a simulation model. Formulations for truss or bar elements, described in Chapter 2 for analysis of truss-type structures, produce only axial strain and stress inside the element (Figure 9.4(a)). Strength assessment of a truss structural member is generally conducted by simply comparing the model-predicted axial stress value with uniaxial tensile strength, determined from material tensile tests, or with compressive strength limit defined based on buckling strength capability of the compressively loaded member. The assessment of member tensile strength capability, based on uniaxial material test result, is a preliminary indication of structural safety. The average tensile strength of members tends to get lower, particularly for brittle and quasi-brittle materials, as the member size gets larger – a phenomenon commonly referred to as "size effect" on material strength (Bažant 2005). Size-adjusted material stress–strain properties can be considered in nonlinear finite element simulation models for mesh-independent strength assessment of structures (Bhattacharjee and Leger 1994).

Simple strength criteria, with or without size effect adjustment, serve the purpose of safety assessment of members that provide resistance to external loads primarily through axial deformation mechanism (e.g. trusses, bars, wire strands, etc.). More general-type structural members may not have a dominant axial stress direction; the internal stress field can be quite complex with multiple stress components acting in different directions at the same stress point. Beam section in Figure 9.4(b), for example, has both normal and shear stresses acting at a point – and both stress components vary over the section of the beam. Figure 9.5 shows a relatively more complex stress field of a plane-stress solid. As discussed with examples in Chapters 3 and 4, three stress components (σ_x, σ_y, and τ_{xy}) are required to describe the stress state inside this

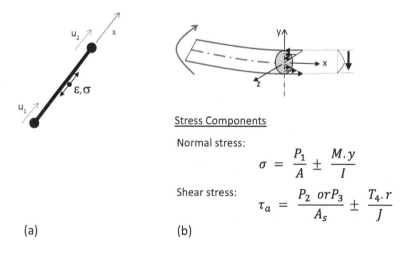

Stress Components

Normal stress:
$$\sigma = \frac{P_1}{A} \pm \frac{M.y}{I}$$

Shear stress:
$$\tau_a = \frac{P_2 \ or P_3}{A_s} \pm \frac{T_4.r}{J}$$

(a)   (b)

FIGURE 9.4   (a) Uniaxial stress–strain state in a truss element; (b) variable normal and shear stresses on a beam section.

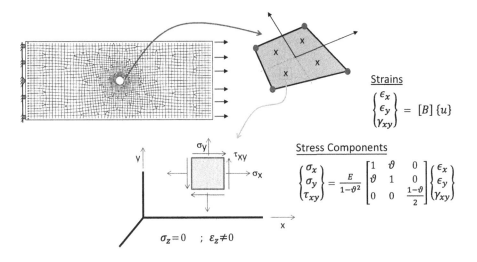

FIGURE 9.5   Strains and stresses at a material point inside a plane-stress solid.

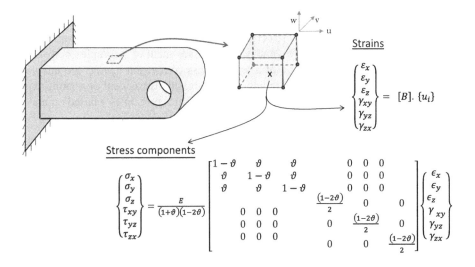

FIGURE 9.6   Strains and stresses at a material point inside a 3D solid.

plane-stress solid. Additional stress components appear in finite element calculations of relatively more complex structural systems. Stress field inside a general 3D solid, without a strong directional bias for geometry or load or boundary conditions (Figure 9.6), produces all six stress components (σ_x, σ_y, σ_z, τ_{xy}, τ_{yz}, and τ_{zx}) inside the material points. A structural shell element, when defined at its mid-plane (Figure 9.7), produces five stress components (σ_x, σ_y, τ_{xy}, τ_{yz}, and τ_{zx}), while the strain and stress in the normal direction of shell's mid-plane are considered negligible. Evidently, stress

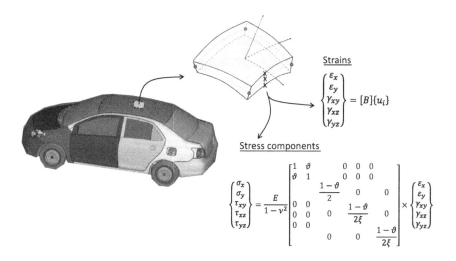

$$\begin{Bmatrix} \varepsilon_x \\ \varepsilon_y \\ \gamma_{xy} \\ \gamma_{xz} \\ \gamma_{yz} \end{Bmatrix} = [B]\{u_i\}$$

Strains

Stress components

$$\begin{Bmatrix} \sigma_x \\ \sigma_y \\ \tau_{xy} \\ \tau_{xz} \\ \tau_{yz} \end{Bmatrix} = \frac{E}{1-v^2} \begin{bmatrix} 1 & \vartheta & 0 & 0 & 0 \\ \vartheta & 1 & 0 & 0 & 0 \\ 0 & 0 & \dfrac{1-\vartheta}{2} & 0 & 0 \\ 0 & 0 & 0 & \dfrac{1-\vartheta}{2\xi} & 0 \\ 0 & 0 & 0 & 0 & \dfrac{1-\vartheta}{2\xi} \end{bmatrix} \times \begin{Bmatrix} \varepsilon_x \\ \varepsilon_y \\ \gamma_{xy} \\ \gamma_{xz} \\ \gamma_{yz} \end{Bmatrix}$$

FIGURE 9.7   Strains and stresses at a material point inside a shell element.

fields inside general solids and structures produce multiple nonzero stress components at measurement points. Obviously, safety assessment of a complex body cannot proceed with the comparison of a single stress component against the uniaxial material tensile or compressive strength value. Strength assessment of a multi-axial stress field requires the use of advanced solid mechanics principles – based on material failure theories discussed in the following Section (9.3).

9.3 STRENGTH ASSESSMENT OF SOLIDS – USE OF MATERIAL FAILURE THEORIES

Figure 9.8 summarizes how material property parameters, extracted from tensile test data, are used in different phases of the finite element simulation exercise. Elastic modulus (E) and Poison's ratio (ν), generated from uniaxial and bi-axial testing of material specimens, are used as input parameters to define material stress–strain relationship matrix, [C], based on Hooke's law. Multi-axial stress components, calculated inside the finite element, are transformed to representative index value ($\bar{\sigma}$) that, in turn, is compared with the uniaxial material strength (S_y or S_t) to assess the safety of body. The transformation of multi-axial stress values to a representative index value requires the use of advanced mechanics theories based on idealized failure characteristics of materials. Environmental and history factors (such as temperature, humidity, stress history, loading rate, etc.) may also need to be considered for a comprehensive assessment of a stressed body. For engineering analysis of simple monotonic static load cases, material failures are usually classified into either brittle failure (fracture) or ductile failure (yielding) type. The most commonly used theories for determining

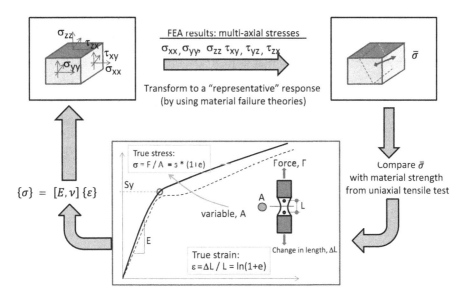

FIGURE 9.8   Process flow for interpretation of finite element stress–strain results.

the failure characteristics of monotonically loaded engineering materials are listed in the following (Budynas 1999):

1. "Principal" stress theory
2. "Principal" strain theory
3. Maximum strain energy theory
4. Maximum shear stress theory
5. Maximum distortion energy theory

1. Principal stress theory: Finite element analysis produces stresses in user coordinate reference system – such as σ_x, σ_y, and τ_{xy} in the 2D element shown in Figure 9.9. Inside the element, stresses on an arbitrary plane can be defined by two components – a normal stress and a shear stress. Interrelation between the normal and shear stresses, for any arbitrary orientation of the internal plane, is defined by a circle – referred to as Mohr's circle of stress in solid mechanics (Popov 1978). The maximum and minimum values of normal stress occur on two orthogonal planes, with zero associated shear stress (Figure 9.9), and are defined by

$$\sigma_{1,2} = \left(\frac{\sigma_x + \sigma_y}{2}\right) \pm \sqrt{\left(\frac{\sigma_x - \sigma_x}{2}\right)^2 + \tau_{xy}^2} \qquad (9.3)$$

where σ_1 is the maximum value of normal stress, referred to as major principal stress; and σ_2 is the minimum value known as minor principal stress. In the 3D stress field, principal stresses are defined by the following equations:

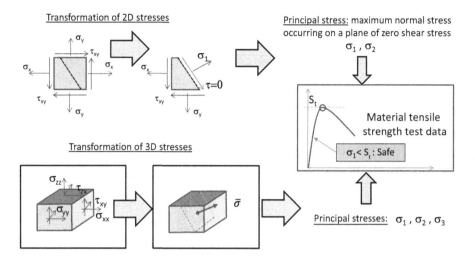

FIGURE 9.9 Principal stress theory for interpretation of stress analysis results (brittle and quasi-brittle materials).

$$\sigma_1 = \frac{I_1}{3} + \frac{2}{3}\left(\sqrt{I_1^2 - 3I_2}\right).cos\varphi$$

$$\sigma_2 = \frac{I_1}{3} + \frac{2}{3}\left(\sqrt{I_1^2 - 3I_2}\right).cos\left(\varphi + \frac{2\pi}{3}\right)$$ (9.4)

$$\sigma_3 = \frac{I_1}{3} + \frac{2}{3}\left(\sqrt{I_1^2 - 3I_2}\right).cos\left(\varphi + \frac{4\pi}{3}\right)$$

where I_1, I_2, and I_3 are called stress invariants, defined by

$$I_{11} = \sigma_{xx} + \sigma_{yy} + \sigma_{zz}$$
$$I_{22} = \sigma_{xx}.\sigma_{yy} + \sigma_{xx}.\sigma_{zz} + \sigma_{yy}.\sigma_{zz} - \tau_{xy}^2 - \tau_{yz}^2 - \tau_{zx}^2$$ (9.5)
$$I_{33} = \sigma_{xx}.\sigma_{yy}.\sigma_{zz} - \sigma_{xx}.\tau_{yz}^2 - \sigma_{yy}.\tau_{zx}^2 - \sigma_{zz}.\tau_{xy}^2 + 2.\tau_{xy}.\tau_{yz}.\tau_{zx}$$

and,

$$\varphi = \frac{1}{3}cos^{-1}\left(\frac{2I_1^3 - 9I_1I_2 + 27I_3}{2\left(I_1^2 - 3I_2\right)^{1.5}}\right)$$ (9.6)

Strength of brittle and quasi-brittle materials is generally defined by the maximum tensile resistance occurring in a uniaxial tensile stest (S_t in Figure 9.9). In multi-axial stress field, the major principal stress is used as the representative indicator of stress state; and the safety status is assessed by comparing it with the material tensile strength (S_t) which may or may not be adjusted for size effect.

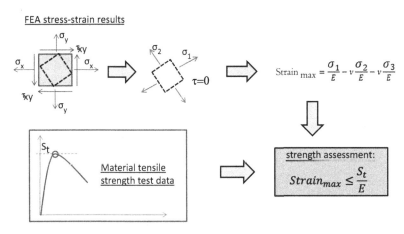

FIGURE 9.10 Principal strain theory for interpretation of stress–strain analysis results (brittle and quasi-brittle materials).

2. <u>Principal strain theory:</u> Similar to the principal stress theory, principal strain theory postulates that failure in brittle and quasi-brittle materials initiates when the major principal strain becomes equal to a pre-determined material tensile resistance defined in term of the material tensile strength (equation 9.7) (Figure 9.10):

$$\varepsilon_1 \leq \frac{S_t}{E} \tag{9.7}$$

In a multi-axial elastic stress field, major principal strain (ε_1) can be calculated from strain components, or from the principal stress values (if those are readily available):

$$\varepsilon_1 = \frac{\sigma_1}{E} - \frac{v}{E}\left(\sigma_2 + \sigma_3\right) \tag{9.8}$$

3. <u>Maximum strain energy theory:</u> In this theory, strength of ductile material is defined by the strain energy per unit volume of uniaxial tensile test specimen that is equal to the area under the stress–strain diagram up to the material yield strength (S_y) (Figure 9.3). Material is assumed to have failed when the strain energy, per unit volume in a finite element calculation point, exceeds the allowable strain energy (as determined from the simple uniaxial tension test):

$$\sum \frac{1}{2}\sigma_{ij}^{T}.\varepsilon_{ij} \leq \frac{S_y^2}{2E} \tag{9.9}$$

4. <u>Maximum shear stress theory:</u> This theory postulates that failure in ductile materials initiates when the maximum shear stress exceeds the critical material shear strength determined from uniaxial tensile test. Mohr's circles for 3D

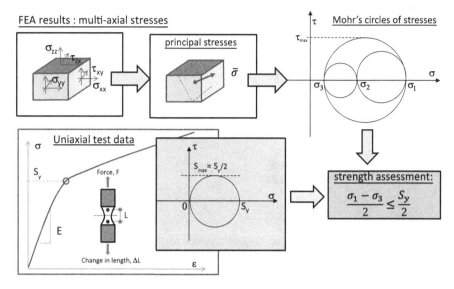

FIGURE 9.11 Maximum shear stress theory for interpretation of stress analysis results (ductile materials).

stress field (Figure 9.11) shows that the maximum shear stress is defined by the following equation (9.10):

$$\tau_{max} = \frac{\sigma_1 - \sigma_3}{2} \tag{9.10}$$

where σ_1 is the major principal stress and σ_3 is the minor principal stress. From the Mohr circle of stress for one-dimensional tensile test of a ductile material specimen, the shear strength of material corresponding to the state of axial stress reaching the yield strength is defined by equation (9.11):

$$S_{max} = \frac{S_y}{2} \tag{9.11}$$

Combining equations (9.10) and (9.11), the criterion for failure initiation in the multi-axial stress field of a ductile material is defined by equation (9.12):

$$\tau_{max} = \frac{\sigma_1 - \sigma_3}{2} \leq \frac{S_y}{2} \tag{9.12}$$

5. Maximum distortion energy theory: General stress state in a three-dimensional body, defined by stress components σ_{xx}, σ_{yy}, etc., can be represented by three principal stresses (σ_1, σ_2, σ_3). The matrix representing the principal stress values can be considered to be made of two matrices – one representing the

average stress state in the body and the other representing the differential stresses in three principal directions as defined in the following equation:

$$
\begin{bmatrix} \sigma_{xx} & \tau_{xy} & \tau_{xz} \\ \tau_{xy} & \sigma_{yy} & \tau_{yz} \\ \tau_{xz} & \tau_{yz} & \sigma_{zz} \end{bmatrix} \equiv \begin{bmatrix} \sigma_1 & 0 & 0 \\ 0 & \sigma_2 & 0 \\ 0 & 0 & \sigma_3 \end{bmatrix}
$$

$$
\equiv \begin{bmatrix} \sigma_{avg} & 0 & 0 \\ 0 & \sigma_{avg} & 0 \\ 0 & 0 & \sigma_{avg} \end{bmatrix} + \begin{bmatrix} \sigma_1 - \sigma_{avg} & 0 & 0 \\ 0 & \sigma_2 - \sigma_{avg} & 0 \\ 0 & 0 & \sigma_3 - \sigma_{avg} \end{bmatrix} \tag{9.13}
$$

where σ_{avg} is the average stress in the body, defined by $\sigma_{avg} = (\sigma_1 + \sigma_2 + \sigma_3)/3$. The differential stresses in three principal directions, represented by the second matrix on the right side of equation (9.13), are generally referred to as distortional stresses $[\sigma']$:

$$
[\sigma'] = \begin{bmatrix} \sigma_1 - \sigma_{avg} & 0 & 0 \\ 0 & \sigma_2 - \sigma_{avg} & 0 \\ 0 & 0 & \sigma_3 - \sigma_{avg} \end{bmatrix} \tag{9.14}
$$

The main assumption in distortion energy theory is that the hydrostatic tension or compression, $[\sigma_{avg}]$, does not cause failure in a ductile material; actual material failure is assumed to initiate when the distortional stresses $[\sigma']$ reach a critical state. Denoting the strains caused by distortional stresses as ε', the strain energy density corresponding to the distortional stress–strain state can be derived as follows (Budynas 1999):

$$
U_d = \frac{1}{2}\{\sigma'\}^T \cdot \{\varepsilon'\} = \frac{1+\vartheta}{6E} \cdot \left[(\sigma_1 - \sigma_2)^2 + (\sigma_2 - \sigma_3)^2 + (\sigma_3 - \sigma_1)^2 \right] \tag{9.15}
$$

Principal stresses in a uniaxial tensile test specimen, at the impending state of yielding, can be defined as follows:

$$
\sigma_1 = S_y, \quad \sigma_2 = \sigma_3 = 0 \tag{9.16}
$$

Substituting the values from equation (9.16) into equation (9.15), critical distortion energy causing material yielding in uniaxial tensile test specimen is given by

$$
U_d^{critical} = \frac{1+\vartheta}{3E} \cdot \left[S_y \right]^2 \tag{9.17}
$$

Maximum distortion energy theory postulates that yielding failure in ductile material initiates when the distortion energy in a 3D stress field, defined by equation (9.15),

FEA results : multi-axial stresses

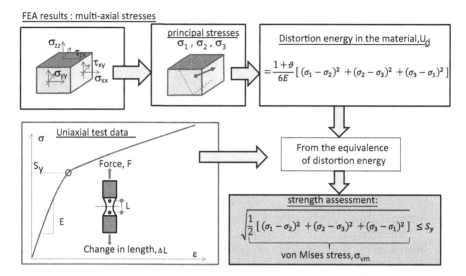

FIGURE 9.12 Maximum distortion energy theory for interpretation of stress analysis results (ductile materials).

reaches the critical material strength defined by equation (9.17), leading to the following criterion for stress assessment (Figure 9.12):

$$\sqrt{\frac{1}{2}\left[\left(\sigma_1-\sigma_2\right)^2+\left(\sigma_2-\sigma_3\right)^2+\left(\sigma_3-\sigma_1\right)^2\right]} \leq S_y \qquad (9.18)$$

The stress function on the left side of equation (9.18) is the well-known von Mises stress (σ_{vm}) that is widely used in strength assessment of ductile materials:

$$\sigma_{vm}=\sqrt{\frac{1}{2}\left[\left(\sigma_1-\sigma_2\right)^2+\left(\sigma_2-\sigma_3\right)^2+\left(\sigma_3-\sigma_1\right)^2\right]} \leq S_y \qquad (9.19)$$

Each of the five failure theories, described in the above, has its special use depending on the physical failure characteristics of materials under consideration. For preliminary strength assessment, based on linear elastic finite element analysis results, principal stress and von Mises stress parameters are widely used for brittle and ductile materials, respectively. Several other material failure theories are also available in the literature for strength assessment of special material cases. Constitutive models for nonlinear material response simulations involve many more theories and material parameters that will be partially discussed in Chapter 12. Material theory manuals of commercially available finite element software products generally provide important details of the implemented formulations that are suitable for simulating specific material failure characteristics.

9.4 POST-PROCESSING OF FINITE ELEMENT STRESS ANALYSIS RESULTS

Solution of matrix equilibrium equations (1.5) produces displacements at nodal degrees of freedom. Strain–displacement relationships, defined by equation (2.6), give strains at internal calculation (integration) points of an element based on the known nodal displacement values. Stress–strain relationships, defined by equation (2.26), provide stresses at the element integration points based on the calculated strain responses. Finite element software packages may save nodal displacements, and element stresses and strains, in a binary database file by default or when requested by the analyst. ABAQUS, for example, saves the nodal and element responses in a database file named with an extension ".ODB". Saved response values generally refer to the user-defined model coordinate system (x–y–z). These finite element stress and strain results can be post-processed to derive the material-specific response parameters, as discussed in Section 9.3. General-purpose post-processing software packages, such as HyperView (Altair.COM) and FEMAP (Siemens.COM), can help with the post-processing of result files saved by many commonly used FEA software packages. For example, ABAQUS result file (.ODB) can be imported into HyperView where an analyst can choose to create contour plots of principal stress, principal strain, or von Mises stress depending on the material strength theory relevant for a specific case study. Structural shell elements, as discussed in Chapter 7 (also summarized in Figure 9.7), produce stress variations through the thickness direction. In graphical mid-plane representation of shell geometry, a user can choose to plot the stress results from one particular layer of through-thickness integration points, or the maximum response occurring among integration points in the thickness direction. User chosen material-failure-theory-specific stress response values, calculated from finite element stress analysis results, can be used in contour plots to identify strength critical areas of a general structural analysis case-study (Figure 9.13). Identification of highly stressed areas, based on linear elastic finite element analysis results, does

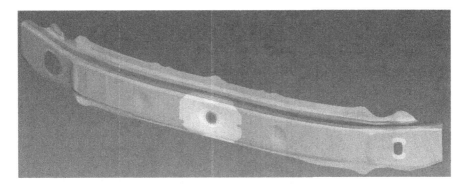

FIGURE 9.13 Contour plot of maximum through-thickness von Mises stress – plotted on the mid-surface of shell element model of an automotive bumper beam (FEA model extracted from a vehicle model – courtesy of NHTSA 2020).

not conclusively indicate safety or failure possibility of a structure. These results provide preliminary indications of critical areas that require further investigations with more rigorous analysis techniques – including the simulation of nonlinear response mechanisms if appropriate.

9.5 STRESS ANALYSIS FOR DURABILITY (FATIGUE LIFE) ASSESSMENT OF STRUCTURES

In monotonic load–displacement response simulations, a simple comparison between the calculated maximum stress response parameter, and the material strength value, is a preliminary indicator of the structural safety margin. When the applied stress exceeds elastic response limit, material can experience partial damage due to yielding or micro-cracking – leaving a permanent distortion in geometry after the external load is removed. A few applications of repeated loading-unloading cycles can accumulate further internal damage; and the material can eventually fail although ultimate material strength is not exceeded in one load cycle (Figure 9.14(a)). This phenomenon is generally referred to as low-cycle fatigue damage of materials. Nonlinear material stress–strain laws are required to predict material damage evolution with finite element simulation models of high-intensity loading events. Most engineering products, however, experience low amplitude fluctuations of elastic stresses and strains during normal operating conditions. An automotive body or chassis component, for example, may experience thousands of low amplitude stress–strain cycles during every-day driving operations. Cumulative damage in material under a high number of loading cycles, for stress–strain fluctuations well within the elastic range of material behavior (Figure 9.14(b)), is generally referred to as high cycle fatigue damage of materials. The terms "durability" and "fatigue life" are often used interchangeably to describe the ability of a component to last for the useful life of the product (for example, 15 years of service or 250,000 km of driving life of a car). Elastic finite element stress–strain analysis results, with high-cycle material fatigue life estimation theories, are routinely used to verify the durability of automotive products (World Auto Steel 2015).

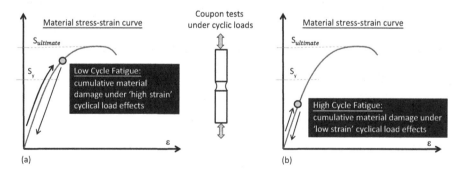

FIGURE 9.14 (a) Material fatigue under high amplitude cyclic stress–strain loading; (b) material fatigue under low-amplitude cyclic stress–strain loading.

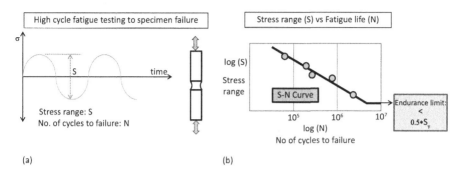

FIGURE 9.15 (a) Constant-amplitude cyclic load testing up to specimen failure; and (b) S–N curve for fatigue life estimation.

Fatigue life of material is usually measured by testing standard specimens under constant amplitude uniaxial load cycles until failure (ASTM E466-15 2021). A series of tests are conducted to determine fatigue lives (number of cycles to failure) for different amplitudes of stress fluctuation. A log–log plot of the stress range (the difference between maximum and minimum values) versus the number of cycles to failure, commonly referred to as S–N curve (Figure 9.15), is considered a property of the specific tested material. When the stress amplitude at a critical location is known from finite element simulation model, the expected fatigue life is determined from the material-specific S–N curve. Like uniaxial material tensile strength, fatigue resistance curve of material is also derived from uniaxial stress testing, although the actual structural components are generally subjected to fluctuations of multi-axial stress state. Consistent with the material failure theories for monotonic loading condition, major principal stress value for brittle and quasi-brittle materials, and von Mises stress value for ductile materials can be used to estimate fatigue life from corresponding material-specific S–N curve.

Stress fluctuation histories in real engineering applications often comprise of variable amplitude random fluctuations. Several empirical and analytical methods are available in the literature for reducing a variable amplitude complex stress history to equivalent blocks of constant amplitude stress cycles (ASTM E1049-85 2021). Partial fatigue damage caused by each constant amplitude stress block is estimated as the ratio between number of applied stress cycles (n_i) to the fatigue life (N_i), corresponding to the specific fluctuation range of stress variation (Figure 9.16). Cumulative fatigue damage of all applied stress fluctuations is calculated from simple arithmetic summation of all fractional contributions:

$$\Sigma \frac{ni}{Ni} \leq 1 \qquad (9.20)$$

Linear superposition rule of cumulative fatigue damage calculation (equation 9.20), known as Miner's rule (Miner 1945), can be used in durability assessment of mechanical components when variable amplitude stress history is known from past-historical measurements. In the early assessment of mechanical designs, external

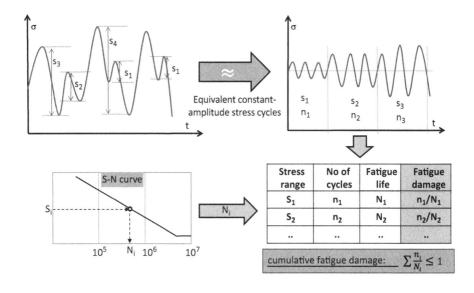

FIGURE 9.16   Durability assessment for variable amplitude stress cycles.

loading events are often represented by few selected critical load cases. For example, fatigue life targets of automotive body structures for normal driving conditions can be pre-defined, based on historical evidence, as follows (World Auto Steel 2015):

Fatigue critical event	Equivalent static loading	Target fatigue life
Pothole impact	3g vertical acceleration of un-sprung mass	200,000 cycles
Vehicle turning	0.7g lateral acceleration of vehicle mass	100,000 cycles
Forward braking	0.8g longitudinal deceleration of vehicle mass	100,000 cycles

Linear-elastic finite element simulation model of automotive body structure is used to identify the critical locations and the corresponding stress values for each of these load cases. Fatigue life at an identified critical location is estimated from the material-specific S–N curve for the stress fluctuation range caused by the relevant loading event. Cumulative fatigue damage for combined load effects at a critical location is estimated from equation (9.20) where "n_i" refers to applicable fatigue load cycles, and "N_i" refers to estimated fatigue life values from material-specific S–N curve. When different critical stress locations arise for each of the external loading conditions, durability assessment reduces to a simple exercise of checking the estimated fatigue life at the critical locations against the pre-defined fatigue life targets set from past product design experiences. Dedicated fatigue life analysis software (such as nCode), coupling the finite element stress analysis results with material-specific S–N curve, can be used for durability assessment of mechanical designs.

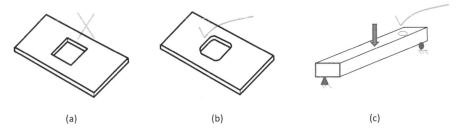

(a)   (b)   (c)

FIGURE 9.17   (a) Hole with undesirable sharp corners; (b) hole with smooth corners to reduce stress concentration effects; and (c) hole in low stressed zone of a structural member.

When a design does not meet pre-defined fatigue life target, engineering solution may consider design alternatives of (i) load input modifications (if possible), or (ii) material modification (to improve fatigue life), or (iii) structural design modifications to reduce internal stresses (by changes to part geometry and thickness). A good rule of thumb is that 10% decrease in stress will double the fatigue life of a component. In other words, if the estimated fatigue life is at 50% of desired fatigue life, decreasing the stress level by 10% would allow the part to achieve full 100% of the durability life. In design for durability, the use of good engineering practice is perhaps more critical than the prediction of stress response with reliable finite element simulation technique. When engineered properly, a component can usually meet the target fatigue life at no cost or weight penalty. The basic principle in design for durability is to eliminate or minimize the effects of stress raisers – without waiting to discover local stress concentration issues during late-stage finite element simulation exercises. As discussed in Section (4.1), and illustrated in Figure 4.2, a simple circular hole in a uniform axial stress field can raise the local stress concentration value to 3-times the nominal stress value. As per equation (4.3), local stress concentration value increases as the hole shape in Figure 4.2 becomes narrower ($a < b$), thereby leading to a lower fatigue life for the component. If a hole is essential in a stressed component design for special functional reasons, a smooth circular hole is much more preferable to a square or rectangular hole. However, when a non-circular hole is required for specific reasons, the corners of the hole must be produced with smooth radii (Figure 9.17(a)) to reduce stress concentration values around the corners. When there is an option, cutouts or holes in a member must be located in low-stressed areas to reduce the impact on durability life of structural component. In addition to geometric discontinuity in mechanical component design, discrete joints in multi-component product assemblies also act as stress raisers leading to fatigue life concerns for engineered products. Spot welds, for example, show poor fatigue resistance when subjected to tensile loading mode (Figure 9.18(a)); better fatigue resistance is achieved when the weld joint is engineered to transfer shear loading (Figure 9.18(b)). Smooth stress transfer at welded joints is further improved by using structural adhesives in between the mating surfaces of structural sheet metals (Dow Automotive Systems 2021).

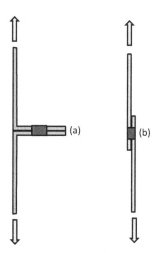

FIGURE 9.18   (a) Tensile peeling load on a welded joint; and (b) shear loading on a welded joint.

9.6   STRUCTURAL SAFETY ASSESSMENT AND QUALITY OF FEA STRESS RESULTS

As evident from the discussions in Sections 9.2–9.5, structural performance assessment for strength and durability, in early design phases, relies on the stress analysis results from finite element simulation models. Quality of the stress analysis models, thus, has a direct impact on the quality of engineering decisions made with respect to strength and durability performance of a proposed structural design configuration. Finite element stress analysis results need to be used with caution while predicting the strength and fatigue life of engineered products. In general, finite element simulation models can estimate the locations of high stressed areas in a structure, but the magnitude of predicted stress can be affected by mesh quality and severity of geometric discontinuity in the model. The simulation model of a uniformly loaded plate, with a smooth circular hole at the center (Figure 4.2), has

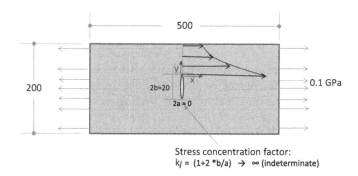

FIGURE 9.19   Indeterminate stress concentration at the tip of a sharp geometric discontinuity in the uniformly stressed plate.

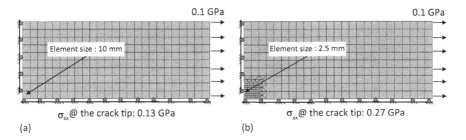

FIGURE 9.20 Mesh-dependent prediction of stress response at the crack tip of a uniformly stressed plate.

predicted the internal stress distribution accurately – producing the local stress concertation effects at the crest of the circular hole consistent with the result of the analytical solution (equation 4.3). When the circular hole is replaced by a narrow crack at the center of plate (Figure 9.19), analytical stress prediction at the tip of sharp geometric discontinuity becomes indeterminate. Finite element simulation models, however, always predict some stress value at the points of discontinuity because of the homogenization of the stress singularity over a finite size element (Figure 9.20). As the finite element mesh is refined, the predicted stress response keeps rising without convergence to a stable predictable response (Figure 9.20). A similar mesh-dependent stress result has been obtained in the vicinity of the rigid boundary constraint of the in-plane bending problem in Figure 3.19(b). It is, therefore, important to verify that the stress response predicted by a finite element simulation model is reliable for use in strength-based criteria of structural integrity assessment. The general rule of practice is to conduct stress analyses with two finite element models – one having double the mesh density compared to the other. If the stress results from two models are very close (within 10%), the predicted stress result can possibly be used with some confidence for structural safety assessment. Close predictions of stress results with two different mesh density models, however, do not mean that the predicted results are accurate. The magnitude of stress results may be affected by the inherent limitations of the finite element formulations (as discussed in Section 3.10). Local stress responses, predicted by finite element simulation models, often need to be supplemented with past experiences of the analyst to reach meaningful conclusions on the safety assessment of structures.

9.7 STRESSES AT POINTS OF DISCONTINUITY: STRESS INTENSITY FACTOR

As evident from the discussions presented in Section 9.6, linear elastic stress analysis results are not objective at the vicinity of singular stress points. Theory of fracture mechanics provides an alternative method for the integrity assessment of solids with cracks. Considering a local coordinate system x–y with origin at the tip of a sharp

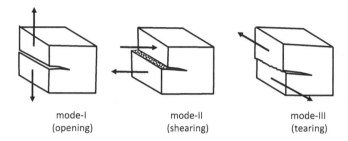

mode-I	mode-II	mode-III
(opening)	(shearing)	(tearing)

FIGURE 9.21   Fracture modes in a general 3D solid.

crack in Figure 9.19, the indeterminate state of elastic stress at the crack tip can be defined by the following expression:

$$\sigma_x = \frac{K_I}{\sqrt{2\pi y}} \qquad (9.21)$$

where the term K_I defines the intensity of stress at the vicinity of crack tip – commonly known as the stress intensity factor (Irwin 1957). As per the theory of fracture mechanics, the strength of a cracked body is determined by the magnitude of stress intensity factor. Crack traversing a normal stress field will have unstable growth when K_I reaches a critical material strength value called Fracture Toughness (K_{Ic}). Toughness value changes depending on the mode of crack tip deformation (Figure 9.21). ASTM Standard (ASTM E1820-20ae1 2021) can be followed to determine the fracture toughness of material specimens for mode-I (opening) crack. Alternatively, reference values for preliminary assessment of material fracture resistance can be obtained from ASM Handbook (ASM International 1997).

Stress intensity factor at the tip of a crack inside a general solid depends on crack length, geometric configuration of cracked body, the external load, and boundary conditions. Analytical and empirical expressions for calculation of stress intensity factors in simple geometric and loading configurations are available in Tada et al. (2000). More complex structural design cases require the use of finite element simulation models to calculate the stress intensity factor. Several alternative numerical techniques are available for the extraction of stress intensity factor values from finite element simulation models, such as crack-tip singular element formulations (Zienkiewicz and Taylor 1991), energy release rate method (Zehnder 2012), and J-integral method (Rice 1968). ABAQUS software includes a special routine for the extraction of stress intensity factors from J-integral calculations (Dassault Systems 2020b). A simpler approximate method for estimation of stress intensity factor involves the use of equation (9.21) directly with the finite element stress analysis results. For the center crack problem of uniformly stressed plate (Figure 9.19), using the stress value σ_x at a distance of 2.5 mm from the crack-tip of finite element model (Figure 9.20(b)), equation (9.21) gives the following estimated value of K_I:

$$K_I = \sigma_x.\sqrt{2\pi y} = 0.1578\sqrt{2\pi * 2.5} \approx 0.625 \, \text{GPa}\sqrt{\text{mm}} \qquad (9.22)$$

Analytical solution for stress intensity factor at the tip of central crack of uniformly loaded large plate gives (Broek 2012):

$$K_{theoretical} = \sigma_{nomial} \cdot \sqrt{\pi * b} = 0.10 * \sqrt{\pi * 10} \approx 0.560 \text{ GPa}\sqrt{\text{mm}} \qquad (9.23)$$

The difference between stress intensity factor values in equations (9.22) and (9.23) is less than 12%. The calculation procedure, based on the local finite element stress response value (equation 9.22), provides an approximate value for the stress intensity factor since the stress values, extrapolated at element nodes, are not very accurate. Alternatively, the stress intensity factor can also be calculated directly from standard finite element models based on the elastic energy release rate concept. Elastic energy stored in the plate with 20 mm long crack at the center is four times the energy stored in the quarter model of plate shown in Figure 9.20(b)):

$$U_0 = 4 * 0.5985 = 2.3941 \text{ kN} - \text{mm} \qquad (9.24)$$

Re-analyzing the model with an extended crack length of 25 mm (2.5 mm extension of each tip), strain energy stored in the model is given by

$$U_1 = 4 * 0.600532 = 2.4021 \text{ kN} - \text{mm} \qquad (9.25)$$

Change in strain energy of the system per unit area of crack extension is given by

$$R = \frac{\Delta U}{\Delta a * 1} = \frac{2.4021 - 2.3941}{5 * 1} = 0.0016 \text{ kN/mm} \qquad (9.26)$$

Stress intensity factor for the rate of strain energy change, R, is given by (Broek 2012):

$$K_I = \sqrt{E * R} = \sqrt{200 * 0.00161} = 0.566 \text{ GPa}\sqrt{\text{mm}} \qquad (9.27)$$

The predicted value from the strain energy release rate method (equation 9.27) is very close to the theoretical value (equation 9.23) – with 1% difference between the two. Standard finite element simulation models, thus, provide reasonable estimates of the stress-intensity factor at a crack tip subjected to the tensile opening mode of loading. Similar calculation procedures can be used to predict stress intensity factor values for other crack opening modes as well (Figure 9.21). Like the stress-based criterion for cumulative fatigue damage calculations, fatigue crack growth in a solid can also be calculated based on the fluctuation of the stress intensity factor:

$$\frac{da}{dN} = C.(\Delta K)^m \qquad (9.28)$$

Equation (9.28) is known as Paris law (Paris and Erdogan 1963), where "a" is crack length, "N" the number of load cycles, "ΔK" the range of stress intensity factor variation, and C,m are material constants determined from experiments. Finite element model results for ΔK can be used with equation (9.28) to determine the fatigue crack growth rate in a body.

9.8   PRACTICE PROBLEMS: ASSESSMENT OF STRUCTURAL STRENGTH AND DURABILITY

PROBLEM 1

From the load-deflection analysis results of the automotive roof problem in Problem-3, Section 7.10, generate a contour plot of maximum von Mises stress values corresponding to the applied pressure load of 0.15 kN. Assuming material yield strength of 0.12 GPa, determine the factor of safety against yielding of material at the vicinity of load application area.

PROBLEM 2

Vehicle drive over a hypothetical durability test track generates the variable amplitude stress history of Figure 9.22(b) – measured at a critical body structural component. For example calculations, Table in Figure 9.22(c) shows the constant amplitude stress ranges and number of cycles equivalent to one week of vehicle driving condition. Using the S–N curve from Figure 9.22(d), estimate the fatigue life of the relevant structural component.

PROBLEM 3

Figure 9.23 shows uniform in-plane stress and boundary constraints on a plate with hexagonal hole at the center. Calculate the stress intensity factor at point "A" of the hexagonal hole.

PROBLEM 4

Figure 9.24 shows a plate with hole – subjected to in-plane uniform traction (0.01 GPa) on the upper edge. Calculate the maximum stress occurring on the perimeter of the circular hole.

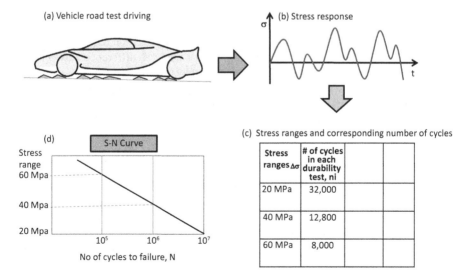

FIGURE 9.22   Cumulative fatigue damage calculation for variable amplitude loading.

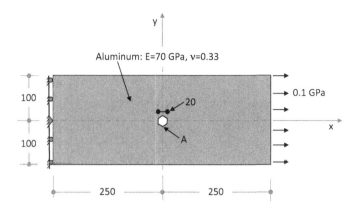

FIGURE 9.23 Uniform stress and boundary constraints on a plate with a hexagonal hole at the center.

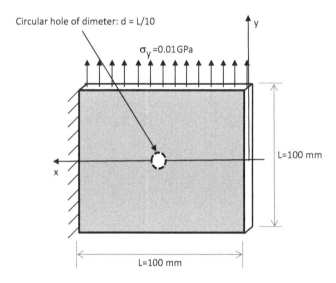

FIGURE 9.24 Hole at the center of a unit thickness steel plate subjected to uniform in-plane traction on the upper boundary (assume $E = 200$ GPa, $\nu = 0.3$).

10 Vibration Frequency Analysis of Structures with FEA Model

SUMMARY

Linear elastic finite element analysis, discussed so far in Chapters 1–9, has considered time-independent load effects – represented by static equilibrium state between applied load and stiffness-based deformation resistance of structures. Vibration response of a structure refers to dynamic oscillations of system responses (stresses, displacements, etc.) under external perturbation effects. The metrics for structural vibration, represented by cycle per unit time or the time period taken for one complete cycle of response variation, depend on the stiffness and mass properties of a given structure. The vibration frequency (or period of structural vibration), representing the dynamic property of a structure, and its relative relationship with the dynamic characteristics of applied load, define the amplitude of linear dynamic response of a structure. The study of vibration frequency is, thus, a very important part of the engineering development process for civil, mechanical, and aerospace engineering products and structures. The basic dynamic equilibrium state, between applied load and corresponding system resistances (representing the deformation and inertia characteristics of a deformable spring-mass system), is established in Section 10.1 based on Newton's second law of motion. The free-vibration response of the single-degree-of-freedom (SDOF) spring-mass system, induced by an initial perturbation to the static rest state, is analyzed in Section 10.2 – eventually leading to the important fundamental relationship among stiffness, mass, and vibration frequency properties of the system. Section 10.3 is devoted to the analytical descriptions of forced vibration response and resonance behavior of SDOF linear elastic system. The effect of internal energy loss mechanism, represented by the addition of a damping term in the description of SDOF system, is also analyzed in this section. The relative relationship between structural vibration property and the dynamic load characteristic, defining the relevance of static versus dynamic response analysis techniques, and the effect of damping on the overall dynamic amplification of system response, are graphically demonstrated based on SDOF system solutions. The use of frequency separation concept, to define targets for subsystem designs, is discussed with reference to a hypothetical automotive system example in Section 10.4. Analytical techniques to estimate the vibration frequency and mode shapes of relatively more complex systems, having uniformly distributed system and mass properties, are developed in Section 10.5.

The basic definitions of SDOF vibration characteristics are extended to multi-degree-of-freedom (MDOF) system property definitions in Section 10.6

– developing the analytical formulations to calculate the vibration frequencies from stiffness and mass property matrices of MDOF systems. Mass matrix calculation of finite elements, not discussed in earlier chapters, is discussed in Section 10.7. Numerical techniques for the calculations of MDOF mode shapes and frequencies, as implemented in finite element software packages, are reviewed in Section 10.8. Relative efficiencies of different numerical techniques, for the analysis of large MDOF systems, are also discussed in that section. Software-based analysis of modal frequencies and mode shapes of finite element models is discussed with ABAQUS-specific options in Section 10.9. Finally, practice problems for modal frequency analysis of MDOF finite element models are presented in Section 10.10.

10.1 INTRODUCTION – DYNAMIC RESPONSE OF STRUCTURES

Newton's second law of motion states that force (F) acting on a free rigid body causes a change in momentum that is proportional to the force applied, and the equation of motion for a given mass value (m) is as follows:

$$F = m * \frac{\partial\left(\dot{u}\right)}{\partial t} \tag{10.1}$$

Writing the rate of momentum change as acceleration of the body, equation (10.1) is re-written as follows:

$$F = m * \frac{\partial^2 u}{\partial t^2} = m * \ddot{u} \tag{10.2}$$

Figure 10.1 shows the direct equilibrium between externally applied force (F) and inertia resistance of a free rigid body ($m*\ddot{u}$). When the body mass is attached to a flexible structural system, represented by a spring in Figure 10.2, the dynamic motion

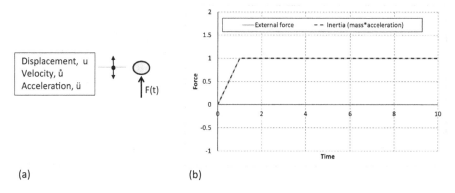

(a)   (b)

FIGURE 10.1 (a) Dynamic response of a unit mass under externally applied load ($F(t)$); and (b) force time histories (external force = inertia resistance).

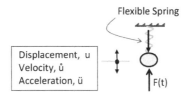

Flexible Spring

Displacement, u
Velocity, u̇
Acceleration, ü

F(t)

FIGURE 10.2   Dynamic response of a flexible spring-mass system under externally applied load F(t).

of mass leads to spring deformation that is resisted by the stiffness property of structure. System resistance to mechanical deformation can be defined by using the structural stiffness properties as described in Section 1.4 (Figure 1.7(b) and equation 1.1). State of equilibrium between the externally applied force and the combined system resistances derived from inertia and stiffness properties of the flexible spring-mass system of Figure 10.2 can be expressed by

$$F(t) = k.u + m * \ddot{u} \tag{10.3}$$

where u is spring deformation and $\ddot{u}$ is the acceleration of mass – both time-dependent responses of the spring-mass system. If the external force $F(t)$ is removed at a time, spring force $(k.u)$ and inertia force $(m \cdot \ddot{u})$ must equilibrate each other leading to free-vibration response of the system – discussed in the following Section 10.2. Forced vibration response of the system, including the effect of system energy losses due to damping, will be discussed in Section 10.3.

10.2   VIBRATION FREQUENCY OF A SINGLE DEGREE OF FREEDOM SPRING-MASS SYSTEM

Figure 10.3(a) shows a mass (m), attached to an elastic spring of stiffness (k). If the mass is pulled from its rest position by a unit displacement $(u = -1)$ and released, the unbalanced spring force on the mass will cause an upward acceleration of the mass. As the mass returns towards the original rest position $(u = 0)$, velocity of the mass gradually increases, and the spring force gradually reduces – eventually becoming zero when the mass returns to its rest position. Assuming no energy losses in the ideal elastic spring-mass system (Figure 10.3(b)), the elastic deformation energy, stored in the spring at stretched state $(u = -1)$, gradually converts to kinetic energy of the mass as it returns towards the initial un-stressed position. The velocity of mass, gained during the retraction phase of spring from deformation state of $u = -1$ to initial stress-free state $u = 0$, will keep driving the mass further upward and that motion will cause progressive compression in the spring until the spring compresses to a state of $u = +1$, and the mass reaches zero velocity state $(\dot{u} = 0)$ due to the spring resistance opposing its upward motion. The compressed spring from that position will start the downward return motion of mass with gradual conversion of spring elastic energy to kinetic energy of mass. This repeated cycle of energy conversion between spring deformation and mass kinetic motion will continue un-interrupted resulting in a

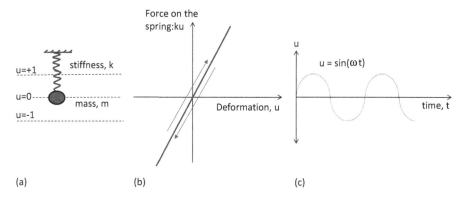

FIGURE 10.3   (a) Single degree of freedom (SDOF) spring-mass system; (b) linear elastic resistance-to-deformation response of the spring; and (c) harmonic motion of the mass.

cyclic harmonic motion of the system (Figure 10.3(c)) – commonly known as the free vibration response of a system:

$$u(t) = \sin(\omega t) \tag{10.4}$$

where ω (= $2\pi f$) is the cyclic motion of the mass and f is the frequency of vibration per unit time. The self-equilibrating state of free vibration response, between the spring resistance force and mass inertia force, can be described by re-writing equation (10.3) as follows:

$$k.u + m * \ddot{u} = 0 \tag{10.5}$$

Substituting the expression (10.4) into equation (10.5) gives:

$$\left(k - \omega^2 m\right) * \sin(\omega t) = 0 \tag{10.6}$$

Equation (10.6) will be satisfied for any value of ω and t. This means that the multiplication term in the parenthesis $(k-\omega^2 m)$ must be zero – leading to the following frequently used relationships among stiffness (k), and mass (m), and vibration characteristics (frequency f, period of vibration, T, and angular velocity ω) of the spring-mass system:

$$\omega^2 = \frac{k}{m} \rightarrow \omega = 2\pi f = \frac{2\pi}{T} = \sqrt{\frac{k}{m}} \tag{10.7}$$

The terms f and T are commonly referred to as natural vibration frequency and fundamental period of vibration of the SDOF spring-mass system. The fundamental vibration property, represented by ω, f or T, defines the dynamic response characteristic of a system when subjected to a time-dependent dynamic force function. The relative relationship, between time-domain characteristics of force function and the responding spring-mass system, is a critical factor defining the intensity of a system's dynamic response to the applied external force. Section 10.3 in the following discusses the forced vibration response – which is an important aspect of structural design for dynamic load effects.

10.3 FORCED VIBRATION RESPONSE AND RESONANCE OF STRUCTURES

Figure 10.4(a) shows a general spring-mass system subjected to a time-dependent force F(t). An example time history of external force is described by the sine harmonic function in Figure 10.4(b). Assuming linear elastic response of spring without any energy losses due to damping effects, the dynamic equilibrium equation (10.3) can be rewritten for the sine harmonic load case as follows:

$$m * \frac{\partial^2 u}{\partial t^2} + k.u(t) = F_o \sin(\bar{\omega} t) \tag{10.8}$$

where F_o is the amplitude, and $\bar{\omega}$ the angular velocity of harmonic load function ($\bar{\omega} = 2\pi \bar{f}$; $\bar{f}$ being the frequency of applied load). Dividing both sides of equation (10.8) by the spring stiffness term, k:

$$\frac{m}{k} * \frac{\partial^2 u}{\partial t^2} + u(t) = \frac{F_o}{k} \sin(\bar{\omega} t) \tag{10.9}$$

Substituting ω^2 for the ratio between spring stiffness k and mass m, equation (10.9) is re-written in the following form:

$$\frac{1}{\omega^2} * \frac{\partial^2 u}{\partial t^2} + u(t) = \frac{F_o}{k} \sin(\bar{\omega} t) \tag{10.10}$$

A particular solution of second-order partial differential equation (10.10) can be expressed in term of the applied loading function multiplied by response amplitude G (equation 10.11):

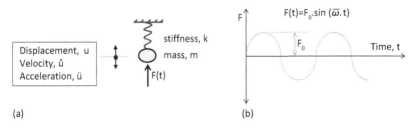

(a)   (b)

FIGURE 10.4 (a) Dynamic response parameters of a spring-mass system under time-dependent applied load $F(t)$; and (b) example of a time-dependent loading history.

$$u_p = G\sin(\bar{\omega}t) \tag{10.11}$$

Equation (10.11) represents one part of the displacement response that is in-phase with the loading function. Substituting expression (10.11) into equation (10.10), and after re-arranging the terms, the response amplitude G is obtained as follows:

$$G = \frac{F_0}{k}\frac{1}{1-\left(\dfrac{\bar{\omega}}{\omega}\right)^2} = \frac{F_0}{k}\frac{1}{1-\beta^2} \tag{10.12}$$

where β is the ratio between $\bar{\omega}$ and ω. A second solution of the second-order partial differential equation (10.10), known as the complimentary solution, is given by the following function:

$$u_c = A\sin(\omega t) + B\cos(\omega t) \tag{10.13}$$

Equation (10.13) represents the free vibration response of the spring-mass system without the presence of external force function. Combining equations (10.11), (10.12), and (10.13), total dynamic response of the spring-mass system is given by the following equation:

$$u(t) = A\sin(\omega t) + B\cos(\omega t) + \frac{F_0}{k}\frac{1}{1-\beta^2}\sin(\bar{\omega}t) \tag{10.14}$$

Considering a special initial condition of $u(0) = \dot{u}(0) = 0$, unknown terms A and B in equation (10.14) are obtained as follows:

$$A = \frac{-F_0}{k}\frac{\beta}{1-\beta^2}, \qquad B = 0 \tag{10.15}$$

Substituting the expressions from equation (10.15) into equation (10.14), the time-domain response of spring-mass system, for the harmonic force function shown in Figure 10.4(b), is given by the following equation:

$$u(t) = \frac{F_0}{k}\frac{1}{1-\beta^2}(\sin\bar{\omega}t - \beta\sin\omega t) \tag{10.16}$$

The first multiplication term (F_0/k) in equation (10.16) represents the static displacement response corresponding to load amplitude F_0; second term $(1/(1-\beta^2))$ is a dynamic amplification factor applied to that static response; the first term inside the parenthesis $(\sin\bar{\omega}t)$ represents the response function in-sync with the external force function described in Figure 10.4(b), and the second term inside the parenthesis $(\beta.\sin\omega t)$ represents the free vibration response of the system with an angular velocity of ω defined by equation (10.7). The parameter β, as defined earlier, is the ratio between frequency of applied loading $(\bar{f})$ and the natural vibration frequency of spring-mass system (f). As evident from equation (10.16), when the frequency of applied external force approaches the natural vibration frequency of system, the dynamic amplification factor $(1/(1-\beta^2))$ approaches infinity – implying that the response of elastic system amplifies indefinitely under the action of external cyclic loading. This phenomenon of dynamic structural response is commonly known as the "resonance" of structure when $f = \bar{f}$. Most built-up structures, however, experience energy losses due to hysteretic load-deformation response as shown in Figure 10.5. This response mechanism is generally represented by adding a velocity-dependent resistance component to the linearized deformation resistance function – leading to the following modified form of the dynamic equilibrium equation (10.8):

$$m * \frac{\partial^2 u}{\partial t^2} + c.\frac{\partial u}{\partial t} + k.u(t) = F_o \sin(\bar{\omega}t) \qquad (10.17)$$

The term "c" in equation (10.17) is commonly known as the damping resistance of the system. Diving both sides of equation (10.17) by the spring stiffness term "k":

$$\frac{m}{k} * \frac{\partial^2 u}{\partial t^2} + \frac{c}{k}.\frac{\partial u}{\partial t} + u(t) = \frac{F_o}{k}\sin(\bar{\omega}t) \qquad (10.18)$$

Introducing $\omega^2 = k/m$, and $\xi = c/2\sqrt{k.m}$, equation (10.18) can be re-written in the following form:

$$\frac{1}{\omega^2} * \frac{\partial^2 u}{\partial t^2} + \frac{2\xi}{\omega}.\frac{\partial u}{\partial t} + u(t) = \frac{F_o}{k}\sin(\bar{\omega}t) \qquad (10.19)$$

The term ξ is generally referred to as the damping ratio. Free vibration response of a damped system diminishes with time due to the presence of damping. Ignoring the free-vibration response part, the dynamic response of damped system can be defined by the following load-dependent time function (Clough and Penzien 1975):

$$u(t) = \frac{F_0}{k}.D.\sin\left(\bar{\omega}t - tan^{-1}\frac{2\xi\beta}{1-\beta^2}\right) \qquad (10.20)$$

where "D" is the dynamic amplification factor for damped spring-mass system defined in equation (10.21):

$$D = \frac{1}{\sqrt{\left(1-\beta^2\right)^2 + \left(2\xi\beta\right)^2}} \qquad (10.21)$$

Figure 10.6 shows graphical representation of the factor D for different damping ratio values. When the structural vibration frequency is very high compared to the frequency of applied loading function ($\beta \approx 0$), or in other words, when the

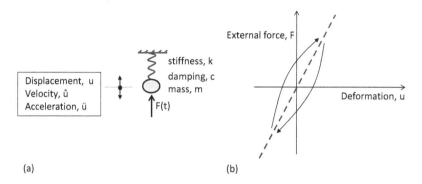

(a)   (b)

FIGURE 10.5   (a) General dynamic characteristics of a spring-mass-damper system; and (b) hysteresis in force–deformation response of the spring.

time-domain variation of loading is very slow compared to the vibration period of structure, the dynamic amplification factor value is close to "1". This means that the system response is well predicted by the static load–deflection analysis irrespective of the damping behavior of structure. On the other end, when ($\beta > 2$), load frequency is high and the structure is relatively flexible with a low frequency of vibration, the dynamic amplification factor is well below "1" irrespective of the damping value. In that situation, inertia of the system dominates the response to dynamically applied load. At resonant frequency ($\beta = 1$), the value of "D" becomes infinite for zero damping value ($\xi = 0.0$) – as expected for an undamped system. Figure 10.6 also shows that for highly damped systems ($\xi > 0.5$), the value of D approaches "1" for frequency ratio in the range of 0–1, implying that the system response can be predicted by simple static load–deflection analysis. However, for lightly damped common structural systems (buildings, towers, bridges, automotive body structures, etc.), the damping ratio tends to be well below 10% ($\xi < 0.1$) which implies that "D" will reach a very high value (>5) when the applied load frequency happens to be in the vicinity

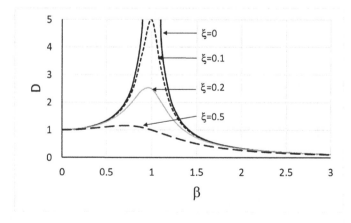

FIGURE 10.6   Dynamic amplification factor for damped systems.

of natural vibration frequency of the structure. A good part of the design process of many structures, thus, involves management of the structural response through separation of the structural vibration frequency from the input load frequency. Section 10.4 presents a discussion of this topic.

10.4 FREQUENCY SEPARATION AND DESIGN TARGETS FOR STRUCTURES

Design of structures, when subjected to cyclic dynamic loads, is generally concerned with the resonant mode of structures, i.e. vibration modes that amplify input forces. As demonstrated with the dynamic amplification curves in Figure 10.6, alignment between the structural vibration frequency and the input load frequency leads to large amplification of the response for lightly damped systems. From a structural design point of view, the issue is how to design a structure by avoiding resonance with input excitations. In automotive product engineering, for example, the target design frequencies for suspension and body structure are set apart from the known input frequencies of powertrain operation and road loads (Figure 10.7). Design targets for natural vibration frequencies of the body structure, in bending and torsional deformation modes, are set in the range of 25–35 Hz, largely to avoid resonance with suspension motions, and also to stay away from powertrain operation and road input frequencies in the range of 100–500 Hz. Dynamic load effects on the structure are significantly reduced by keeping the natural vibration frequencies far apart from the applied loading frequencies (Figure 10.6). This same idea of frequency separation is also used in the earthquake resistant design of civil engineering structures (BSSC 2020), where earthquake load is defined as a function of the structural vibration period; and the design process is driven with the objective of minimizing the dynamic load effect on the structure. Calculation of the structural vibration frequencies, and the adjustment of mass and stiffness properties to tune the frequency values, are of critical importance in all structural engineering disciplines. Sections 10.5 and 10.6 in the following present commonly used analytical techniques for calculating the structural vibration frequencies. More powerful finite element model-based calculation

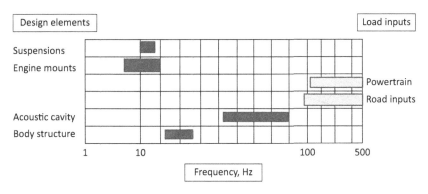

FIGURE 10.7   Example of frequency separation: automotive system design.

procedures, for large and complex multi-degree-of-freedom (MDOF) structural systems, are presented next in Section 10.7.

10.5　VIBRATION MODE SHAPE AND FREQUENCY OF SDOF STRUCTURES

Implicit assumption for the SDOF in Figure 10.3 is that the spring-mass system can vibrate in only one mode – producing axial expansion and contraction in the spring. The vibration frequency (and period of vibration) is calculated from equation (10.7) by using discrete mass and spring stiffness property values. This analysis method can be used to determine the vibration characteristics of elementary structural members that can be reduced to an equivalent SDOF system. Figure 10.8(a) shows a beam member supporting a discrete mass (M) at its mid-span. Any transient perturbation to the mass element is expected to cause up-and-down vibrational motion of the beam-mass system. Assuming that the mass of beam element is negligible compared to the lump mass (M), the vibration frequency of beam-mass system can be calculated from equation (10.7) by using the effective mass value ($M_{eff} = M$) and the bending stiffness of beam that resists the vertical vibrational motion of mass element. The relevant bending stiffness of simply supported beam, of length L and flexural stiffness EI, can be calculated by measuring the lateral deflection "δ" for an arbitrary lateral load "P" at midspan (Figure 10.8(b)):

$$K_{eff} = \frac{P}{\delta} = \frac{48EI}{L^3} \qquad (10.22)$$

Frequency of vibration of the idealized SDOF beam-mass system can be readily calculated by using the effective mass value (M_{eff}) and the bending stiffness of beam "K_{eff}" from equation (10.22).

This above calculation procedure for SDOF vibrating system can be extended to calculate the vibration frequency of a simply supported beam of uniform flexural stiffness "EI" and distributed mass per unit length "m" (Figure 10.9(a)). The vibration mode shape of prismatic beam is described by the following sine harmonic function:

$$v(x) = \sin\left(\frac{\pi x}{L}\right) \qquad (10.23)$$

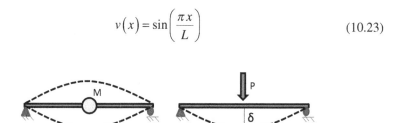

(a)　　　　　　　　　　(b)

FIGURE 10.8　(a) Lateral vibration of a simply-supported beam-mass system; and (b) bending deformation shape of the beam under a lateral load at mid-span.

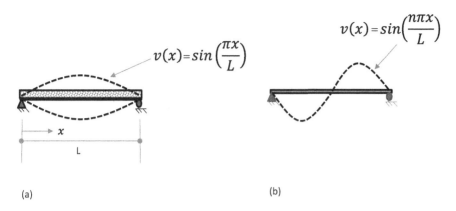

$$v(x) = sin\left(\frac{\pi x}{L}\right)$$

$$v(x) = sin\left(\frac{n\pi x}{L}\right)$$

(a)   (b)

FIGURE 10.9   (a) Lateral vibration of a simply-supported beam of uniform flexural rigidity (EI) and mass per unit length (*m*); (b) higher vibration mode shape of the beam.

Following the Euler–Bernoulli's beam deflection theory (equation 6.58), the internal bending moment (*BM*) in the beam corresponding to the vibration mode shape of equation (10.23) is given by

$$BM(x) = EI\frac{d^2v}{dx^2} = -\frac{\pi^2}{L^2}.EI.\sin\left(\frac{\pi x}{L}\right) \qquad (10.24)$$

Bending stiffness of the beam, defined by the lateral load resistance corresponding to the bending response of equation (10.24), is calculated as follows:

$$K_{eff} = \int_0^L p(x)dx = \int_0^L \frac{d^2BM(x)}{dx^2}dx = -\frac{\pi^4}{L^4}.EI.\int_0^L \sin\left(\frac{\pi x}{L}\right)dx = 2.\frac{\pi^3}{L^3}.EI \qquad (10.25)$$

Effective total mass of beam participating in the sine harmonic vibration mode of equation (10.23) is calculated as follows:

$$M_{eff} = m\int_0^L \sin\left(\frac{\pi x}{L}\right)dx = \frac{2}{\pi}.mL \qquad (10.26)$$

Substituting the effective stiffness and mass values from equations (10.25) and (10.26) into equation (10.7), the vibration frequency response for a simply supported prismatic beam is given by the following equation:

$$\omega = 2\pi f = \frac{2\pi}{T} = \left(\frac{\pi}{L}\right)^2\sqrt{\frac{EI}{m}} \qquad (10.27)$$

For example calculations, substituting *EI* = 1540980 kN-mm², *L* = 600 mm and *m* = 0.0083 kg/mm, equation (10.27) gives ω = 0.3728 rad/ms. If the effective stiffness of beam bending is approximately calculated by equation (10.22), simulating the bending effect of a concentrated vertical load at the midspan, $K_{eff} = 48EI/L^3 = 0.34244$ kN/mm, the bending frequency response of beam, with the effective mass value from equation (10.26), can be calculated from equation (10.7) as, $\omega = 0.3182$ rad/ms. This approximate value is within 7% of the theoretical value of 0.3728 rad/ms. Approximation of the vibration frequency value, by using simplified assumptions for stiffness and mass values, provides valuable insights during the preliminary design phase of structures.

Extending the half-sine harmonic description of beam vibration mode (Figure 10.9(a)) to higher degree vibration mode shapes (Figure 10.9(b)), a generalized definition for the vibration modes of a simply supported prismatic beam can be given by modifying equation (10.23) as follows:

$$v(x) = \sin\left(\frac{n\pi x}{L}\right); \quad where \quad n = 1, 2, 3....etc. \tag{10.28}$$

The vibration properties of simply supported beam corresponding to higher degree mode shapes are defined by modifying equation (10.27) as follows:

$$\omega_n = 2\pi f_n = \frac{2\pi}{T_n} = \left(\frac{n\pi}{L}\right)^2 \sqrt{\frac{EI}{m}}; where\, n = 1, 2, 3....etc. \tag{10.29}$$

Evidently, the lowest vibration frequency of system is given by $n = 1$ (corresponding to the shape shown in Figure 10.9(a). All other frequency values (for $n > 1$), corresponding to relatively more complex mode shapes, will be higher than the first frequency of vibration. The lowest frequency value is generally referred to as the fundamental frequency of vibration of an MDOF system. And the vibration mode shape, corresponding to the lowest frequency value, represents the lowest energy mode of the structure. Equation (10.29), thus, presents frequency solutions for the prismatic beam structure by reducing the distributed stiffness and mass properties to effective stiffness and mass values of SDOF systems representing different vibration mode shapes. The methodology, described thus far, represents a very basic but powerful method for calculating the vibration frequencies of structures, having distributed mass and stiffness properties, by using appropriate functions for vibration mode shapes. Considering the simply supported flat rectangular plate of dimensions $a \times b$ (Figure 10.10), and distributed mass per unit area $= m$ (m = material density * plate thickness), the fundamental period of plate vibration is given by

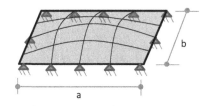

FIGURE 10.10 Vibration mode shape of a simply supported rectangular plate.

$$\omega = \sqrt{\frac{1}{m} * \frac{E.t^3}{12(1-v^2)}} * \pi^2 * \left(\frac{1}{a^2} + \frac{1}{b^2}\right) \qquad (10.30)$$

Equations (10.7), (10.27), and (10.30), defining the vibration characteristics of elementary structural examples, provide basic analytical tools to check the results of more complex finite element models (to be discussed later in this chapter). For example, assuming the overall bending stiffness target of automotive body structure to be 10 kN/mm (Figure 10.11), and the effective mass in bending vibration mode given by equation (10.26) as $M_{eff} = 2/\pi*$body structure mass (364 kg) = 231.5 kg, the vibration frequency target in bending mode of structural vibration is obtained from equation (10.7) as follows:

$$f = \frac{1}{2\pi} * \sqrt{\frac{10}{231.5}} \frac{cycles}{ms} * 1000 = 33 \text{ Hz} \qquad (10.31)$$

Bending frequency, calculated in equation (10.31), serves as a reference value to guide the development of body structure design with finite element analysis models.

10.6   VIBRATION FREQUENCIES OF MDOF SYSTEMS

Figure 10.12 shows two vibration mode shapes (ϕ_1 and ϕ_2) of an arbitrary MDOF structure, and the time–domain function for the vibration mode shapes are defined as:

$$\{u_i\} = \{\varphi_i\} sin(\omega_i t) \qquad (10.32)$$

where $\{\phi_i\}$ describes the vibration mode shape of system (i = 1,2 …etc.), and ω_i is the angular velocity of vibration ($\omega_i = 2\pi f_i$, where f_i is cyclic frequency) corresponding to mode shape $\{\phi_i\}$. Re-writing equation (10.5), equilibrium state during free-vibration response of an MDOF system can be defined with the following matrix and vector terms:

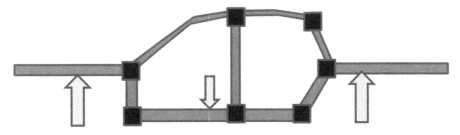

FIGURE 10.11   Schematic representation of an automotive body structure subjected to bending mode of deformation.

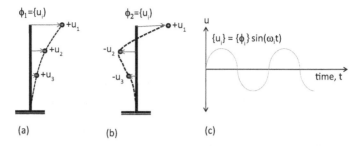

FIGURE 10.12   (a) Vibration mode-1; (b) vibration mode-2; and (c) time function of the vibration mode response.

$$[k].\{u\} + [m].\{\ddot{u}\} = 0 \qquad (10.33)$$

where $[k]$ is the stiffness matrix, $[m]$ the mass matrix, $\{u\}$ the displacement response vector and $\{\ddot{u}\}$ the vector of accelerations of nodal DOF. Substituting expression from equation (10.32) into (10.33):

$$\left[[k].\{\varphi_i\} - \omega_i^2[m].\{\varphi_i\}\right]\sin(\omega_i t) = 0 \qquad (10.34)$$

Since equation (10.34) is true for any value of time function $\sin(\omega_i t)$, the following condition must be satisfied by the stiffness and mass property matrices of the MDOF system:

$$[k].\{\varphi_i\} - \omega_i^2[m].\{\varphi_i\} = 0 \qquad (10.35)$$

Equation (10.35) resembles the standard eigenvalue problems encountered in many engineering mechanics problems, where λ_i ($= \omega_i^2$) is the eigenvalue and ϕ_i is the eigenvector. Properties of eigenvector dictate that mode shapes be orthogonal to one-another, i.e.:

$$\{\varphi_i\}^T.\{\varphi_j\} = \delta_{ij} \qquad (10.36)$$

where δ_{ij} is the Kronecker delta ($\delta_{ij} = 1$ when $i = j$ and $\delta_{ij} = 0$ when $i \neq j$). Structural vibration frequency analysis problem of equation (10.35) can be analytically described by the following polynomial of degree "n" where "n" is the number of DOF in the system:

$$determinant\left|[k] - \omega_i^2[m]\right| = 0 \qquad (10.37)$$

Solutions of equations (10.37) can be analytically calculated for 2 or 3 DOF. Special numerical techniques are required for large systems. Commonly used solution techniques focus on determining the mode shapes (i.e. eigenvectors), ϕ_i, that are subsequently used to calculate the frequency values. Pre-multiplying equation (10.35) with $\{\phi_i\}^T$, and after rearranging the terms:

$$\omega_i^2 = \lambda_i = \frac{\{\varphi_i\}^T [k] \cdot \{\varphi_i\}}{\{\varphi_i\}^T [m] \cdot \{\varphi_i\}} \qquad (10.38)$$

The terms defined in equation (10.38) are known as Rayleigh's quotient having the following properties (Bathe 1996):

$$\lambda_1 \leq \lambda_2 \leq .... \leq \lambda_n \qquad (10.39)$$

With positive definite[k] and [m] matrices, equation (10.38) readily provides the vibration frequency values of MDOF structure provided that the mode shape vectors $\{\phi_i\}$ are known. The minimum value of λ_i providing the lowest vibration frequency is known as the fundamental vibration frequency of structure. The pair of eigenvalue λ_i and corresponding eigenvector ϕ_i is commonly known as eigenpair. Eigenvectors are occasionally normalized to represent the following condition:

$$\{\varphi_i\}^T \cdot [m] \cdot \{\varphi_i\} = [I] \qquad (10.40)$$

where [I] is a diagonal matrix of unit values. The numerator on the right side of equation (10.38) represents the effective stiffness value of structure (structural resistance) against mode shape ϕ_i, and the denominator represents the effective mass value participating in that mode of vibration. Stiffness matrix of the structure, [k], is calculated by using the standard finite element formulation given in equation (2.44). Calculation of mass matrix [m] from the finite element model of a structure is described in the following Section 10.7. Section 10.8 will introduce the unique numerical techniques required for the calculation of vibration mode shapes $\{\phi_i\}$ and corresponding frequencies ω_i.

10.7 CALCULATION OF SYSTEM MASS MATRIX FOR MDOF SYSTEMS

Discrete lump mass elements at nodal DOF can be directly assembled into the system mass matrix. Mass matrix calculation for distributed mass systems requires special numerical procedures. Using the same displacement interpolation functions (alternatively known "shape functions"), described earlier for calculation of deformation inside finite elements, the acceleration response at a material point inside a finite element ($\ddot{u}$) can be calculated from the nodal acceleration responses ($\ddot{u}_i$) as follows:

$$\{\ddot{u}\} = [H_i]\{\ddot{u}_i\} \qquad (10.41)$$

Exact definition of the matrix $[H_i]$ in equation (10.41) will depend on the type of finite elements used to model the structure. For example, considering the 2D solid element of Figure 3.8(a), the acceleration responses in two orthogonal directions at a material point inside the element can be calculated from the nodal acceleration values as follows:

$$\begin{Bmatrix} \ddot{u} \\ \ddot{v} \end{Bmatrix} = \begin{bmatrix} H_1 & 0 & H_2 & 0 & H_3 & 0 & H_4 & 0 \\ 0 & H_1 & 0 & H_2 & 0 & H_3 & 0 & H_4 \end{bmatrix} \begin{Bmatrix} \ddot{u}_1 \\ \ddot{v}_1 \\ \ddot{u}_2 \\ \ddot{v}_2 \\ \ddot{u}_3 \\ \ddot{v}_3 \\ \ddot{u}_4 \\ \ddot{v}_4 \end{Bmatrix} \tag{10.42}$$

where H_i (i = 1, 2, etc.) are the iso-parametric shape functions defined in Equations (3.65). Adding the inertia resistance term, the virtual work of equation (2.39) can be re-written as:

$$\int \{\bar{\varepsilon}\}^T \{\sigma\} dV + \int \{\bar{u}\}^T \{\rho \ddot{u}\} dV = \bar{U}^T \left[\int [B]^T \{\sigma\} dV + \int [H_i]^T \{\rho \ddot{u}\} dV \right] \tag{10.43}$$

where ρ is the density of material. Substituting equation (10.41) into (10.43), the equilibrium state between internal resistances and external forces can be expressed by re-writing equation (2.43) as follows:

$$\left[\int [B]^T.[C].[B].dV \right].\{u_i\} + \left[\rho.\int [H_i]^T.[H_i].dV \right].\{\ddot{u}_i\} = \{P\} \tag{10.44}$$

where $\{u_i\}$ and $\{\ddot{u}_i\}$ are displacements and accelerations at the nodal DOF. The second integral term on the left side of equation (10.44) represents the mass matrix of finite element:

$$[m] = \left[\rho.\int [H_i]^T.[H_i].dV \right] \tag{10.45}$$

Equation (10.45) produces a positive definite mass matrix formulation that is essential for the calculation of frequency values from equation (10.38). Mass matrix derived from numerical integration of equation (10.45) is commonly referred to as consistent mass matrix since the same displacement interpolation functions are used for the interpolation of acceleration response. For the simple case of two DOF truss element in Figure 2.11, equation (10.45) provides the following consistent mass matrix definition:

$$[m] = \frac{\rho A L}{6} \begin{bmatrix} 2 & 1 \\ 1 & 2 \end{bmatrix} \tag{10.46}$$

where "ρ" is material density, "A" the cross-sectional area of truss element, and "L" the length of member. Taking summation of the terms in each row, and lumping the value at diagonal position will lead to the following definition of diagonal mass matrix:

$$[m] = \frac{\rho A L}{2} \begin{bmatrix} 1 & 0 \\ 0 & 1 \end{bmatrix} \tag{10.47}$$

Lumped mass matrix definition in equation (10.47) shows half of the element mass effective at each of the two translational DOF. Consistent mass matrix for square 2D solid element of Figure 3.16(a), corresponding to 4 translational DOF in the x-direction only, is obtained from equation (10.45) with the use of shape functions from Equations (3.65):

$$[m] = \frac{\rho A t}{36} \begin{bmatrix} 4 & 2 & 1 & 2 \\ 2 & 4 & 2 & 1 \\ 1 & 2 & 4 & 2 \\ 2 & 1 & 2 & 4 \end{bmatrix} \tag{10.48}$$

where "ρ" is material density, "A" the plan view area of 2D solid, and "t" the thickness of element. By using the row summation technique, the diagonal lumped mass matrix is obtained as:

$$[m] = \frac{\rho A t}{4} \begin{bmatrix} 1 & 0 & 0 & 0 \\ 0 & 1 & 0 & 0 \\ 0 & 0 & 1 & 0 \\ 0 & 0 & 0 & 1 \end{bmatrix} \tag{10.49}$$

Effective mass at each DOF of 4-node square element turns out to be one-quarter of the element mass. Lumped mass matrix formulation with equal distribution of element mass at corner nodes, however, is not applicable for higher order elements and for elements with rotational degrees of freedom. For the transverse shear and bending deformation modes of a 2D beam element (Figure 6.21, Equations 6.75), the consistent mass matrix from equation (10.44) is calculated as:

$$[m] = \frac{\rho A L}{420} \begin{bmatrix} 156 & 22L & 54 & -13L \\ 22L & 4L^2 & 13L & -3L^2 \\ 54 & 13L & 156 & -22L \\ -13L & -3L^2 & -22L & 4L^2 \end{bmatrix} \tag{10.50}$$

where "ρ" is material density, "A" the cross-sectional area of beam element, and "L" the length of member. The diagonal lumped mass matrix of this beam element is defined, based on engineering intuition of beam deformation behavior, as follows:

$$\begin{bmatrix} m \end{bmatrix} = \frac{\rho AL}{2} \begin{bmatrix} 1 & 0 & & \\ & & 0 & 0 \\ 0 & \dfrac{L^2}{12} & 0 & 0 \\ 0 & 0 & 1 & 0 \\ 0 & 0 & 0 & \dfrac{L^2}{12} \end{bmatrix} \tag{10.51}$$

where 1st and 3rd diagonal terms represent half of beam mass effective at each of the transverse DOF; and the second and fourth diagonal terms represent the mass moment of inertia for spinning motion of half-length of beam about each end:

$$\int_0^{L/2} x^2 . (\rho A).dx = \frac{\rho AL}{2} . \left(\frac{L^2}{12} \right) \tag{10.52}$$

Various techniques of lumped mass matrix formulations have been extensively documented by Cook et al. (1989) and Zienkiewicz and Taylor (1991). Both mass matrix formulations, consistent and lumped, have their merits and demerits. There is no standard procedure to assess the accuracy of results obtained by one or the other method. The lumped mass matrix is a convenient numerical tool at the expense of some accuracy. Consistent mass matrices tend to be accurate for flexural problems when modeled with beam and shell elements. In nonlinear impact simulations, e.g. crashworthiness analysis of automotive structures, lumped mass matrices are commonly used for computational efficiency as well as for greater numerical stability against spurious oscillations. Finite element software packages may selectively use lumped mass matrix formulations for computational efficiency reasons. ABAQUS software, for example, uses lumped mass matrix formulations for linear truss and solid elements, and consistent mass matrix for higher order solids, beams, and shell elements.

In some structural analysis models, physical mass properties could be lumped to selected translational DOF, while the rotational DOF of the system is assumed to be massless. Vibration analysis problem of equation (10.35) can be solved by reducing the full $n \times n$ system to a reduced set containing "m" master DOF that have nonzero mass properties. Dividing the equations (10.35) into to subsets of master DOF (with mass properties) and massless slave DOF:

$$\begin{bmatrix} k_{mm} & k_{ms} \\ k_{sm} & k_{ss} \end{bmatrix} . \begin{Bmatrix} \varphi_m \\ \varphi_s \end{Bmatrix} - \omega_i^2 \begin{bmatrix} m_{mm} & 0 \\ 0 & 0 \end{bmatrix} . \begin{Bmatrix} \varphi_m \\ \varphi_s \end{Bmatrix} = \begin{Bmatrix} 0 \\ 0 \end{Bmatrix} \tag{10.53}$$

where "m" refers to the master DOF (to be retained), "s" refers to the slave DOF (to be reduced), and $[k_{sm}] = [k_{ms}]^T$. Using the bottom part of equations (10.53), the transformation relationship between reduced and full system can be written as follows:

$$\begin{Bmatrix} \varphi_m \\ \varphi_s \end{Bmatrix} = [Z]\{\varphi_m\}; \quad where \, [Z] = \begin{Bmatrix} I \\ -k_{ss}^{-1}.k_{sm} \end{Bmatrix} \tag{10.54}$$

The reduced stiffness and mass matrices are defined by using the transformation matrix $\{Z\}$:

$$[k]_{m \times m} = [Z]^T [k][Z] \qquad [m]_{m \times m} = [Z]^T [m][Z] \tag{10.55}$$

Reduced stiffness and mass matrices, defined in equation (10.55), can be used in equation (10.35) in lieu of full system matrices for vibration frequency analysis of the structure. Condensation of the massless DOF, commonly known as Guyan reduction, is frequently used in earthquake response analysis of tall building frames.

As discussed in Section 1.4, a structure with un-constrained rigid body modes will not produce a positive definite stiffness matrix $[k]$. A work-around for this issue is to shift the eigenvalue problem of equation (10.35) by applying an arbitrary shift "α" as shown in the following:

$$[[k] + \alpha.[m]].\{\varphi_i\} - (\omega_i^2 + \alpha)[m].\{\varphi_i\} = 0 \tag{10.56}$$

This "shifted" system equations can be solved by one of the numerical techniques described in the following Section 10.8. The first vibration frequency, predicted by solving equations (10.56), is the one closest to the applied shift value (α).

10.8 NUMERICAL CALCULATION OF VIBRATION MODE SHAPES AND FREQUENCIES OF MDOF SYSTEMS

Assuming that system property matrices [k] and [m] in equation (10.35) are of $n \times n$ size, there exist "n" number of frequencies and corresponding mode shapes for the system. Several different numerical techniques exist, with varying degrees of efficiency, for solving the eigenvalue problem of equation (10.35). Effectiveness of solution techniques to be used depends on the relative ratio between the number of desired mode shapes "p" to the size of the overall problem "n" ($p \leq n$). When all mode shapes and frequencies are required for a small size problem, Jacobi transformation method provides effective solution to the Eigenvalue problem. In that method, a transformation matrix [P] is developed through iterations to diagonalize the stiffness and mass matrices of the system:

$$\begin{aligned} [P]^T [k][P] &= diagonal[k_{ii}] \\ [P]^T [m][P] &= diagonal[m_{ii}] \end{aligned} \tag{10.57}$$

The columns of matrix [P] represent the eigenvectors (i.e. mode shapes), and the determination of eigenvalues becomes a straight-forward operation given by

$$\lambda_i = \omega_i^2 = k_{ii}\Big/m_{ii} \qquad (10.58)$$

A systematic trial-and-error approach is followed to find the transformation matrix [P] that progressively leads the system property matrices [k] and [m] to diagonal forms. Householder-QR method is another matrix transformation method that initially transforms the matrices [k] and [m] to tri-diagonal forms, and eventually produces diagonal property matrices in the subsequent steps. Detail numerical steps for the derivation of transformation matrices, with convergence checks of successive iterations, can be found in Bathe (1996).

When a small number of mode shapes and frequencies are required for a large system ($p \ll n$), vector iteration method provides a very efficiency solution strategy. Starting with a trial vector $\{x_1\}_k$ for mode shape $\{\phi_1\}$, an improved prediction for the vector is obtained by solving the following re-arranged form of equation (10.35):

$$\left[k\right].\left\{x_1\right\}_{k+1} = \omega_1^2\left[m\right].\left\{x_1\right\}_k \qquad (10.59)$$

where ω_1^2 is the value predicted from equation (10.38) by using the trial vector $\{x_1\}_k$. Successive iterations with equations (10.38) and (10.59) eventually lead to a converged eigenpair solution of ω_1 and $\{\phi_1\}$. Iterations for second eigenpair follow the same iterative scheme, but with added interim step to enforce orthogonality condition between mode-1, $\{\phi_1\}$, and the next trial vector $\{x_2\}_k$:

$$\left\{x_2\right\}_k = \left\{x_2\right\}_k - \left[\left\{\varphi_1\right\}^T.\left\{x_2\right\}_k\right].\left\{\varphi_1\right\} \qquad (10.60)$$

Actual numerical implementation, however, does not need to follow the sequential extraction of mode shapes one by one. A set of trial vectors can be used simultaneously to progressively improve all predictions in each iteration step. Numerical implementation of such a multi-vector trial scheme is commonly referred to as "subspace" iteration method (Bathe 1996). It is a good analysis practice to use a larger number of trial vectors "q" for producing a good set of "p" mode shapes ($p < q \ll n$). The method works efficiently when only a handful of eigenpairs are desired, e.g., 2 or 3 mode shapes are generally desired in dynamic analysis of tall building frame structures. However, a much larger number of modal frequency values need to be checked in vibration-sensitive structural designs, such as in automotive body structures. Computational efficiency of standard subspace iteration method goes down significantly when "p" becomes high ($p > 20$). An accelerated form of subspace iteration method is sometimes constructed by extracting a limited number of modes in one step, and then by extracting additional mode steps in subsequent steps from the solution of shifted eigenvalue problem (equation 10.56). Lanczos transformation method, producing tri-diagonal forms of system property matrices ([k] and [m]), is often used in conjunction with subspace iteration method to extract large number of eigenpairs efficiently. An important consideration in iterative solution of eigenvalue problem is the verification with Sturm sequence check that all eigenpairs within a frequency target range have been calculated (Bathe 1996). General-purpose finite element software packages tend to offer many of the above referenced alternative eigensolver options to be explicitly selected by the analyst when setting up a modal analysis problem.

10.9   VIBRATION FREQUENCY ANALYSIS WITH ABAQUS

Finite element analysis models, described in earlier chapters for static load–deflection analysis, remain the same for stiffness matrix calculations in vibration analysis models. Material property descriptions need additional data inputs for density values, with the use of data block identifier *DENSITY as shown in the following:

*MATERIAL, NAME=*mat-1*
*ELASTIC
210,0.3
*DENSITY
7.8e-6

The specified material density value is used, with equation (10.45), for the calculations of element mass matrices [m]. Discrete mass values can be assigned to selected nodes by defining virtual elements with TYPE=MASS:

*ELEMENT, TYPE=MASS, ELSET=*aName*
ID no, nodal ID
ID no, nodal ID
ID no, nodal ID
.....

Actual mass value to be assigned to the translational DOF of nodes selected in mass element descriptions are defined with *MASS data block identifier:

*MASS, ELSET=*aName*
a numerical mass value

Discrete lumped mass values are combined with element mass matrices, calculated from finite element analysis model, to get the overall mass matrix [m]. An ABAQUS analysis step for vibration mode analysis can be constructed by using the following commands:

*STEP
*FREQUENCY
p, fmin, fmax, λ
* END STEP

where "p" is the number of desired mode shapes from the analysis model. This field can be left blank if the maximum frequency of interest ($fmax$) is specified and the evaluation of all the eigenvalues in the given range is desired. The optional shift parameter (λ) in ABAQUS analysis is specified in the unit of (f^2). Calculated modal frequency values (f_i) are saved by ABAQUS in standard output data file (*.dat). Mode shape data are saved in the general binary database output file (*.ODB). When no specific eigensolver routine is selected, ABAQUS uses the LANCZOS method by-default for eigenpair extraction. However, a user may optionally select the subspace iteration method, if desired, by using the following syntax:

*STEP
*FREQUENCY, EIGENSOLVER=SUBSPACE
$p,, fmax, \lambda, q$
* END STEP

The number of trial vectors (q), if omitted, is internally set by ABAQUS as the minimum of ($2p$ and $p + 8$). ABAQUS normalizes the mode shapes, by default, so that the largest displacement or rotation value in each reported vector is unity. However, a user may optionally specify the mass normalization method (equation 10.40) to be used in eigenproblem solution:

*STEP
*FREQUENCY, EIGENSOLVER=SUBSPACE, NORMALIZATION=MASS
p
* END STEP

10.10 PRACTICE PROBLEMS: VIBRATION ANALYSIS OF STRUCTURES

PROBLEM 1
Figure 10.13 presents geometry, material, and boundary condition data for a thin-wall C-section beam. Using SDOF simplified modal analysis concepts, calculate by

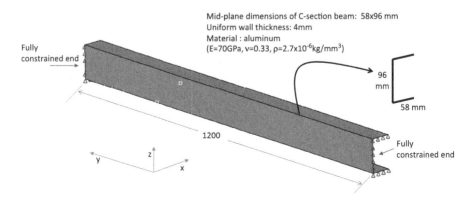

FIGURE 10.13 Properties of a thin-wall C-section beam.

hand the vibration frequencies of the member for: (1) lateral vibration mode in the x-direction, (2) vertical mode in the z-direction, and (3) torsional mode about the y-axis. Build a finite element model of the member using shell elements; and conduct modal frequency analysis for the first five modes of vibration. Compare the hand calculation results (frequency and mode shapes) with the finite element model results

PROBLEM 2
Roof panel of an automotive body structure is made of 0.65 mm thick steel panel – spot welded @ approximately 60 mm spacing along the edges to the perimeter body structure frame. Geometry data of the roof panel mid-surface is available in the file Roof_Panel.step. Prepare a finite element analysis model of the roof panel assuming fixed boundary conditions at the spot weld locations; calculate the lowest vibration frequency of finite element shell model; and plot the corresponding mode shape. Using equation (10.30), what will be the vibration frequency of an equivalent simply supported flat panel having approximate panel dimensions shown in Figure 10.14? What are the reasons for differences between FEA and hand calculation results?

PROBLEM 3
Re-analyze finite element analysis model of Problem-2 assuming 1.0 mm thick aluminum roof panel ($E = 70$ GPa, $\nu = 0.33$, $\rho = 2.7 \times 10^{-6}$ kg/mm³). Compare the fundamental vibration frequency responses of 2 alternative material choices (steel versus aluminum). What is the relative weight ratio of aluminum versus steel design?

PROBLEM 4
Re-analyze the roof panel finite element model, without support constraints at spot-weld locations, by using the frequency shifting technique described by equation (10.56). Verify the vibration frequency result of shifted model against that of base model.

PROBLEM 5
Beam structure in Figure 6.30 is made of steel (density, $\rho = 7.8e^{-6}$ kg/mm³). Assume an additional lumped mass of 10 kg supported at point C. Calculate the fundamental vibration frequency of the system; and verify the result with hand calculations.

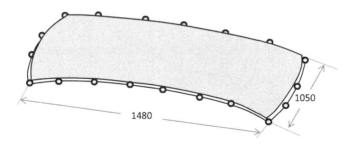

Mid-surface geometry of a vehicle roof panel: Roof_Panel.step
Uniform sheet metal thickness: 0.65 mm
Material : steel (E=210GPa, v=0.3, ρ=7.8x10⁻⁶ kg/mm3)

FIGURE 10.14 Properties of a vehicle roof panel (Roof panel extracted from vehicle FEA model: courtesy of NHTSA 2020).

11 Linear Dynamic Response Analysis of Structures

SUMMARY

Structural response to cyclic dynamic loads and the management of dynamic response through frequency separation have been discussed in Chapter 10. However, structures are also frequently subjected to non-cyclic short-duration dynamic events. This chapter is dedicated to the analysis of structural response for such non-cyclic dynamic load events. Duhamel integral formulation, to predict the elastic dynamic response of SDOF systems to external impulse loads, is introduced in Section 11.1. This analysis technique provides very useful conclusions on the range of dynamic amplifications that a system can experience when subjected to impulse events of arbitrary duration. The concept of design response spectra, based on the envelope of peak dynamic responses of linear elastic systems, to pre-defined single loading function, is discussed in Section 11.2. Duhamel integration method, although very powerful for predicting the linear dynamic response of SDOF systems, is not an efficient method for predicting the time history response of MDOF systems. The alternative direct integration techniques, for time-domain dynamic response simulations, are introduced in Sections 11.3–11.6. Section 11.4 specifically focuses on the accuracy and stability aspects of implicit time integration method. Relative efficiencies of direct integration versus modal superposition methods, for linear elastic dynamic response prediction, are discussed in Sections 11.5 and 11.6. Explicit time integration method, which is more potent for nonlinear dynamic response analysis, is introduced in Section 11.7. ABAQUS-specific commands for dynamic response analysis of finite element models are reviewed in Section 11.8 followed by the presentation of practice problems in Section 11.9.

11.1 LINEAR ELASTIC RESPONSE OF SDOF SYSTEMS TO IMPULSIVE LOADING

Figure 11.1 shows a general dynamic loading $F(\tau)$ acting on an SDOF spring-mass system. Re-arranging the terms of equation (10.1), the effect of dynamic force $F(\tau)$, acting for a short time duration, $d\tau$, can be expressed as follows:

$$d\dot{u}(\tau) = \frac{1}{m} * F(\tau).d\tau \qquad (11.1)$$

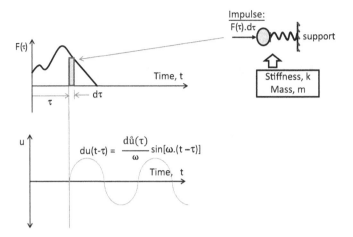

FIGURE 11.1   Impact force acting on a flexible SDOF spring-mass system.

where $d\dot{u}(\tau)$ is the change in velocity of mass (m) caused by the impulse $[F(\tau). d\tau]$. Vibration response of the spring-mass system at a time (t), following the application of infinitesimal impulse, is given by (Clough and Penzien 1975)

$$du(t-\tau) = \frac{d\dot{u}(\tau)}{\omega} \sin\left[\omega.(t-\tau)\right] \qquad (11.2)$$

Substituting the expression from equation (11.1) into equation (11.2), displacement of the spring-mass system due to short-duration impulse effect is given by

$$du(t-\tau) = \frac{1}{\omega m} F(\tau).\sin\left[\omega.(t-\tau)\right].d\tau \qquad (11.3)$$

where ω is the natural vibration frequency of spring-mass system (defined by equation 10.7). Integrating both sides of equation (11.3) over the duration of impulse (τ: 0 to t), displacement response of the system, for a general dynamic load of duration (t) and for the initial state of $u(0) = 0$, is given by

$$u(t) = \frac{1}{\omega m} \int_0^t F(\tau).\sin\left[\omega.(t-\tau)\right].d\tau \qquad (11.4)$$

Equation (11.4) is commonly known as Duhamel integral – a powerful expression for calculating the undamped linear elastic dynamic response of an SDOF system when subjected to a general dynamic loading $F(\tau)$. With known system properties ω and m, elastic dynamic response $u(t)$ can be calculated, for any given force function $F(\tau)$, by conducting step-by-step numerical integration of the expression on the right side of equation (11.4).

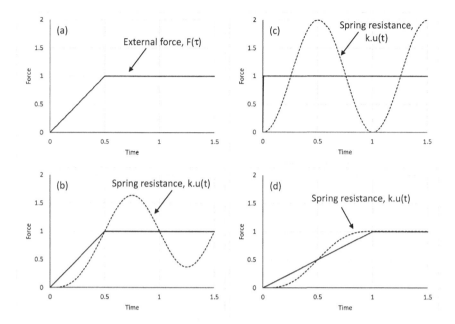

FIGURE 11.2 (a) External dynamic load $F(\tau)$; (b) spring force response to dynamic external load; (c) response to a sudden impact force; (d) response to a slow external load.

An example case study is considered here assuming an SDOF with mass, $m = 1$, and stiffness, $k = 39.4784$, such that the period of vibration of the system comes out to be, $T = 1$. Figure 11.2(a) shows a dynamic force function reaching to unit peak amplitude over a time duration of 0.5. Numerical integration of equation (11.4) produces the displacement response of the mass $u(t)$, that, when multiplied with spring stiffness k, gives the dynamic spring force response (shown by the dotted line in Figure 11.2(b)). The dynamic amplification of the spring response, thus, happens to be 1.64 when the force function ramps up to unit peak amplitude at half time of the period of system vibration. Figure 11.2(c) shows the predicted spring force response when the external load is applied suddenly over an infinitesimally short duration of time. Peak spring force reaches a value of 2 times the amplitude of applied external force, thus, defining a limit for the maximum dynamic amplification that can occur when a sudden impact force is applied on a system. On the other hand, if the external force is applied slowly, over a time span equal to or larger than the period of vibration of the system (Figure 11.2(d)), spring force response oscillates about the applied external force trend line before eventually approaching the value of applied external force. This implies that, in the limit case of very slowly applied force (time to peak force > period of structural vibration), system response can be predicted by solving the static equilibrium equation for applied peak force (equation 1.1). Figure 11.3 summarizes the peak dynamic responses, obtained from equation (11.3), for the external force function with variable time-to-peak values. Dynamic amplification of system response diminishes as the time-to-peak of externally applied force becomes

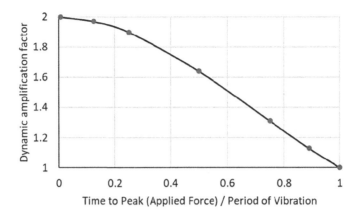

FIGURE 11.3 Dynamic amplification of structural response as a function of the ratio between time-to-peak force and vibration period of system.

longer than the period of vibration of the structure. For shorter duration impact events ($t < T$), the Duhamel integral approach provides useful information about the dynamic amplification of a system response for external loading function of any general shape.

11.2 RESPONSE SPECTRUM OF LINEAR DYNAMIC SYSTEMS

Dynamic amplification of a system response, as discussed in Section 11.1, is determined by the relative relationship between rate of loading and the period vibration of structure. In engineering design practices, generic external loading functions are often pre-defined, such as earthquake loading functions for building design in specific geographic locations, and road input loads for automotive system designs. Linear dynamic response of a system, to the pre-defined loading function, can be predicted by using the Duhamel integral technique described in Section 11.1. Figure 11.4(a) shows an example force function of trapezoidal shape (solid line spanning a time duration of $t = 1.5$). The predicted force responses of two SDOF spring-mass

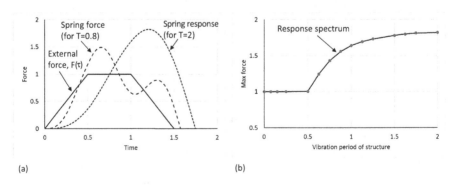

(a)   (b)

FIGURE 11.4 (a) External force and example response histories; (b) response spectrum (envelop of maximum responses) as a function of vibration period of structure.

systems, having period of vibration values $T = 2$ and $T = 0.8$, are shown by the dotted lines in the same plot. Similar response plots can be generated, for the same force function, but with many possible values of the structural vibration period (T). Taking the peak values from such response histories, Figure 11.4(b) shows an envelope of the peak force values for different values of the structural vibration period – generally known as response spectrum in structural design discipline. Like this example of spring force response spectrum, similar envelopes can also be generated for acceleration, velocity, or displacement responses of the system. From a pre-defined response spectrum, the expected peak dynamic response can be estimated easily during preliminary design iterations of a structure.

In practical engineering applications, engineers are often interested to know the maximum response amplitude of a system for standard loading functions. The response spectra method, described in the above for SDOF systems, becomes very useful to predict the peak response of MDOF systems as well. Vibration periods of an MDOF structure can be calculated, by using the techniques presented in Chapter 10, when system property matrices, $[k]$ and $[m]$, are known from a finite element model of the system. Peak response for each of the known modal period of vibration can be calculated from the response spectrum – assuming each mode responds independently as an SDOF system. As it is evident from Figure 11.4(a), peak responses of different modes (with different vibration period values) will occur at different times. Superposition of the modal response values to generate the combined system response requires special considerations (Tedesco et al. 1999, Chopra 2017). One commonly used combination technique is to take the square root of the sum of the squares of all relevant modal peak response values:

$$Max\,response = \sqrt{\sum_{i=1}^{n}\left[x_i\right]^2} \qquad (11.5)$$

where x_i ($i = 1\ldots n$) are the peak response values for individual mode shapes of an MDOF system. A good estimate of the peak structural response can be produced by considering a small number of structural vibration modes when the system response is dominated by a handful of low-frequency modes (such as earthquake and wind load effects on bridges and tall buildings). However, a pre-defined design response spectrum may not always represent case-specific general dynamic load scenarios; thus, requiring time-domain response analysis of structures in many engineering development projects. Numerical integration of equation (11.4), as demonstrated with elementary case studies, can predict time history of dynamic response for SDOF systems. The method is, however, not convenient when numerous mode shapes may need to be considered to predict solutions for large complex systems. More specialized numerical technique, based on direct step-by-step integration of the system equilibrium equations (10.17), is preferred because of the generality of the method that can be easily adapted to nonlinear problems as well when needed (to be discussed in Chapter 12). Sections 11.4–11.6 in the following review different aspects of the numerical solution techniques for time-domain response history analysis of structures.

11.3   TIME-DOMAIN ANALYSIS OF DYNAMIC STRUCTURAL RESPONSE

Following the developments from Chapter 10, dynamic equilibrium of a damped structure at a time (t) can be expressed as follows:

$$k.u(t) + c.\dot{u}(t) + m.\ddot{u}(t) = F(t) \tag{11.6}$$

Solution of dynamic equilibrium equation (11.6), as an analytical function of time (t), is difficult to achieve for general structural systems. Instead, solutions of discretized finite element models can be derived by considering system equilibrium at discrete time steps. Re-writing equation (11.6), equilibrium after a small time increment, Δt, can be expressed as follows:

$$k.u(t + \Delta t) + c.\dot{u}(t + \Delta t) + m.\ddot{u}(t + \Delta t) = F(t + \Delta t) \tag{11.7}$$

The derivation of actual solution at time ($t + \Delta t$) requires certain assumptions about the variations of displacement, velocity, and acceleration responses over the time step Δt. As one of the simplest forms of numerical approximation, taking the average of acceleration responses $\ddot{u}(t)$ and $\ddot{u}(t + \Delta t)$ over the time step Δt, kinematic relations among displacement, velocity, and acceleration can be defined as follows:

$$\dot{u}(t + \Delta t) = \dot{u}(t) + \frac{\Delta t}{2}\left[\ddot{u}(t) + \ddot{u}(t + \Delta t)\right] \tag{11.8}$$

$$u(t + \Delta t) = u(t) + \dot{u}(t).\Delta t + \frac{1}{4}[\ddot{u}(t) + \ddot{u}(t + \Delta t)](\Delta t)^2 \tag{11.9}$$

Assuming that the stiffness remains un-changed during the time interval Δt, resistance to deformation can be expressed as follows:

$$k.u(t + \Delta t) = R(t) + k.\Delta u \tag{11.10}$$

where $R(t)$ is the internal resistance of the structure to deformations, calculated from the stress responses of all finite elements by using equation (2.42). Combining equations (11.7)-(11.10), and after re-arranging the terms:

$$\left(\frac{4}{\Delta t^2}m + \frac{2}{\Delta t}c + k\right)\Delta u = F(t + \Delta t) - R(t) + m.\left(\frac{4}{\Delta t}\dot{u}(t) + \ddot{u}(t)\right) + c.\left(\dot{u}(t)\right) \tag{11.11}$$

Equation (11.11), with known quantities on the right-hand side, can be solved to determine the incremental displacement response Δu occurring over the time step Δt.

The dynamic response of an undamped spring-mass system (with $T = 0.8$, $m = 1$, $k = 61.685$), for the trapezoidal force function of Figure 11.4(a), is analyzed with

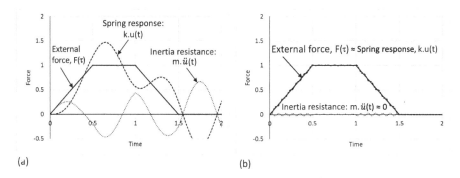

FIGURE 11.5 (a) Dynamic responses of a flexible SDOF spring-mass system; (b) force responses of a very stiff system.

step-by-step application of equation (11.11); and the time histories of external force and internal resistances (for both inertia and spring deformation) are shown in Figure 11.5(a). Evidently, relative contributions of spring resistance and inertia components vary with time. The system resistance components oscillate with time while maintaining overall equilibrium with the externally applied force. When the spring stiffness is increased by a factor of 100 ($k = 6168.5$), the mechanical resistance dominates the overall system response; and the inertia component of resistance mechanism becomes relatively very small (Figure 11.5(b)). In the limit case of relatively negligible contribution from the inertia resistance, solution of the dynamic equilibrium of forces (equation 11.6) converges to the solution obtained by considering pure static equilibrium state (equation 1.1), thus proving the validity of the step-by-step numerical integration formulation presented in equation (11.11). Equations (11.8) and (11.9) have used average value of accelerations at time "t" and "$t + \Delta t$", to estimate the velocity and displacement response changes over time step Δt. Evidently, many other assumptions can be made to forecast the response variations over that small discrete time step Δt. Accuracy and stability of various commonly used numerical integration techniques are discussed in Section 11.4.

11.4 NUMERICAL INTEGRATION PARAMETERS FOR TIME-DOMAIN ANALYSIS OF STRUCTURES

Assumptions made to approximate the variation of dynamic motions (i.e. of displacement, velocity, and acceleration), over the times step Δt, have led to numerical integration techniques of many different names in the field of structural dynamics. Section 11.3 presented time-domain solution method using the assumption of average acceleration response over the time step Δt. A more general form of response variation can be considered by re-writing the kinematic relations of equations (11.8) and (11.9) as follows:

$$\dot{u}(t + \Delta t) = \dot{u}(t) + \left[(1 - \gamma)\ddot{u}(t) + \gamma.\ddot{u}(t + \Delta t)\right].\Delta t \qquad (11.12)$$

$$u(t+\Delta t) = u(t) + \dot{u}(t).\Delta t + \left[\left(\frac{1}{2} - \beta\right)\ddot{u}(t) + \beta.\ddot{u}(t+\Delta t)\right](\Delta t)^2 \qquad (11.13)$$

where parameters β and γ provide a general framework for defining response variations over time step Δt; and the method is commonly known as Newmark integration method (Newmark 1959). The average acceleration method, presented in Section 11.3, is a special form of the Newmark method when integration parameters are defined as $\beta = 1/4$, $\gamma = 1/2$. Linear acceleration method, another widely used assumption in structural dynamics calculations, is derived from the Newmark method with parametric definitions of $\beta = 1/6$, $\gamma = 1/2$. A special form of linear acceleration method is the Wilson θ method that assumes linear acceleration trend applies over a time span from "t" to $(t + \theta.\Delta t)$ where $\theta \geq 1$ (Bathe and Wilson 1976). An enhanced form of Newmark's average acceleration method has been proposed by Hilber et al. (1977) to include a numerical damping parameter α in the step-by-step calculation of structural dynamic response by defining the numerical integration parameters as follows:

$$\beta = \frac{(1-\alpha)^2}{4}; \quad \gamma = \left(\frac{1}{2} - \alpha\right) \qquad (11.14)$$

The value of α is usually defined in the range of $(-1/3 \leq \alpha \leq 0)$. This method degenerates to Newmark's average acceleration method for $\alpha = 0$. The constant or linear acceleration assumptions, as used in the above integration techniques, are in fact simplifications of infinite Taylor series expressions for continuous time-domain response functions (Modak and Sotelino 2002). Modak's T-method uses higher order Taylor series expressions for displacement, velocity, and acceleration variations in the time domain. The dynamic equilibrium equation is derived by using a weighted-residual approach, and a recursive integration technique is derived with 9 numerical parameters after truncating the Taylor series expansions. Since this method allows a wide range of values for the nine parameters, it provides the opportunity for optimization of the parameters to make the algorithmic error minimum while keeping it unconditionally stable and second-order accurate. The optimal form of this generalized method provides higher numerical accuracy compared to other integration methods currently available in finite element software packages.

The direct step-by-step integration methods, commonly known as implicit time integration methods, enforce system equilibrium at discrete time steps $(t, t + \Delta t, \ldots$ etc.) – thus providing stable solution over the time domain. Accuracy of the solution is somewhat affected by the assumptions made about the nature of response variation over time step. However, the most critical parameter affecting the accuracy of solution appears to be the size of time step (Δt) selected for discrete solution steps. For demonstration purpose, an SDOF spring-mass system, stretched to an initial deformation state of $u = 1$ at time $t = 0$, and then released to undergo free vibration, is considered for step-by-step time-domain analysis (Figure 11.6). The period of vibration of the system is calculated as follows:

$$T = \frac{1}{f} = 2\pi.\sqrt{\frac{m}{k}} = 2\pi * \sqrt{\frac{21.54}{0.34}} = 50\,ms \qquad (11.15)$$

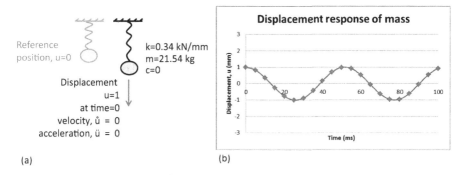

FIGURE 11.6 (a) Spring-mass system stretched to an initial displacement of $u = 1$ at time = 0; (b) calculated displacement response with time step $\Delta t = T/10$.

Upon release of the mass from displaced state of $u = 1$, the spring-mass system is expected go through simple harmonic free-vibration oscillations with an amplitude of "1". The response of the system can be calculated, by using the step-by-step implicit integration of equation (11.11), with any assumed value of time step Δt. The time history of predicted response with $\Delta t = T/10 = 5$ ms, shown in Figure 11.6(b), shows harmonic response with a peak amplitude of $0.99999 \approx 1$. Figure 11.7 shows the error in predicted peak amplitude as a function of the time step used in numerical calculations. Evidently, the accuracy of predicted response goes down for time steps $\Delta t > T/10$. The general analysis practice is to use a time step smaller than one 10th of the important vibration period of a structure.

11.5 TIME DOMAIN ANALYSIS OF MDOF SYSTEMS

The time-domain dynamic equilibrium equation of SDOF system (equation 11.6) can be re-written for an MDOF system by replacing the system property terms m, k and c with corresponding property matrices, and by replacing the dynamic response parameters u, $\dot{u}$ and $\ddot{u}$ with corresponding nodal response vectors:

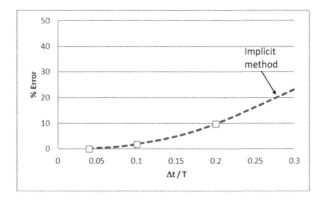

FIGURE 11.7 Effect of time step on solution error.

$$[k].\{u(t)\} + [c].\{\dot{u}(t)\} + [m].\{\ddot{u}(t)\} = \{F(t)\} \qquad (11.16)$$

Stiffness matrix of the structure, $[k]$, is calculated by using the standard finite element formulation given in equation (2.44); and the mass matrix $[m]$ is also calculated from the finite element model of a structure as described in Section 10.7. In many general structural analysis cases, without the presence of distinct damping elements, the damping matrix is usually defined proportional to stiffness and mass property matrices:

$$[c] = a.[m] + b.[k] \qquad (11.17)$$

Damping matrix, defined in equation (11.17), is often known as Rayleigh damping matrix; and the multiplication parameters "a" and "b" are known as Rayleigh damping coefficients. Since the stiffness and mass matrices are reduced to diagonal forms by modal transformation, the damping matrix defined by equation (11.17) is also reduced to diagonal form with same modal transformation. Pre-multiplication of equation (11.17) with the transpose of a mode shape vector, $\{\phi_i\}^T$, and post multiplication with $\{\phi_i\}$, will provide the following expression for damping:

$$\{\varphi\}_i^T [c]\{\varphi\}_i = a.\{\varphi\}_i^T [m]\{\varphi\}_i + b.\{\varphi\}_i^T [k]\{\varphi\}_i = (a + b.\omega_i^2) \qquad (11.18)$$

where ω_i is the vibration frequency and $\{\phi\}_i$ is the mode shape vector that is normalized to give $\{\varphi\}_i^T [m]\{\varphi\}_i = 1$. Equating the modal damping definition of equation (11.18) with that of an SDOF system, $c = \xi.2\sqrt{k.m}$, (introduced in equation 10.19):

$$c = \xi_i.2\sqrt{k.m} = \xi_i.2.\omega_i = (a + b.\omega_i^2) \qquad (11.19)$$

Simplification of the equation (11.19) leads to the following relationship between modal damping value (ξ_i) and Rayleigh damping coefficients:

$$\xi_i = \frac{1}{2}\left(\frac{a}{\omega_i} + b.\omega_i\right) \qquad (11.20)$$

Equation (11.20) can be solved to determine the unknown Rayleigh damping coefficients (a and b) with damping factors for any two selected mode shapes ($i = j,k$). The Rayleigh damping matrix $[c]$, thus calibrated for two modal damping targets, can be substituted in dynamic equilibrium equation (11.16) of an MDOF system. Step-by-step time-domain analysis technique for predicting the response of SDOF system, presented earlier, is equally applicable to MDOF systems as well where the system property terms m, k and c in equation (11.11) are substituted with corresponding property matrices, and the dynamic response parameters u, $\dot{u}$ and $\ddot{u}$ are substituted with corresponding nodal response vectors:

$$\left(\frac{4}{\Delta t^2}[m]+\frac{2}{\Delta t}[c]+[k]\right)\{\Delta u\}=\{F(t+\Delta t)\}-\{R(t)\}$$

$$+[m].\left(\frac{4}{\Delta t}\{\dot{u}(t)\}+\{\ddot{u}(t)\}\right)+[c].\{\dot{u}(t)\}$$

(11.21)

Solution of MDOF system equilibrium equations, with Rayleigh damping coefficients calculated for two selected mode shapes (equation 11.20), does not induce uniform damping for all modal frequency values (Figure 11.8). The stiffness proportional damping term ("$b. \omega_i$" in equation 11.20) shows linearly increasing damping for higher frequency values – leading eventually to unrealistically high damping of high-frequency modes. However, where few low-frequency modes dominate the system response, high damping value imposed on high-frequency modes, by Rayleigh damping formulation, does not cause a significant error in overall dynamic response of the MDOF systems. Additional damping values can be added to the diagonal terms of the Rayleigh damping matrix, [c], if discrete damping elements exist in an MDOF system.

Solution of incremental displacement response, $\{\Delta u\}$, from equation (11.21), requires the assembly of all 3 property matrices for the entire system ([k], [m], and [c]). Enforcement of entire system equilibrium at discrete time steps (t, $t + \Delta t$, …etc.) inherently maintains the stability of solution process although the accuracy can vary depending on the time step used for step-by-step calculation of system response in time domain. Evidently, the step-by-step integration of dynamic response with equation 11.21 implicitly computes the responses of all mode shapes present in an MDOF system. The use of a constant time step value, Δt, implies that the modal response values are computed with different degree of accuracy – with higher modes (i.e. modes with shorter time periods) being calculated with lower accuracy according to the analysis results of SDOF systems presented in Figure 11.7. In practical engineering analyses, time step Δt is chosen small enough to accurately calculate the response of the highest mode of interest (equation 11.22):

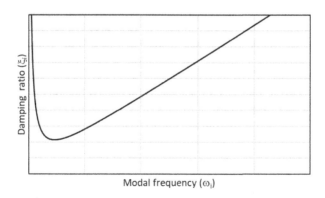

Modal frequency (ω_i)

FIGURE 11.8 Relationship between damping ratio and frequency (for Rayleigh damping).

$$\Delta t = \frac{T_k}{10} \qquad (11.22)$$

where "k" refers to the highest critical mode. Obviously, the identification of critical mode "k" varies depending on loading history and dynamic properties of the system being analyzed. In general engineering practice, the highest significant modal frequency of structure is considered four times the highest significant frequency of a loading function. This conclusion is derived from the observation of dynamic amplification plot in Figure 10.6 that implies that dynamic response of a structure approaches the static response when the ratio of applied loading frequency to the vibration frequency of structure is below 0.25. Fourier series analysis technique can be used to identify the significant frequency content of a loading function. After identifying the highest frequency of interest in a given loading function, the time step for direct integration of system equilibrium equations can be set at ($\Delta t < T_k/10$); where T_k refers to the period of vibration corresponding to four times the critical loading frequency. Although modal frequency analysis of a structure is not a prerequisite of the direct integration method, the finite element analysis model of structure, however, should possess detailed contents to represent up to "kth" mode shapes correctly. In linear elastic system analyses, the large system property matrices ($[k]$, $[m]$ and $[c]$) can be assembled once; however, the calculation steps (defined by equation 11.21) need to be repeated at discrete time steps Δt for the duration of dynamic event. Direct time-domain integration of system equilibrium equations (11.21), without the need for solving large eigenvalue problem, provides one way of calculating the time-domain response of a large system. Modal superposition method, described in the following Section 11.6, can be an efficient alternative technique for linear elastic systems when the response is dominated by few low-frequency modes only.

11.6 MODE SUPERPOSITION METHOD FOR ANALYSIS OF MDOF SYSTEMS

In Section 11.5, dynamic equilibrium equations (11.16) for MDOF system have been solved by direct integration of responses involving all degrees of freedom in the system. When the overall elastic response of a large structure is dominated by few low-frequency modes (such as tall building and long-span bridges), the modal superposition technique is used for the convenience and efficiency of combining few modal response values. Theoretically, the dynamic response at nodal DOF can be calculated by superposing the time-domain responses of multiple mode shapes:

$$\{u(t)\} = [\varphi]\{x_i(t)\} \qquad (11.23)$$

where $[\phi]$ is an $n \times p$ matrix of "p" mode shape vectors, each mode with "n" terms corresponding to "n" DOF in the system ($p \leq n$), and $x_i\{t\}$ are time function of response amplitudes corresponding to mode shapes $i = 1,2 \dots p$ (defined by equation 10.32). Substituting equation (11.23) into equation (11.16), and pre-multiplying both sides with $[\phi]^T$:

$$\left[\varphi\right]^{T}\left[k\right]\left[\varphi\right].\left\{x_{i}\left(t\right)\right\}+\left[\varphi\right]^{T}\left[c\right]\left[\varphi\right].\left\{\dot{x}_{i}\left(t\right)\right\}+\left[\varphi\right]^{T}\left[m\right]\left[\varphi\right].\left\{\ddot{x}_{i}\left(t\right)\right\}=\left[\varphi\right]^{T}\left\{F\left(t\right)\right\} \quad (11.24)$$

Assuming that the damping matrix is defined by the Rayleigh damping coefficients (equation 11.17), orthogonality property of the modal shape vectors, ϕ_i, transforms equations (11.24) into the following set of uncoupled equations in time domain:

$$\omega_{i}^{2}.x_{i}\left(t\right)+2.\xi_{i}.\omega_{i}.\dot{x}_{i}\left(t\right)+\ddot{x}_{i}\left(t\right)=r_{i}\left(t\right) \qquad (11.25)$$

where ω_i is the angular velocity of mode shapes ($i = 1, 2 \dots p$), ξ_i is the modal damping ratio, and $r_i(t)$ is the time function of transformed load function corresponding to mode shape "i". Equation (11.25) is analogous to the dynamic equilibrium of an SDOF system that can be solved in time domain by using the Duhamel integral for undamped systems (equation 11.4), or by using the direct integration technique described in Sections 11.3–11.4. Time-domain response history of overall system is calculated by superposing the modal response histories (equation 11.23). The mode superposition method of dynamic response analysis, thus, involves three distinct steps: (1) numerical solutions to identify mode shapes and frequencies (discussed in Chapter 10); (2) time history solution of decoupled modal responses (equation 11.25); and (3) final superposition of modal responses to derive the overall system response (equation 11.23). The choice between dynamic response history calculation methods, either by using the modal transformation method, or by the direction integration of equilibrium equations of entire system, depends on the relative computational efficiency of two methods. As discussed earlier, direct integration of equations (11.21) requires processing of an entire system of equations – posing significant computational demand. If the response of a system is expected to be dominated by a handful of low-frequency modes (such as earthquake response of tall buildings), few modes of desired frequency range can be extracted with limited computational effort ($p \ll n$), and the dynamic response of few individual modes can be calculated from the uncoupled equations (11.25). The modal superposition method of dynamic response history analysis, however, does not provide any computational benefit for nonlinear systems that will be the topic of discussion in Chapter 12.

11.7   EXPLICIT TIME-DOMAIN ANALYSIS OF MDOF SYSTEMS

Implicit time-domain analysis of structural response requires the assembly and numerical operations with the property matrices [k] and [m] of the entire system. Modal superposition method, which produces a set of decoupled system of equations (11.25) for dynamic response analysis, also requires the assembly of large system property matrices at the beginning to find the modal frequencies and mode shapes. Explicit time-domain analysis technique, on the other hand, provides an alternative numerical method that considers each nodal DOF an independent entity. The time-domain responses at future time step are forecasted explicitly from the past responses of local DOF, without enforcing the equilibrium state over the entire system during the forecasting process. The unbalanced forces produced during the response forecasting process are equilibrated with iterative updates to the local response quantities. The equilibrium state at time (t) of an MDOF system can be expressed by rearranging the terms of equation (11.16) as follows:

$$[m].\{\ddot{u}(t)\} = \{F(t)\} - \{R(t)\} - \{X(t)\} \tag{11.26}$$

where $\{R(t)\}$ refers to system resistance to deformation which is equal to $[k].\{u(t)\}$ for a linear elastic system; and $X(t)$ is the force vector generated by local damping resistance mechanism. Assuming that initial values $\{u(0)\}$ and $\{\ddot{u}(0)\}$ are known for any given system, equations (11.26) can be readily solved to determine the initial acceleration response for any unbalanced force on the right-hand side of equation (11.26). This solution process becomes highly efficient for an undamped system having a diagonal mass matrix – requiring no assemblies of large system property matrices. Once the initial state of equilibrium is established, dynamic responses at a next time step are forecasted based on the current response values of displacement, velocity, and acceleration. Among various explicit forecasting methods, the forward integration method is the simplest one that forecasts the velocity and displacement response at each individual DOF based on the current known responses as follows:

$$\dot{u}(t + \Delta t) = \dot{u}(t) + \ddot{u}(t).\Delta t \tag{11.27}$$

$$u(t + \Delta t) = u(t) + \dot{u}(t + \Delta t).\Delta t \tag{11.28}$$

The displacement values at all DOF, forecasted locally by equations (11.27) and (11.28), are used to calculate strains and stresses in finite elements. And the corresponding system resistance to deformation, $R(t+\Delta t)$, is calculated from integral expression in equation (2.42). The damping resistance $\{X(t)\}$, if included in analysis, is also assembled from local element properties. Equilibrium state at time $(t+\Delta t)$ is enforced by solving the equation (11.26) for updated acceleration response at time $(t+\Delta t)$. The process of explicit forecast of local responses, and subsequent correction of acceleration response to take account of unbalanced force field, can proceed step-by-step to predict the time-domain response of the entire system.

Forward forecasting method, described by equations (11.27) and (11.28), is prone to numerical instability. A relatively more effective explicit integration method is defined based on the central difference theorem where the relationships among displacement, velocity, and accelerations are defined over two times steps as follows:

$$\dot{u}(t) = \frac{1}{2.\Delta t}\left[u(t + \Delta t) - u(t - \Delta t)\right] \tag{11.29}$$

$$\ddot{u}(t) = \frac{1}{(\Delta t)^2}\left[u(t + \Delta t) - 2.u(t) + u(t - \Delta t)\right] \tag{11.30}$$

Substituting the expressions (11.29) and (11.30) into dynamic equilibrium equations (11.16), and after rearranging the terms, the following explicit relationship is obtained to calculate the displacement response at time $(t + \Delta t)$ based on the equilibrium states at (t) and $(t-\Delta t)$:

$$\left(\frac{1}{\Delta t^2}[m] + \frac{1}{2.\Delta t}[c]\right)\{u(t+\Delta t)\}$$

$$= \{F(t)\} - \left([k] - \frac{2}{\Delta t^2}[m]\right)\{u(t)\} - \left(\frac{1}{\Delta t^2}[m] - \frac{1}{2.\Delta t}[c]\right)\{u(t-\Delta t)\} \quad (11.31)$$

Equation (11.31) can be solved at local DOFs provided that mass and damping matrices are of the diagonal form. Explicit calculation of dynamic motion, thus, does not require the assembly and decomposition of a positive definite stiffness matrix which is a fundamental requirement in static load–deflection analysis with finite element models. The very nature of explicit time integration method, which uses local response variables to forecast the future response state, requires that the time step size must not be large. Re-analysis of the free-vibration response of the spring-mass SDOF system (Figure 11.6), using the explicit time integration method, produces solution error much higher than that of implicit method (Figure 11.9). In fact, the solution error increases exponentially as the time step size becomes larger than a quarter of the period of vibration of the system. For stable and accurate response prediction, enough number of calculation points must be used within the period of vibration of the spring-mass system (Figure 11.10), thus defining the following criterion for time step selection – commonly known as the Courant–Friedrichs–Lewy (CFL) law (Courant et al. 1928):

$$\Delta t \le \frac{T}{2\pi} \quad (11.32)$$

Considering the axial stress wave propagation in a one-dimensional member (Figure 11.11), the CFL condition of equation (11.32) can be re-written in the following from (equation 11.33):

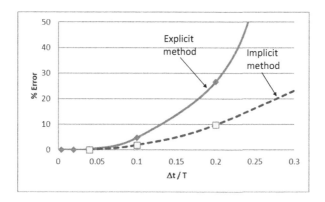

FIGURE 11.9 Solution errors for implicit and explicit integration methods.

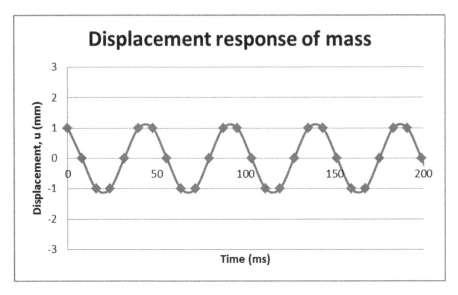

FIGURE 11.10 Stable prediction of the vibration response of spring-mass system using explicit time integration method with $\Delta t = T/2\pi$.

FIGURE 11.11 Axial stress wave propagation in a one-dimensional field.

$$\Delta t \leq \frac{T}{2\pi} \rightarrow \sqrt{\frac{m}{k}} = \sqrt{\frac{\rho A l_c}{\frac{AE}{l_c}}} = \frac{l_c}{\sqrt{E/\rho}} = \frac{l_c}{c} \qquad (11.33)$$

where "c" is the velocity of stress wave propagation in a one-dimensional field. Equation (11.33) establishes a very important relationship between the time step Δt, for finite-difference calculation of response time history, and the finite element size l_c. The stability of explicit method requires that the time step limit, imposed by the size of finite elements (l_c), be satisfied – thus making the method conditionally stable. Physical interpretation of the Courant condition is demonstrated with a one-dimensional wave propagation problem in Figure 11.12. In explicit integration method, local calculations at node "i" assume that the instantaneous effect of impact force $F(t)$ is felt by the mass DOF at node "i" only; and the impact wave does not reach the next node "j" within the same step Δt:

$$c.\Delta t \leq l_c \qquad (11.34)$$

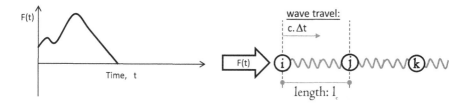

FIGURE 11.12 Impact wave propagation in a one-dimensional field.

where l_c is the distance between adjacent nodes "i" and "j" – commonly known as characteristic length of finite elements. For 1-D finite elements (truss, cable, etc.), l_c is simply equal to the element length (Figures 11.11 and 11.12). For surface-based solid elements, such as the shell in Figure 11.13(a), element characteristic length is often defined with one of the two alternative forms shown in equation (11.35) (LSTC. COM 2021):

$$l_c = \frac{surface\ area}{Max\left(D13,D24\right)} or, \quad l_c = \frac{surface\ area}{Max\left(L_1,L_2,L_3,L_4\right)} \tag{11.35}$$

where D_{13}, D_{24}, L_1, L_2, s..., etc., are geometric dimensions of shell element shown in Figure 11.13(a). The value of l_c is used in equation (11.33) to calculate the critical time step where the stress wave velocity for planar solid is defined by equation (11.36) – in terms of elastic modulus (E), material density (ρ), and Poisson's ratio (ν):

$$c = \sqrt{\frac{1}{\left(1-v^2\right)} \cdot \frac{E}{\rho}} \tag{11.36}$$

For solid elements (Figure 11.13(b)), characteristic length is defined by

$$l_c = \frac{volume}{Max\left(A1,A2,A3,A4,A5,A6\right)} \tag{11.37}$$

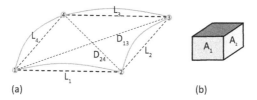

FIGURE 11.13 (a) Geometric dimensions of a shell element; (b) surface areas of a 3D solid element.

where A1, A2 … etc., are surface areas of the solid element. The stress wave velocity in 3D solid is given by

$$c = \sqrt{\frac{(1-v)}{(1+v)(1-2v)} \cdot \frac{E}{\rho}} \qquad (11.38)$$

In complex structural analysis models, comprising of different element types, critical time step limits are checked for all elements, and the smallest of all values defines the time step limit for step-by-step time-domain solution of the MDOF system. For an example shell element size of 5 mm, and for material properties of steel (E = 210 GPa, v = 0.3, and ρ = 7.8e^{-6} kg/mm^3), equation (11.33) limits the explicit integration time step to 0.0009 ms. Despite the very tight limit on allowable time step size, explicit time-domain analysis tends to be very efficient since computations are carried out at local element and nodal DOF levels without the need for assembly and operations of large system property matrices. Local uncoupled analysis process also makes the explicit technique particularly suitable for massive parallel computing processing of highly nonlinear systems. Finite element model preparation for explicit time-domain analysis must control the minimum element size so that the minimum time step is not unduly burdensome for rest of the model analysis. Analysis software packages, such as LS-DYNA (LSTC.COM 2021), allow artificial scaling of mass at local DOF to maintain a minimum time step for efficient execution of the step-by-step analysis of large system models. Mass scaling, however, is not recommended for critical structural elements. Instead, FEA mesh should be revised to meet the time step limit without mass scaling. Alternatively, large complex models can be divided into subdomains; and explicit integration process can proceed independently with different time steps in different subdomains – followed by periodic synchronization of interface responses at larger time steps (Borrvall et al. 2014).

11.8   LINEAR DYNAMIC RESPONSE ANALYSIS WITH ABAQUS

Finite element analysis models, described in earlier chapters for static load and vibration frequency analyses, remain same for time-domain dynamic load analysis of structures. Additional data inputs are required to specify: (1) time integration parameters, and (2) time history of load function. An example analysis step for implicit dynamic analysis in ABAQUS can be described as follows:

*STEP
*DYNAMIC, TIME INTEGRATOR=*HHT-TF*
Δt_0, t_f, Δt_{min} , Δt_{max}
*END STEP

In the above example, selected time integration scheme "*HHT-TF*", which is the default ABAQUS method, refers to the α-integration method proposed by Hilber, Hughes, and Taylor (1977), with slight numerical damping (α = −0.05). Numerical integration parameters β and γ are calculated by using equation (11.14). If desired, time-domain integration parameters can be optionally specified for α, β, and γ as follows:

*STEP
*DYNAMIC, ALPHA=α, BETA=β, GAMMA=γ
Δt_0, t_f, Δt_{min}, Δt_{max}
*END STEP

The value Δt_0 in the above specifies the initial time step to be used by ABAQUS in step-by-step time-domain calculations. ABAQUS has built-in capability to adjust the time step during analysis, based on convergence characteristics of nonlinear solution algorithms. The parameters Δt_{min} and Δt_{max} exert analyst's control on the minimum and maximum values to be considered by ABAQUS during automatic time step adjustments. An analyst can choose fixed time step calculations by optionally defining a parameter, "DIRECT=$value$", in the keyword line *DYNAMIC. The duration of time-domain solution is specified by the input value t_f.

External loads on finite element nodes or surfaces, described by *CLOAD or *DLOAD keywords, can vary in time as defined by a user-defined time function. The following example describes time-dependent concentrated load applications at selected nodes of a finite element model:

*STEP
*DYNAMIC, ALPHA=α, BETA=β, GAMMA=γ
Δt_0, t_f, Δt_{min}, Δt_{max}
*CLOAD, AMPLITUDE=Loading-TH
node1, 3, c1
node2, 3, c2
node3, 3, c3
*END STEP

where, "Loading-TH" refers to a time function of load amplitude that is defined in ABAQUS input data file by using *AMPLITUDE key word:
*AMPLITUDE, name=Loading-TH
0.0, 0.0, 5.0, 3.2, 10.0,0.0, 100.0, 0.0

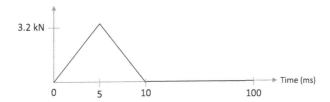

FIGURE 11.14   Time history of an arbitrary load function.

In the above description, time history of load amplitude is described by time and amplitude values, in pairs, with as many pairs as needed to describe the complete time history of applied load. Figure 11.14 graphically describes the load function described in the above example of input data block. The generic load function, thus defined, is multiplied by a factor "c_1", and applied in coordinate direction "3" at node "$node_1$" (per the description of input data presented in the above example). Similarly scaled load functions are also applied at nodes $node_2$, $node_3$, etc. Time history of calculated finite element model responses can be saved in "odb" database file for graphical post-processing of results, at specified time intervals by assigning a numerical value to parameter "FREQUECNY" in *OUTPUT control command:

```
*STEP
*DYNAMIC, ALPHA=α, BETA=β, GAMMA=γ
Δt₀, t_f Δt_min , Δt_max
*CLOAD, AMPLITUDE=Loading_TH
node1, 3, c1
node2, 3, c2
node3, 3, c3
*OUTPUT, FIELD, FREQUENCY=1
*NODE OUTPUT
U
*END STEP
```

In the discussions so far, time-domain dynamic analysis option has been initiated by using the *DYNAMIC command. This analysis option can be repeatedly used to generate target structural responses for many different load functions when needed. In some practical applications, engineers often seek to determine the sensitivity of a structural response parameter to the frequency of applied load (where the geometric distribution and amplitude of loading do not change). Figure 11.15 presents a conceptual description of the frequency response function for floor vibration of a hypothetical automotive body structure subjected to external loads of variable input

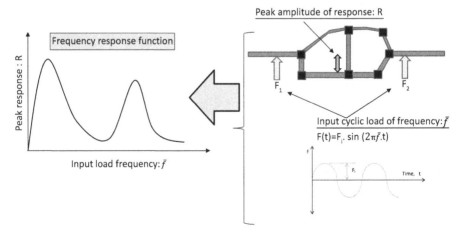

FIGURE 11.15 Example of a frequency response function.

frequency. Such frequency response functions can be efficiently generated in ABAQUS by using the dedicated analysis procedure, *STEADY STATE DYNAMICS. The analysis step for response function generation must be preceded by a frequency analysis step to generate mode shape vectors. The following example lists key ABAQUS command steps for steady-state dynamic analysis procedure:

```
*STEP
*FREQUENCY
, , f_f
*END STEP
*STEP
*STEADY STATE DYNAMICS, Interval=Range, Frequency scale=linear
f_0, f_f, n_steps
*CLOAD
set1, 3, c1
set2, 3, c2
etc.
*OUTPUT, HISTORY
*NODE OUTPUT, NSET=set1
UT, VT, AT
*END STEP
```

Modal frequency and shape data, generated in the first analysis step up to the frequency limit of f_f, are used in the next steady-state dynamic analysis step for generating frequency response function by using the mode superposition technique. Frequency-dependent displacement (UT), velocity (VT), and acceleration (AT) responses, for the node set "*set1*" in the above example, are saved in the database file for subsequent post-processing.

For explicit dynamic analysis, ABAQUS uses the central difference time integration method (Equations 11.29–11.31). Analysis step for explicit dynamic analysis is specified in ABAQUS input data file by inserting the required parameter "EXPLICIT" with *DYNAMIC keyword:

```
*STEP
*DYNAMIC, EXPLICIT, DIRECT USER CONTROL
Δt_0, t_f
*END STEP
```

The optional parameter "DIRECT USER CONTROL" specifies that the analysis is conducted at a fixed user-defined time step of Δt_0; and the time span of dynamic response analysis is defined by t_f. Automatic time step definition, based on element wave characteristics (Equations 11.32–11.38) can be activated by choosing the optional parameters "ELEMENT BY ELEMENT" in lieu of "DIRECT USER CONTROL":

```
*STEP
*DYNAMIC, EXPLICIT, ELEMENT BY ELEMENT
t_f, Δt_max
*END STEP
```

The input parameter Δt_{max} limits the maximum time step that ABAQUS can use from element-by-element time step size calculations. Explicit dynamic analysis can be conducted without direct initial load applications to nodes or elements. Instead, time-domain system response can be calculated starting with an initial velocity state, or with prescribed motion applied to a set of nodes. Following example shows how to define initial velocity to a set of nodes:

*INITIAL CONDITIONS, TYPE=VELOCITY
set_i, DOF_i, v_i
...

where initial velocity v_i is applied to node set set_i at degree of freedom DOF_i. A user can include additional lines of similar data to define initial velocity values for multiple node sets.

11.9   PRACTICE PROBLEMS: DYNAMIC RESPONSE ANALYSIS OF STRUCTURES

PROBLEM 1
Static load–deflection analysis problem of Figure 6.28 is re-introduced in Figure 11.16. Analyze the problem, using implicit time-domain analysis technique, assuming that the externally applied load amplitude follows the time function described in Figure 11.14. Calculate the dynamic amplification factor for downward deflection at monitoring point A (shown in Figure 11.16).

PROBLEM 2

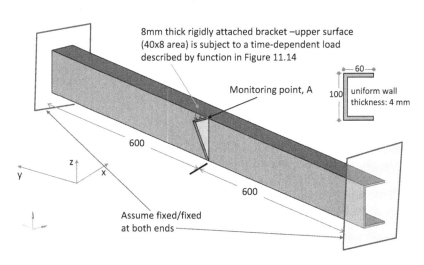

8mm thick rigidly attached bracket –upper surface (40x8 area) is subject to a time-dependent load described by function in Figure 11.14

Monitoring point, A

60

100   uniform wall thickness: 4 mm

600

z

y   x

600

Assume fixed/fixed at both ends

FIGURE 11.16   Load-deflection analysis for a time-dependent load.

FIGURE 11.17 Prescribed downward displacement at the monitoring point A in Figure 11.16

Assuming that the loading in Problem-1 follows sine harmonic function, generate frequency response function for peak downward deflection at monitoring point A (shown in Figure 11.16) in the frequency range of 0.1 to 1 cycle/ms.

PROBLEM 3

Re-analyze the problem, described in Problem-1, for prescribed downward displacement function defined in Figure 11.17 (in lieu of the external loading function applied in Problem-1). Compare the relative computational efficiency of implicit vs explicit time integration techniques for this problem.

12 Nonlinear Analysis of Structures

SUMMARY

Finite element analysis methods, presented in Chapters 1–11 of this book, have been built on four key assumptions: (a) material stress–strain relationship is linear elastic; (b) strain and displacement responses are very small, thus, keeping the first-order derivative relationship between strain and displacement constant throughout the response history; (c) the boundary conditions and inter-body contact conditions do not change over the course of load–displacement history; and (d) applied load vector is independent of the structural displacement response. Assumptions (a), (b) and (c) make the structural property matrices, $[k]$ and $[m]$, constant. The stress-based assessment of structural strength and durability, thus, generally falls in the domain of linear elastic finite element analysis. The vibration frequency analysis problem, discussed in Chapter 10, is a linear eigen solution problem based on the constant structural property matrices $[k]$ and $[m]$. The time-domain load–displacement analysis technique, discussed in Chapter 11, is also a linear elastic finite element analysis method, based on constancy assumptions for system property matrices, boundary conditions, and geometric distribution of time-dependent loading. The buckling load analysis of structures, frequently discussed in finite element literature, is another class of eigenvalue problem where the load capacity is estimated for a small perturbation to the elastic stability of structure. This analysis technique is available in general-purpose finite element software packages. However, this topic is omitted from this reference book since the author has not seen an opportunity to use that technique in the past 25 years of professional practice in civil and automotive structural design. While many of the structural engineering problems get analyzed and solved by using linear elastic finite element simulation methods, practicing engineers do encounter frequent problems where one or more of the linearity assumptions get violated. And in certain engineering problems, for example, in automotive structural design for crashworthiness, nonlinear finite element simulation method is used as the daily analysis tool soon after the development of preliminary design based on empirical design rules and prior engineering experiences.

The volume of current knowledge on nonlinear finite element methods is way beyond the scope of one entire book. This chapter here should be considered as partial introductions to some of the key ideas and techniques that are used in the analysis of practical engineering problems. Section 12.1 starts with brief references to the key sources of nonlinearity that appear in structural analysis problems. The general idea of deriving iterative solution for nonlinear load–displacement problem is introduced in this Section. Section 12.2 summarizes the basic approach of how to include material nonlinear behavior in finite element software implementation. Details of material plasticity formulations are presented for demonstration of the key intricacies that get implemented in constitutive models of finite element software packages. Section 12.3 is devoted to the formulation of geometric nonlinear problems – involving higher order derivative relationships between strain and displacement responses of solids. The basic ideas of contact formulations have been discussed in Chapter 8. Section 12.4 in this chapter provides a brief description of how contact status changes can be included in step-by-step simulation of nonlinear structural response analysis. Section 12.5 describes the integration of nonlinear response mechanisms in step-by-step implicit and explicit dynamic response analysis of structures. Nonlinear response analysis, with deformation-dependent external loading description, is not explicitly discussed in this book. However, it is understood that such changes can be integrated into general nonlinear analysis steps when required. Section 12.6 is devoted to the discussion of material fracture propagation in finite element simulation model – that has experienced huge amount of research contributions over the past few decades. Structural form simulation, a very specialized nonlinear structural engineering problem, is discussed in Section 12.7 followed by Section 12.8 presenting a set of practice problems for nonlinear analysis.

12.1   SIMULATION OF NONLINEAR FORCE-DEFORMATION RESPONSE OF STRUCTURES

Linear elastic finite element analysis technique, described thus far in Chapters 1–11, rests on two critical assumptions: (1) material undergoes infinitesimal small deformation resulting into a linear elastic stress–strain relationship matrix $[C]$; and (2) deformation response does not effectively change the geometry of structure, thus keeping the strain–displacement relationship matrix, $[B]$, constant during the analysis step. The constancy assumptions for element property matrices, $[C]$ and $[B]$, lead to the calculation of a system stiffness matrix $[k]$ (equation 2.44) that is independent of the deformation response. Linear elastic solution of the static equilibrium state, described by equation (2.43), uses two additional assumptions: (3) boundary conditions embedded into the displacement vector $\{u\}$ do not change, and (4) load vector $\{P\}$ is independent of the structural deformation response. These four assumptions, together, imply that a scalar multiple of the load vector $\{P\}$ will generate same scalar multiple of the displacement response $\{u\}$. Equilibrium equations (2.43), for linear elastic systems, are solved with one single step of load application – producing the displacement response $\{u\}$, the elastic strain response $\{\varepsilon\} = [B]\{u\}$, and the stress response $\{\sigma\} = [C]\{\varepsilon\}$. Assumptions of material and geometric linearity, as well as the constancy of load distribution and boundary conditions, serve the purpose of many engineering analysis problems. However, one or more deviations from the linearity assumptions may appear in many structural design situations when the analyst is required to consider nonlinear analysis techniques for safety assessment of structures.

Nonlinear response of elementary structural members, for standard load and boundary conditions, can be occasionally predicted with analytical techniques. Figure 12.1 shows the ultimate load capacity calculation of a simply supported beam

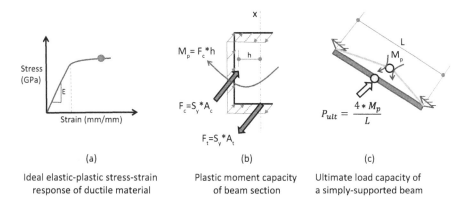

(a)

Ideal elastic-plastic stress-strain
response of ductile material

(b)

Plastic moment capacity
of beam section

(c)

Ultimate load capacity of
a simply-supported beam

FIGURE 12.1   Ultimate strength analysis of a beam based on the concept of plastic moment capacity of beam section.

assuming elastic-plastic material behavior. However, the progressive evolution of nonlinear response mechanisms in complex structures is often very difficult to visualize for analytical formulation. Finite element analysis, with step-by-step load applications, provides an effective method to predict the nonlinear response mechanism of structures. Analysis can start with usual linear elastic model, with a small initial load application (F_0 in Figure 12.2), providing preliminary estimate for displacement response u_i. This displacement response can be used to predict local strain and stress responses using the preliminary assumption of linear elastic material response. However, when locally calculated stresses exceed the elasticity limit of material response, a nonlinear stress–strain relationship model can be invoked to define the local resistance values. The integrated overall structural resistance, corresponding to the first step of calculated displacement response (point A in Figure 12.2), may show an imbalance with the externally applied force F_0. Newton-Raphson iteration method provides a solution to this problem with the prediction of an incremental displacement response for the unbalanced force value (equation 12.1):

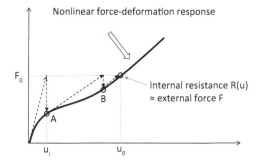

FIGURE 12.2   Nonlinear force–deformation response of a system.

$$K^i.\left(\Delta u\right)^{i+1} = F_0 - R^i \tag{12.1}$$

where K^i is the tangential stiffness matrix after iteration i, R^i is the internal resistance of structure corresponding to displacement response u_i, and $(\Delta u)^{i+1}$ is the correction to displacement response for the unbalanced force:

$$u^{i+1} = u^i + \left(\Delta u\right)^{i+1} \tag{12.2}$$

The internal resistance of structure can be updated for the new displacement response u^{i+1}, and equation (12.1) can be solved repeatedly until the unbalance force on the right-hand side becomes negligible. In stable structural configurations, the use of tangential stiffness matrix K^i in iterative calculations, demonstrated graphically in Figure 12.2, generally leads to a quick convergence to the target solution point (F_0, u_0). The above procedure of iterative solution to a discrete load step can be repeated for a second load step, and so on. For computational efficiency reasons, in large structural system analyses, the stiffness matrix can be updated only once in the first iteration of a load step, and the subsequent iterations in the same load step can be conducted by re-using the same stiffness matrix – a method generally referred to as modified Newton–Raphson method. In the above discussion, nonlinear material behavior has been cited as the reason for stiffness matrix change. Large geometric configuration change may also lead to the change of stiffness properties of a structure. Same can happen if the boundary conditions change due to part-to-part contacts. And the fourth source on nonlinearity, as noted earlier, can appear when the external loading condition changes due to changes in structural deformation response.

The step-by-step linearization of nonlinear structural response, presented by equation (12.1), is a mere extension of the standard solution technique used in linear elastic finite element analysis models (equation 2.43). The success of this analysis technique depends on the positive definite property of tangent stiffness matrix K^i in equation (12.1) which in indicated in Figure 12.2 by monotonically increasing structural resistance against deformation response. However, failure zone localization in some structures, at or after the peak response point, may lead to post-peak softening response (Figure 12.3). The tangent stiffness property of structure is no-longer positive definite at the post-peak state – meaning that standard technique of tangential stiffness matrix decomposition and inversion cannot be used to predict the nonlinear load–deformation response history beyond the peak resistance point (B). A solution to this problem can be obtained by controlling the displacement response at one or more control points – solution technique commonly known as displacement control method. Structural DOF with pre-defined non-zero displacement responses are separated from other DOF in the system equilibrium setup:

$$\begin{bmatrix} K_a & K_{as} \\ K_{sa} & K_s \end{bmatrix}^i . \begin{Bmatrix} \Delta u_a \\ \Delta u_s \end{Bmatrix}^i = \begin{Bmatrix} \Delta f_a \\ \Delta r_s \end{Bmatrix} \tag{12.3}$$

where suffix "s" refers to the DOF under specified displacement control, and suffix "a" to all other non-zero DOF. Pre-defined displacement increment at control DOF is

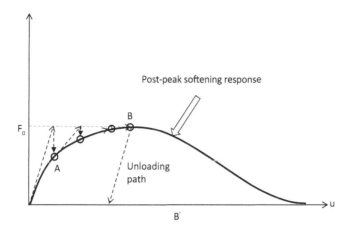

FIGURE 12.3   Post-peak softening response of a structure.

identified by Δu_s and the corresponding unknown reaction forces by Δr_s. The term Δf_a refers to zero or nonzero externally applied forces at un-constrained DOF in the system. Partitioning of the system equilibrium equations allows the calculation of unknown displacement responses by solving the following reduced system of equations (12.4):

$$\left[K_a\right]^i \cdot \left\{\Delta u_a\right\}^i = \left\{\Delta f_a\right\} - \left[K_{as}\right]^i \cdot \left\{\Delta u_s\right\} \tag{12.4}$$

Solution of equation (12.4) provides the first approximation to unknown deformation responses Δu_a that can be used to find out the unbalance between external and internal forces, and the iterative improvement to predicted solution can proceed by using the Newton-Raphson method (or by using its modified form). The displacement control method, presented in the above, helps to predict nonlinear structural responses, beyond the peak resistance point, provided that the controlled displacement response increases monotonically. However, special situations exist where the load–displacement response may not progress monotonically. Figure 12.4 shows a schematic shear test setup for notched concrete beam that happens to show snap-back in load–deflection response (Arrea and Ingraffea 1981). Neither the step-by-step load control nor the displacement control analysis techniques can numerically simulate such structural responses. Successful numerical techniques have been implemented in academic research to predict the snap-back behavior by using specially derived monotonic response function, such as the crack tip displacement, to adjust the externally applied loads or displacements during incremental analyses (Bhattacharjee 1993).

   Not all four nonlinear response mechanisms, listed earlier in this section, occur simultaneously in a structural analysis problem. Nonlinear material and geometric responses tend to be the commonly encountered problems followed by contact-related nonlinearity. Section 12.2 in the following summarizes the basic approaches of how material nonlinear response is handled in finite element formulations.

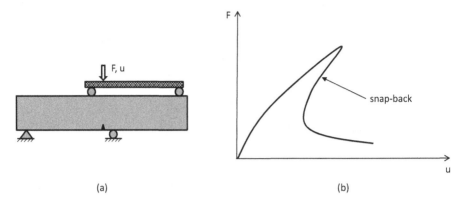

(a)   (b)

FIGURE 12.4 (a) Single notch shear test of a beam made of quasi-brittle concrete; (b) snap-back in force-displacement response.

12.2 NONLINEAR MATERIAL MODELS FOR FINITE ELEMENT SIMULATION OF STRUCTURES

Material stress–strain relationship, alternatively known as constitutive model, defines the material elasticity matrix $[C]$, that is used to calculate the stiffness property of structures, and to calculate the stress responses from strains. In linear elastic finite element simulations, both displacement and strain values are assumed to be very small, thus retaining the initial reference coordinate system to define finite element strain and stress responses (equations 2.5 and 2.25). Hooke's law, using the principle of superposition of stress–strain responses, relates the total strain responses to total stress responses through equation (2.18). This linear relationship, between stresses and strains, often changes as the deformation increases (Figure 12.5). Similar to the piece-wise linearization of load–deformation response of structures, material stress–strain law in nonlinear finite element analysis can also be expressed in the following incremental form:

$$\{\Delta\sigma\}^i = [C]^i . \{\Delta\varepsilon\}^i \tag{12.5}$$

where $[C]^i$ is the instantaneous elasticity matrix at a given state of deformation in the material. In step-by-step nonlinear load–deformation analysis of structures, instantaneous elasticity matrix $[C]^i$ can be used, in lieu of the initial elasticity matrix $[C]$, to define the tangential stiffness matrix $[K]^i$ in equation (12.1). The total stress response, at the newly deformed state of material, is calculated by adding the incremental stress values to those of the previous equilibrium state:

$$\{\sigma\}^i = \{\sigma\}^{i-1} + \{\Delta\sigma\}^i \tag{12.6}$$

These updated stress values are used to calculate the internal resistance of structure $\{R\}^i$ in equation (12.1), and the standard iterative correction for unbalanced forces

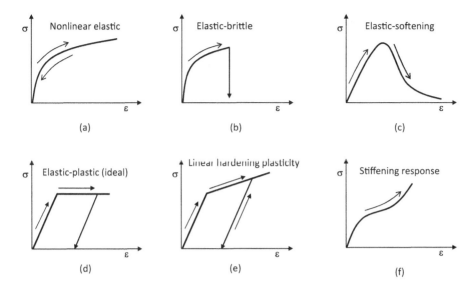

FIGURE 12.5   Nonlinear stress–strain responses of solids.

can continue as usual. Finite element formulations for multi-axial nonlinear stress response, however, require special techniques to define the tangential elasticity matrix $[C]^i$.

One of the commonly used material models is the plasticity-based formulation for simulation of ductile material behavior. Figure 12.6 schematically shows a nonlinear stress–strain response curve, where material resistance, after initial yield point (S_y), increases with increasing permanent deformation (plastic strain) in the material – commonly known as hardening plastic behavior of ductile metals. The incremental strain response $\Delta \varepsilon^i$ in equation (12.5) comprises of two parts – an elastic (recoverable) part, $\Delta \varepsilon_e$, and a permanent deformation (plastic) part, $\Delta \varepsilon_p$ (equation 12.7):

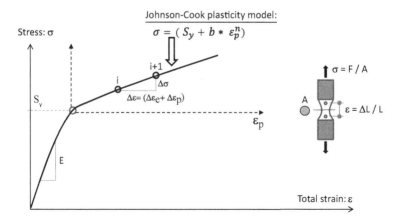

FIGURE 12.6   Hardening plasticity response of a ductile metal specimen.

$$\{\Delta\varepsilon\}^i = \{\Delta\varepsilon_e\}^i + \{\Delta\varepsilon_p\}^i \tag{12.7}$$

For pure uniaxial stress–strain case, relationship between incremental stress and strain is defined as follows:

$$\left(\Delta\sigma\right)^i = E^i.\left(\Delta\varepsilon\right)^i = E.\left(\Delta\varepsilon - \Delta\varepsilon_p\right)^i \tag{12.8}$$

where "E^i" refers to tangent elastic modulus of material, and "E" the Young's modulus that relates incremental stress to the elastic component of incremental strain. The relationship between incremental stress and plastic strain is defined by using a material parameter "H", commonly known as strain hardening parameter:

$$\left(\Delta\varepsilon_p\right)^i = \left(\Delta\sigma\right)^i / H^i \tag{12.9}$$

Combining equations (12.8) and (12.9), the following definition is obtained for the tangent modulus E^i:

$$E^i = \frac{E.H^i}{E + H^i} = E\left(1 - \frac{E}{E + H^i}\right) \tag{12.10}$$

With known hardening parameter value H^i from nonlinear uniaxial stress–strain test data, equation (12.10) gives the tangent modulus, E^i, which is used to define incremental stress–strain relationship and tangential stiffness properties of one-dimensional finite elements (truss, cable, etc.). The value of H^i, and thereby, the incremental stress–strain relationship and stiffness property matrices can be updated during the iterative calculations of nonlinear structural responses. Evidently, an ideal elastic-plastic material law (Figure 12.5(d)) can be generated by setting $H^i \approx 0$; and a linear hardening model (Figure 12.5(e)) can be generated by using a constant value of H^i. Plastic hardening may occur only in the direction of straining without affecting other directions – known as kinematic hardening of material. Alternatively, hardening can be assumed to occur iso-tropically. Depending on the prior understanding of specific material behavior, a user of general-purpose finite element software packages can choose either of the hardening options, kinematic or isotropic, with material plasticity models.

The accumulated plastic strain over a loading duration indicates the degree of permanent deformation in the material – which is also used to assess the material integrity at finite element calculation points:

$$\{\varepsilon_p\}^i = \{\varepsilon_p\}^{i-1} + \{\Delta\varepsilon_p\}^i \tag{12.11}$$

In one-dimensional finite element models (truss, cable, etc.), strains and stresses are calculated in local element direction, thus explicitly defining the direction of plastic strain in the same axis direction. The calculation of tangent elasticity matrix, $[C]^i$ in equation (12.5), is achieved with straight forward substitution of initial Young's modulus, E, with the tangent modulus E^i calculated from equation (12.10). In multi-directional stress fields, with no apparent member axis direction, special analytical assumptions are needed to define a comprehensive plasticity model: (a) an initial yield function (σ, ε); (b) a plastic flow rule defining the direction and magnitude of plastic strain; (c) material hardening rule; and (d) the tangent (incremental) elasticity matrix $[C]^i$. The von Mises stress function, defined in equation (9.19), is one possible option to define a yield function in multi-axial stress field:

$$Y = \left[\sqrt{\frac{1}{2}\left[\left(\sigma_1 - \sigma_2\right)^2 + \left(\sigma_2 - \sigma_3\right)^2 + \left(\sigma_3 - \sigma_1\right)^2 \right]} - S_y \right] \leq 0 : elastic \qquad (12.12)$$

where σ_1, σ_2 and σ_3 are principal stresses calculated at a finite element integration point, and S_y is the yield strength of material determined from uniaxial material test specimen. For hardening plasticity, a general form of yield function can be written as

$$Y\left(\sigma, \kappa\right) = 0 \qquad (12.13)$$

where κ is a material hardening parameter. For plastic deformation occurring on the yield surface, derivative of yield function gives:

$$\frac{\partial Y}{\partial \sigma_1}.\Delta\sigma_1 + \frac{\partial Y}{\partial \sigma_2}.\Delta\sigma_2 + ... + \frac{\partial Y}{\partial \kappa}.\Delta\kappa = 0 \qquad (12.14)$$

Using vector notation, equation (12.14) can be rewritten as:

$$\left\{ \frac{\partial Y}{\partial \sigma} \right\}^T .\left\{\Delta\sigma\right\} + \frac{\partial Y}{\partial \kappa}.\Delta\kappa = 0 \qquad (12.15)$$

When a stress state falls outside the yield function, plastic deformation is expected in the material. In general constitutive models, a dedicated plastic potential function, Q is defined as a function of stresses (σ) and hardening parameter (κ):

$$Q\left(\sigma, \kappa\right) = 0 \qquad (12.16)$$

The incremental plastic strain vector $\{\Delta\varepsilon_p\}$ is defined as (after dropping the iteration index "i" for simplicity):

$$\left\{\Delta\varepsilon_p\right\} = \Delta\lambda.\left\{ \frac{\partial Q}{\partial \sigma} \right\} \qquad (12.17)$$

where $\Delta\lambda$ is proportionality constant (to be determined). Equation (12.17) defines the plastic strain increments in a multi-directional stress field. A constitutive model using

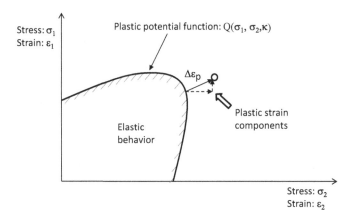

FIGURE 12.7 Plastic strain potential function in a 2D strain field.

same analytical functions for Y and Q is known as associated plasticity model – commonly used for ductile metals. When Y and Q use different analytical functions, the constitutive model is known as non-associated plasticity model – generally used for soil and granular materials. For graphical simplicity, an example plastic potential function is shown in Figure 12.7 for a two-dimensional stress field. In hardening plasticity models, the plastic strain increment, defined by equation (12.17), grows in normal direction to the original surface with components occurring in the original directions of σ_1 and σ_2. A three-dimensional plastic potential function will include a third dimension to this two-dimensional model.

The total strain increment in equation (12.7) can now be rewritten as

$$\{\Delta\varepsilon\} = \left[C\right]^{-1}.\{\Delta\sigma\} + \Delta\lambda.\left\{\frac{\partial Q}{\partial\sigma}\right\} \tag{12.18}$$

where $[C]$ is the standard elasticity matrix of material. Defining the work hardening parameter κ (in Equations 12.13 and 12.16) as the amount of plastic work done during plastic deformation, incremental plastic work is given by

$$\Delta\kappa = \{\sigma\}^{T}.\{\Delta\varepsilon_{p}\} \tag{12.19}$$

Substituting equation (12.17) into equation (12.19) gives:

$$\Delta\kappa = \{\sigma\}^{T}.\left\{\frac{\partial Q}{\partial\sigma}\right\}.\Delta\lambda \tag{12.20}$$

Substituting equation (12.20) into equation (12.15) gives:

$$\left\{\frac{\partial Y}{\partial\sigma}\right\}^{T}.\{\Delta\sigma\} + \frac{\partial Y}{\partial\kappa}.\{\sigma\}^{T}.\left\{\frac{\partial Q}{\partial\sigma}\right\}.\Delta\lambda = 0 \tag{12.21}$$

Equations (12.18) and (12.21) can now be combined to write the following expression:

$$
\left\{ \begin{matrix} \Delta\varepsilon \\ 0 \end{matrix} \right\} = \left[\begin{matrix} [C]^{-1} & \left\{ \dfrac{\partial Q}{\partial \sigma} \right\} \\[2ex] \left\{ \dfrac{\partial Y}{\partial \sigma} \right\}^{T} & \dfrac{\partial Y}{\partial \kappa} \cdot \{\sigma\}^{T} \cdot \left\{ \dfrac{\partial Q}{\partial \sigma} \right\} \end{matrix} \right] \cdot \left\{ \begin{matrix} \Delta\sigma \\ \Delta\lambda \end{matrix} \right\}
\tag{12.22}
$$

Inverting the relationship (12.22), the following relationship is obtained between incremental stresses and strains (Zienkiewicz and Taylor 1991):

$$
\{\Delta\sigma\} = \left([C] - [C] \cdot \frac{\left\{ \dfrac{\partial Q}{\partial \sigma} \right\} \cdot \left\{ \dfrac{\partial Y}{\partial \sigma} \right\}^{T} \cdot [C]}{\left\{ \dfrac{\partial Y}{\partial \sigma} \right\}^{T} \cdot [C] \cdot \left\{ \dfrac{\partial Q}{\partial \sigma} \right\} - \dfrac{\partial Y}{\partial \kappa} \cdot \{\sigma\}^{T} \cdot \left\{ \dfrac{\partial Q}{\partial \sigma} \right\}} \right) \cdot \{\Delta\varepsilon\}
\tag{12.23}
$$

The incremental stress–strain relationship matrix, defined by the terms inside the parenthesis on the right-hand side of equation (12.23), can be used to calculate tangent stiffness matrix and incremental stress responses in finite element calculations. Stress–strain curve generated from one-dimensional material test specimen is used to define the basic material parameters embedded in the incremental formulation. The post-yield stress–strain data can be described with piece-wise linearized steps, or can also be described with an analytical function. Johnson–Cook model (Johnson and Cook 1983), a commonly used isotropic plasticity formulation, describes the post-yield hardening response with an analytical expression (Figure 12.6):

$$
\sigma = \left(S_y + b * \varepsilon_p^{n} \right)
\tag{12.24}
$$

In ABAQUS model input files, the Johnson–Cook plasticity model can be chosen by a user by using the following data inputs:

*MATERIAL, NAME=*mat1*
*ELASTIC
E, ν
*PLASTIC, HARDENING=JOHNSON COOK
S_y, b, n

where E and ν are elastic material properties (Young's modulus and Poisson's ratio, respectively); and S_y, b and n are Johnson–Cook material parameters that are generally calculated by fitting equation (12.24) to a uniaxial test data set. A step-by-step nonlinear static analysis of the two-member truss assembly problem of Figure 2.18 can be constructed for a load amplitude of 960 kN by using the following input commands in an ABAQUS input file:

*STEP
*STATIC
1,10,0.1,2.0
*CLOAD
2, 2,-960.0
*END STEP

The option for linear elastic small displacement – small strain analysis, keyword "PERTURBATION", has been excluded in the above analysis step definition. In this example, a concentrated load of 960 kN is applied at node ID # 2 in negative direction of coordinate ID #2 (y-direction). The step parameters, described after the keyword *STATIC, specifies that the first analysis step will apply 1/10th of the load; the total load will be applied in 10 steps; and during automatic load adjustments, ABAQUS is allowed to use a minimum step of 0.1 times the initial load step, but no more than 2 times of that initial step. Figure 12.8 shows predicted load–deflection responses of the system for both linear elastic material assumption and for Johnson–Cook material plasticity model ($S_y = 0.43$ GPa, $b = 0.824$, and $n = 0.51$). The consideration of material hardening plasticity has led to a hardening structural resistance against increased deformation of the structure. In this simple system model of two truss members, material nonlinearity has caused a very large downward displacement response at the load application point – making the geometric configuration of structure very different from the initial state (Figure 12.9). Large geometry change requires consideration of geometric nonlinearity in finite element simulations – the subject of discussion in next Section 12.3. It should be noted, however, that large geometry change can also occur without causing nonlinear material response.

As noted earlier, isotropic plasticity model is relevant for ductile metal behavior only. An analyst will need to choose an appropriate constitutive model depending on the prior knowledge about nonlinear response mechanism of the specific material being analyzed. Numerous material constitutive models have been proposed in the literature to simulate nonlinear response mechanisms of different material types. The

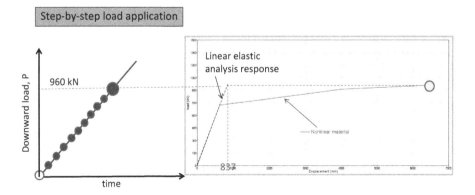

FIGURE 12.8 Elastic versus nonlinear load–deformation responses of two-member truss assembly.

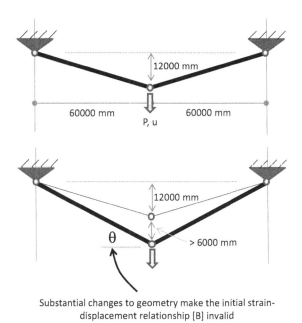

Substantial changes to geometry make the initial strain-
displacement relationship [B] invalid

FIGURE 12.9 Nonlinear geometric response of a two-member truss assembly.

list of available options has become so large over the decades that a comprehensive review of potential choices is beyond the scope of a limited size book. Descriptions of few selected constitutive models for softening, creep and visco-elastic type material behavior are available in Bathe (1996), Lemaitre and Chaboche (1994), and Zienkiewicz and Taylor (1991). General-purpose finite element simulation software packages include many commonly used material constitutive models. Based on prior knowledge, an analyst can choose a constitutive model appropriate for the material case study with known values of relevant model parameters.

12.3 SIMULATION OF LARGE DEFORMATION RESPONSE – NONLINEAR GEOMETRIC PROBLEMS

Large geometric changes, with or without material nonlinear response, means that the strain–displacement relationship matrix [B] (equation 2.6) does not remain constant. Using the initial reference coordinate system, the Green-Lagrange strain components for a nonlinear large deformation problem can be defined by extending the definitions from equations (2.1) and (2.2) into the following forms (Fung 1965):

$$\varepsilon_x = \frac{\partial u}{\partial x} + \frac{1}{2}\left[\left(\frac{\partial u}{\partial x}\right)^2 + \left(\frac{\partial v}{\partial x}\right)^2 + \left(\frac{\partial w}{\partial x}\right)^2\right]$$

$$\gamma_{xy} = \frac{\partial u}{\partial y} + \frac{\partial v}{\partial x} + \left[\frac{\partial u}{\partial x} \cdot \frac{\partial u}{\partial y} + \frac{\partial v}{\partial x} \cdot \frac{\partial v}{\partial y} + \frac{\partial w}{\partial x} \cdot \frac{\partial w}{\partial y}\right]$$

(12.25)

Other components of strain (ε_y, ε_z, γ_{yz}, γ_{zx}) can also be defined in similar formats of equations (12.25). Evidently, the strain components can be separated into two terms:

$$\{\varepsilon\} = \{\varepsilon_0\} + \{\varepsilon_L\} \tag{12.26}$$

where ε_0 refers to the linear first-order terms in equations (12.25), and ε_L to the second-order nonlinear terms. Using the standard nomenclature for strain–deformation relationships, described in earlier chapters, equation (12.26) can be re-written as:

$$\{\varepsilon\} = \left([B]_0 + [B]_L\right)\{u\} = [\bar{B}]\{u\} \tag{12.27}$$

where, nonlinear strain–deformation relationship matrix comprises of two parts:

$$[\bar{B}] = [B]_0 + [B]_L \tag{12.28}$$

$[B]_0$ refers to the first-order linear components of strains, defined by Equations (2.5) and (2.6), and $[B]_L$ is a new term referring to the higher order terms in equations (12.25). The quadratic nonlinear strain terms can be conveniently expressed as follows:

$$\varepsilon_L = \frac{1}{2}\begin{bmatrix} \theta_x^T & 0 & 0 \\ 0 & \theta_y^T & 0 \\ 0 & 0 & \theta_z^T \\ 0 & \theta_z^T & \theta_y^T \\ \theta_z^T & 0 & \theta_x^T \\ \theta_y^T & \theta_x^T & 0 \end{bmatrix}\begin{Bmatrix} \theta_x \\ \theta_y \\ \theta_z \end{Bmatrix} = \frac{1}{2}[A]\{\theta\} \tag{12.29}$$

where, matrix [A] is of 6 × 9 size; and the terms in $\{\theta\}$ are defined as follows:

$$\theta_x^T = \begin{bmatrix} \frac{\partial u}{\partial x} & \frac{\partial v}{\partial x} & \frac{\partial w}{\partial x} \end{bmatrix}; \quad \theta_y^T = \begin{bmatrix} \frac{\partial u}{\partial y} & \frac{\partial v}{\partial y} & \frac{\partial w}{\partial y} \end{bmatrix}; \quad \theta_z^T = \begin{bmatrix} \frac{\partial u}{\partial z} & \frac{\partial v}{\partial z} & \frac{\partial w}{\partial z} \end{bmatrix} \tag{12.30}$$

The vector $\{\theta\}$ in equation (12.29), thus, can be defined in terms of nodal displacements and finite element shape functions (defined in earlier Chapters 2–7):

$$\{\theta\} = [G]\{u\} \tag{12.31}$$

The differential form of equation (12.31) leads to the following relationship:

$$\{d\theta\} = [G]\{du\} \tag{12.32}$$

Re-writing equation (12.29) in differential form leads to the following relationship (Zienkiewicz and Taylor 1991):

$$\{d\varepsilon_L\} = \frac{1}{2}[dA]\{\theta\} + \frac{1}{2}[A]\{d\theta\} = [A]\{d\theta\} \qquad (12.33)$$

Combining equations (12.32) and (12.33):

$$\{d\varepsilon_L\} = [B]_L\{du\} \qquad (12.34)$$

where $[B]_L$ is a deformation-dependent relationship matrix for higher order strain components:

$$[B]_L = [A][G] \qquad (12.35)$$

The complete nonlinear strain–deformation relationship (equation 12.28), is thus, known from the first-order relationship matrix $[B]_0$ (equation 2.5) and the higher order relationship, $[B]_L$ defined by equation (12.35).

The step-by-step solution of a nonlinear response, described by equation (12.1), can be re-written as a minimization of difference between external force vector $\{F\}^i$ and the internal resistance vector $\{R\}^i$ as follows:

$$\Psi = \{F\} - \int [\bar{B}]^T.\{\sigma\}.dV = 0 \qquad (12.36)$$

where $[\bar{B}]$ refers to the strain–displacement relationship matrix at the new deformed state of a structure; and $\{F\}$ to the external load vector at that instance. The variational form of equation (12.36) can be written as:

$$d\Psi = \int [d\bar{B}]^T.\{\sigma\}.dV + \int [\bar{B}]^T.\{d\sigma\}.dV = [K]^i.\{du\}^i \qquad (12.37)$$

where $[K]^i$ is the tangential stiffness matrix of the system and $\{du\}^i$ is the incremental displacement response for the unbalance between external force and internal resistance. The tangent stiffness matrix, $[K]^i$, is defined by two terms in equation (12.37): (i) the first term arising from the resistance offered by existing stresses to the geometry change $[d\bar{B}]$; and (ii) the second term arising from the resistance offered by the current geometry to stress change $\{d\sigma\}$. The change in stress response of a finite element can be defined as follows:

$$\{d\sigma\} = [C]^i.\{d\varepsilon\} = [C]^i[\bar{B}]\{du\} = [C]^i([B]_0 + [B]_L)\{du\} \qquad (12.38)$$

where $[C]^i$ is the incremental stress–strain relationship matrix discussed in Section (12.2). In the case of linear elastic material behavior, $[C]^i$ will simply be equal to the initial elasticity matrix. Using the expression for stress increment from equation

(12.38), the associated stiffness contribution in equation (12.37) can be defined as follows:

$$\int \left[\bar{B}\right]^T.\{d\sigma\}.dV = \int \left(\left[B\right]_0^T + \left[B\right]_L^T\right).\left[C\right]^i.\{d\varepsilon\}.dV \qquad (12.39)$$

Re-writing equation (12.27) in differential form, and using the incremental nonlinear strain definition from equation (12.34):

$$\{d\varepsilon\} = \left[\bar{B}\right]\{du\} = \left(\left[B\right]_0 + \left[B\right]_L\right)\{du\} \qquad (12.40)$$

Substituting expression (12.40) into equation (12.39), finite element resistance against stress change is defined as follows:

$$\begin{aligned}
&\left(\int\left(\left[B\right]_0^T.\left[C\right]^i.\left[B\right]_0\right).dV + \int\left(\left[B\right]_0^T.\left[C\right]^i.\left[B\right]_L\right).dV\right.\\
&\left.+\int\left(\left[B\right]_L^T.\left[C\right]^i.\left[B\right]_L\right).dV + \int\left(\left[B\right]_L^T.\left[C\right]^i.\left[B\right]_0\right).dV\right).\{du\}\\
&=\left(\left[K\right]_0 + \left[K\right]_L\right).\{du\}
\end{aligned} \qquad (12.41)$$

where,

$$\left[K\right]_0 = \int\left(\left[B\right]_0^T.\left[C\right]^i.\left[B\right]_0\right).dV$$

$$\left[K\right]_L = \int\left(\left[B\right]_0^T.\left[C\right]^i.\left[B\right]_L\right).dV + \int\left(\left[B\right]_L^T.\left[C\right]^i.\left[B\right]_L\right).dV + \int\left(\left[B\right]_L^T.\left[C\right]^i.\left[B\right]_0\right).dV$$

The first term in equation (12.41), $[K]_0$, refers to the stiffness matrix for small deformation–small strain system response, and the second term , $[K]_L$, to the nonlinear large deformation response of the system. The complete definition of tangential stiffness matrix $[K]^i$ in equation (12.37) requires an evaluation of the first resistance term associated with stress resistance to the geometry change:

$$\int\left[d\bar{B}\right]^T.\{\sigma\}.dV = \left[K\right]_\sigma.\{du\} \qquad (12.42)$$

Re-writing equations (12.28) and (12.35) in derivative forms, and combining the expressions:

$$\left[d\bar{B}\right] = \left[dB\right]_L = \left[dA\right]\left[G\right] \qquad (12.43)$$

Substituting expression (12.43) into equation (12.42):

$$\left[K\right]_\sigma.\{du\} = \int\left[G\right]^T.\left[dA\right]^T.\{\sigma\}.dV \qquad (12.44)$$

The product term $[dA]^T.\{\sigma\}$ inside equation (12.44) can be expressed as (Zienkiewicz and Taylor 1991):

$$\left[dA\right]^T.\{\sigma\} = \begin{bmatrix} \sigma_x.I_3 & \tau_{xy}.I_3 & \tau_{xz}.I_3 \\ \tau_{xy}.I_3 & \sigma_y.I_3 & \tau_{yz}.I_3 \\ \tau_{xz}.I_3 & \tau_{yz}.I_3 & \sigma_z.I_3 \end{bmatrix}.\{d\theta\} = \left[M\right].\{d\theta\} \tag{12.45}$$

where I_3 is a 3×3 identity matrix. Combining the equations (12.32) and (12.45):

$$\left[dA\right]^T.\{\sigma\} = \left[M\right].\{d\theta\} = \left[M\right].\left[G\right].\{du\} \tag{12.46}$$

Finally, combining equations (12.44) and (12.46), following definition is obtained for the stiffness matrix associated with current stress resistance to geometric changes:

$$\left[K\right]_\sigma = \int \left[G\right]^T.\left[M\right].\left[G\right].dV \tag{12.47}$$

The tangential stiffness matrix, $[K]^i$, in equation (12.37) is thus defined by combining all resistance mechanisms:

$$\left[K\right]^i = \left[K\right]_\sigma + \left[K\right]_0 + \left[K\right]_L \tag{12.48}$$

where stiffness component $[K]_\sigma$ is defined by equation (12.47), and the other two terms, $[K]_0$ and $[K]_L$, by equations (12.41).

Green-Lagrange strain components, defined by equations (12.25), as well as all subsequent derivations presented in the above, refer to the initial reference coordinate system defined in a standard finite element model description. The incremental nonlinear structural response calculation process, referring to the original coordinate reference system, is generally known as the total Lagrangian method. The stress values, calculated by equation (12.38), and referring to the same initial coordinate reference system, are known as second Piola–Kirchoff stresses. By virtue of deformation gradient calculations based on iso-parametric formulations of finite elements, discussed in earlier Chapters 2–7, both Green–Lagrange strain and second Piola–Kirchoff stress values are invariant to rigid body rotations of a structural component. This important invariant feature makes these strain and stress definitions particularly suitable in finite element calculations. However, these response measurements, with respect to the original reference system, are not easily useful in engineering interpretations. This ambiguity related to interpretation of engineering stress–strain calculations is illustrated with a truss element example in Figure 12.10. During large deformation response of a structural assembly, element A-B moves to a new deformed position defined by position A′–B′. Stress values, calculated from Green-Lagrange strains, referring to the original reference coordinate system, are shown by values σ_x and σ_y. Understandably, these values are not directly meaningful to assess the strength and integrity of the member. A more meaningful interpretation of the results can be

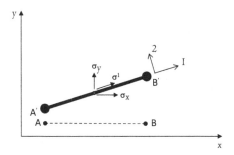

FIGURE 12.10 Coordinate reference systems (*x*–*y* vs. 1–2) for strain and stress definitions during large deformation response.

made if the stress values refer to the post-deformation local axis system (1–2) – generally referred to as true or Cauchy stresses. This analysis objective can be achieved by updating the element geometric reference system after each small incremental deformation response of the system; and then using the latest configuration geometry in next iteration of calculations. This continuously updated calculation process, referred to as updated Lagrangian method, leads to true strain and stress responses when calculations are conducted with very small load steps.

The consideration of geometric nonlinearity can be achieved in an ABAQUS simulation model by including the nonlinear geometric option (NLGEOM) in analysis step description:

```
*STEP, NLGEOM
*STATIC
1,15,0.1,2.0
*CLOAD
2, 2,-1440.0
*END STEP
```

Figure 12.11 shows load–deformation response of problem described in Figure 12.9 including both material and geometric nonlinear response mechanisms. As expected,

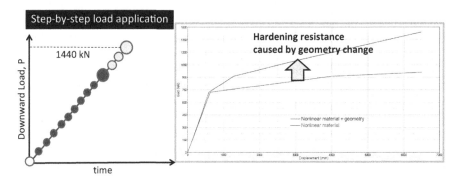

FIGURE 12.11 Load–deformation response of two-member truss assembly – including geometric nonlinearity.

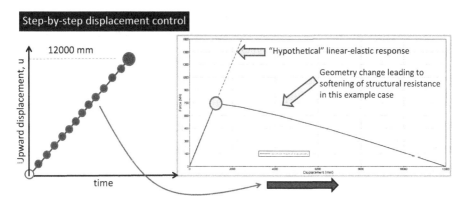

FIGURE 12.12   Post-peak nonlinear load–deformation response prediction with displacement control analysis.

downward deflection of load application point contributes to steeper geometric orientation of the truss members, thus, increasing the structural resistance to downward deflection of load application point. Nonlinear geometric formulation has reproduced the hardening behavior of structure in this special example case study. However, when the loading direction is reversed in this particularly simple structural configuration, the geometric stiffness of structure diminishes with increasing upward deflection until the geometric assembly becomes nearly flat. As discussed in Section 12.1, the post-peak softening response prediction requires special incremental step-by-step analysis, such as by prescribing displacement controls at critical DOF:

```
*STEP, NLGEOM
*STATIC
1,30,0.1,2.0
*BOUNDARY, TYPE=DISPLACEMENT
2, 2,12000.0
*END STEP
```

Figure 12.12 shows step-by-step displacement control analysis results of the two-member truss assembly with a prescribed upward displacement motion at node # 2.

12.4   NONLINEARITY ARISING FROM CHANGES TO INTER-BODY CONTACTS

Body-to-body interface condition may change, from no-contact to contact or vice-versa, during the progression of load–deformation analysis of structural assemblies. Any change in the interface contact state leads to changes in stiffness matrix of the overall structural assembly, and to the interface stress distributions. Changes of the interface contact condition in finite element simulation models are simulated by tracking the relative motion between nodes and element surfaces. Section 8.5 has presented a detail discussion on two alternative numerical techniques, namely Lagrangian multiplier and penalty methods, for simulation of interface contacts. Either of these two methods can be incorporated in step-by-step nonlinear finite element simulation models. Penalty method tends to be the preferred choice in step-by-step simulation of nonlinear response of large structural systems because of improved stability and overall computational efficiency. The success of iterative nonlinear simulation depends on the proper definition of contact stiffness values. Special formulation techniques are often required for the simulation of contact response between soft and hard bodies. A low penalty stiffness typically results in better convergence of the solution, while the higher stiffness keeps the interface penetration at an acceptable level as the contact pressure builds up. Nonlinear contact stiffness formulations, based on gap closure and interface sliding state, generally offer improved stability to simulation models. Finite element simulation software packages often use predefined formulations to set contact stiffness values based on material properties of contacting bodies, provided contact check algorithm is activated with appropriate user input instructions. Contact-related nonlinearity may or may not occur in conjunction with material and geometric nonlinear response mechanisms, However, if appropriate, all three nonlinear mechanisms can be included in a simulation model. A finite element analysis model input data must include specifications for which body parts be checked for potential contact condition changes during the incremental step-by-step simulation process. Specific input data syntax, for activation of contact consideration in ABAQUS simulation models, has been described in Section 8.5. It is important to take care of good quality finite element model preparation with no initial intersections and minimum interface penetrations (Figure 8.8), particularly for simulation of large deformation problems. Initial positions of part surfaces must be placed with adequate offset, taking account of the part thickness values, to minimize initial interface penetrations. During the simulation of multipart shell models, interface contact calculations using hypothetically scaled down (70–90%) values of part thickness often provide stable solution results.

12.5 NONLINEAR DYNAMIC RESPONSE ANALYSIS OF STRUCTURES

Step-by-step simulation technique of nonlinear structural response, as described earlier in this chapter, is naturally adaptable to time-domain dynamic response analysis techniques – that inherently rely on step-by-step simulation process to take account of complex load-time histories. The implicit time-domain simulation technique, described by equation (11.21), requires that the system property matrices $[k]$, $[m]$, and $[c]$ be updated frequently as necessitated by the nonlinear mechanisms arising in a system. Similarly, in explicit time-domain solutions, described by equation (11.31), nonlinear response mechanisms are taken care of during the calculations of local element and nodal response variables. In both cases, implicit and explicit, time step size can be affected by the nonlinear response mechanisms occurring in the system. Time step selection criteria, such as those of Equations (11.22) and (11.33), are used to

initiate the analysis process. However, there is no universally usable analytical criterion to adapt the time step size as the nonlinear mechanisms emerge in a system.
Finite element software packages utilize various internal ad-hoc criteria to adjust the
time step size as nonlinearity emerges in a finite element simulation model.
Calculation of the energy error in overall system response provides an indication of
the quality of results. When the energy error fails to meet a certain threshold limit
within a preset number of iterations, software may scale down the time step size to
meet the energy error criterion. In explicit time-domain calculations, the penalty
stiffness at contact interfaces (defined by equation 8.8), can set the time step size
during step-by-step nonlinear dynamic response calculations. Finite element software package may resort to the use of artificial mass scaling technique, discussed in
Section 11.7, to maintain a user-specified minimum time step during nonlinear
response simulation of structures. ABAQUS data input syntax for a nonlinear implicit
dynamic analysis problem is provided in the following:

*STEP, NLGEOM
*DYNAMIC
$\Delta t_0, t_f, \Delta t_{min}, \Delta t_{max}$
*END STEP

The input values for Δt_0, t_f, $\Delta t_{min,}$ and Δt_{max} define time step control parameters as
discussed earlier in Section 11.8. The optional parameter "NLGEOM" turns on the
nonlinear geometric analysis method in this analysis step. Simulation of nonlinear
material behavior can be activated by using the appropriate constitutive model in
material input data under the keyword "*MATERIAL". Constitutive models in simulation software packages generally include options to describe the strain rate sensitivity of material stress–strain response. For example, the rate sensitivity of the
John–Cook hardening plasticity model is generally expressed with a modified form
of equation (12.24) (Figure 12.13) (equation 12.49):

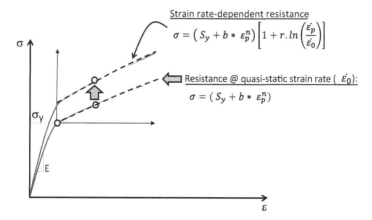

FIGURE 12.13   Strain-rate sensitivity of material stress–strain response.

$$\sigma = \left(S_y + b * \varepsilon_p^n \right) \left[1 + r.\ln\left(\frac{\dot{\varepsilon}_p}{\dot{\varepsilon}_0} \right) \right] \qquad (12.49)$$

where "r" is a material parameter defining the strain rate sensitivity of yield strength; $\dot{\varepsilon}_p$ the instantaneous strain rate; and $\dot{\varepsilon}_0$ the reference strain rate defining the quasi-static stress–strain response. This description in equation (12.49) is one of many different possible formulations for rate-dependent constitutive models. A user of finite element simulation software packages should consult the software-specific material theory manual to identify strain rate dependent constitutive models relevant for the material being analyzed. Users also need to consult the software manuals to identify critical output data that can be used for assessing the reliability of nonlinear simulation results.

12.6 MATERIAL FAILURE SIMULATION IN NONLINEAR FINITE ELEMENT ANALYSIS

Structural safety assessment often involves the simulation of material fracture initiation and propagation in a continuum finite element model. This objective is often achieved by allowing the stress resistance at finite element integration points to degrade or disappear during step-by-step simulation of nonlinear structural response (Figure 12.14). Ductile material failures are often simulated when the accumulated plastic strain at finite element integration points (equation 12.11) exceeds a predefined material failure strain. Sustained resistance during large plastic deformation of ductile materials helps with the evolution of yielding zone beyond artificial

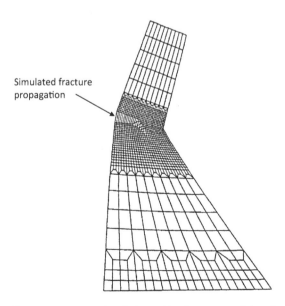

Simulated fracture propagation

FIGURE 12.14 Fracture response simulation with plane stress finite element model of a concrete gravity dam (Bhattacharjee 1993).

inter-element boundaries, thereby, preventing the emergence of mesh-dependent sudden localization of failure at discrete finite element integration points. Strength-based failure simulation for brittle and quasi-brittle materials generally leads to mesh-dependent response predictions with uncertain practical application value. Extensive research work has been pursued since 1990s to produce mesh-objective fracture simulation results by making the softening part of quasi-brittle material stress–strain response (Figure 12.5(c)) a function of the fracture energy dissipation and finite element size parameter (Bhattacharjee and Leger 1994). Successful numerical simulation case studies have been documented for both static and dynamic load cases (ABAQUS Example Manual 2020). Alternative numerical techniques to simulate discrete crack propagation in solids, using linear and nonlinear fracture mechanics criteria, have also been developed over the past decades where crack propagation involves continuous updates to the finite element mesh to represent the geometric profile of discrete crack and crack-tip. While research use of discrete crack simulation has been very active over many decades, the practical use of it in everyday engineering decision process remains very limited.

12.7   COMPUTATIONAL METHODS FOR STRUCTURAL FORM SIMULATION

General-purpose finite element simulation method starts with the assumption that geometric design data for parts and structures exist to build the initial configuration model. However, engineering development process often requires that the forming process of initial geometry itself be also assessed. For example, geometric shapes of thin sheet metal parts are often formed by punching flat metal blanks to the desired shape – a process involving large geometric deformation with significant plastic straining of the original metal blanks (Altan and Tekkaya 2012). Sheet metal forming simulation helps with the formability assessment of a desired geometric shape – starting from flat blanks of cold or hot raw materials (Figure 12.15). Explicit

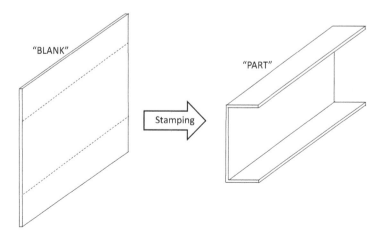

FIGURE 12.15   Stamping of metal blank to a desired part geometry.

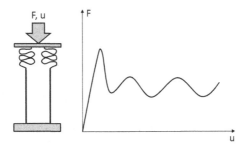

FIGURE 12.16   Progressive folding of a thin-wall metal tube under axial compressive load.

time-domain analysis technique, described in Chapter 11, with geometric and material nonlinear finite element formulations, provides a very useful tool for upfront formability assessment, thus, helping to predict if excessive thinning and fracture may appear during actual forming of geometric shapes. Computational simulation of metal stamping process is an essential part of modern automotive product development process for reducing the time spent on traditional trial-and-error approach used in building and adjusting the physical punch and die (Andersson 2004). Crashworthiness simulation of automotive body structures also falls in the domain of explicit finite element simulation where thin-wall structures are engineered to achieve a preferred post-crush geometry (Figure 12.16).

There exists, however, a special class of structural engineering problem where initial geometric form of structure is unknown. Design of tensioned fabric structures, generally, starts with a few known geometric points, such as the cable attachment points in Figures 3.2 and 7.11. Engineering analysis task involves the finding of stable geometric form that will possess uniform acceptable tension in the deployed configuration of cable-fabric assembly. As illustrated earlier with Figure 12.9, resistance to out-of-plane forces is produced by the geometric configuration of prestressed truss members. Similar resistance mechanism is produced by membranes in three-dimensional space with bi-directional curvature at each point on the surface. The greatest stiffness of a prestressed fabric surface is achieved in an anticlastic surface, i.e. a surface with opposing curvatures at any point (producing a negative Gaussian curvature) (Lewis 2008, 2013). Analysis techniques for structural form finding rely on iterative discovery of a geometric shape without using the stiffness-based finite element formulations. The computation starts with an initial fictitious surface configuration, represented by a mesh of simple geometric elements (cable and membranes). The unbalanced force at each node is calculated based on preset design stress values in the connecting elements. Depending on the magnitude and direction of unbalanced force, the node positions are incrementally adjusted in successive iterations. Several iterations are generally required to find a geometry with overall system equilibrium. Boundary configurations, defined with initial key points and cable configuration, have a significant impact on the success of iterative form-finding analysis. Computational form-finding, driving the discovery of detailed geometric configuration of a surface, for given boundary constraints and target prestress values in the component elements, does not strictly fall within the domain of standard deformation-based finite element method discussed in this book. The form-finding algorithm basically iterates on geometric configuration to reach an equilibrium state between assumed stress field and external forces and boundary constraints. Although not explicitly stated in current research publications, the structural form-finding technique can be considered an iterative solution of the geometric nonlinear problem with predefined stress functions for discrete finite element domains (representing membranes and cables). Obviously, the merger of two simulation techniques, displacement versus stress function-based approach, has not occurred in the currently available commercial software packages.

12.8 PRACTICE PROBLEMS: NONLINEAR RESPONSE ANALYSIS OF STRUCTURES

PROBLEM 1
Re-analyze the steel bumper beam problem #4 in Chapter 7 (Figure 7.16), assuming nonlinear material behavior with Johnson-Cook hardening plasticity model defined by

$$S_y = \left(0.43 + 0.824 * \varepsilon_p^{0.51} \right) GPa \tag{12.50}$$

Conduct a step-by-step analysis by applying the pressure given in Figure 7.16. Plot the contour of maximum plastic strain in the beam structure.

PROBLEM 2
Figure 12.17 shows a nonlinear spring-mass system impacting on a rigid wall. Conduct hand calculations to predict the acceleration response of the mass by using both implicit and explicit time integration methods; and compare the results.

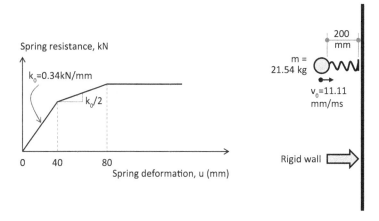

FIGURE 12.17   Impact of a nonlinear spring-mass system on the rigid wall.

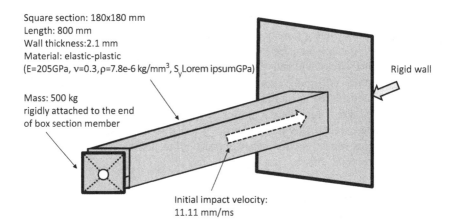

Square section: 180x180 mm
Length: 800 mm
Wall thickness: 2.1 mm
Material: elastic-plastic
(E=205GPa, v=0.3, ρ=7.8e-6 kg/mm^3, S_y Lorem ipsumGPa)

Rigid wall

Mass: 500 kg
rigidly attached to the end
of box section member

Initial impact velocity:
11.11 mm/ms

FIGURE 12.18 Simulation of the axial impact of a thin-wall box-section member on the rigid wall.

PROBLEM 3

Figure 12.18 shows a thin-wall box-section member carrying a rigidly attached mass of 500 kg at one end, and it is impacting on a rigid wall at the other end with an initial velocity of 11.11 mm/ms. Average crush strength of a box section beam is approximately defined by the following empirical equation (Mahmood and Paluzsny 1981):

$$P_{avg} = 19.796(t)^{1.86} \cdot (b)^{0.14} \cdot (S_y)^{0.57}$$ (12.51)

where "b" is the side dimension of square section in mm, "t" the wall thickness in mm, and S_y the material yield strength in GPa. Conduct a nonlinear finite element simulation of the impact problem; assess the average dynamic crash strength from simulation results, and compare that with the quasi-static strength predicted by the empirical equation (12.51). Assume elastic-plastic material behavior.

PROBLEM 4

Re-analyze Problem-3 by using the Johnson–Cook hardening plasticity model defined by equation (12.50); and compare the simulation results with that of Problem-3.

References

ABAQUS Example Manual, Seismic Analysis of a Concrete Gravity Dam, 2020, https://help.3ds.com/2020/English/DSSIMULIA_Established/SIMACAEEXARefMap/simaexa-c-concretedam.htm?ContextScope=all&id=3b047001b85d4d5ba54899be58c0cea7#Pg0.

Ahmad, S., Irons, B.M., and Zienkiewicz, O.C., "Analysis of thick and thin shell structures by curved elements", *International Journal of Numerical Methods in Engineering*, 2: 419–451, 1970.

Altan, T. and Tekkaya, E., *Sheet Metal Forming: Fundamentals*, ASM International, 2012.

Altair University, HyperWorks 12.0 Student Edition, 2020. https://altairuniversity.com/free-altair-student-edition/.

Altair University, 2D Meshing, 2014, https://www.altairuniversity.com/wp-content/uploads/2014/02/2Dmeshing.pdf.

Andersson, A., "Comparison of sheet-metal-forming simulation and try-out tools in the design of a forming tool", *Journal of Engineering Design*, 15(6): 551–561, 2004.

ANSI, American National Standards Institute, The Initial Graphics Exchange Specification (IGES), 1996, https://webstore.ansi.org/.

Arrea, M., and Ingraffea, A.R., "Mixed-mode crack propagation in mortar and concrete", Department of Structural Engineering Report 81-13, Cornell University, 1981.

ASM International, ASM Handbook Volume 20: Materials Selection and Design, 1997, https://www.asminternational.org/search/-/journal_content/56/10192/06481G/PUBLICATION, 2021.

ASTM E8/E8M, Standard Test Methods for Tension Testing of Metallic Materials, 2021, https://www.astm.org/Standards/E8.

ASTM E132-17, Standard Test Method for Poisson's Ratio at Room Temperature, 2021, https://www.astm.org/Standards/E132.htm.

ASTM E466-15, Standard Practice for Conducting Force Controlled Constant Amplitude Axial Fatigue Tests of Metallic Materials, 2021, https://www.astm.org/Standards/E466.htm.

ASTM E1049-85, Standard Practices for Cycle Counting in Fatigue Analysis, 2021, https://www.astm.org/Standards/E1049.

ASTM E1820-20ae1, Standard Test Method for Measurement of Fracture Toughness, 2021, https://www.astm.org/Standards/E1820.htm.

Bathe, K.J., *Finite Element Procedures*, Prentice Hall Inc., 1996.

Bathe, K.J. and Wilson, E.L., *Numerical Methods in Finite Element Analysis*, Prentice-Hall, 1976.

Bažant, Z.P., *Scaling of Structural Strength*, 2nd edition, Elsevier Butterwortg-Heinemann, 2005.

Belytschko, T., Lin, J., and Tsay, C.S., "Explicit algorithms for nonlinear dynamics of shells", *Computer Methods in Applied Mechanics and Engineering*, 42: 225–251, 1984.

Beta-CAE.com, META Post-Processor, https://www.beta-cae.com/meta.htm.

Bentley.com, STAAD Pro – 3D Structural Analysis and Design Software, https://www.bentley.com/en/products/brands/staad.

Bhattacharjee, S.S., *Smeared Fracture Analysis of Concrete Gravity Dams for Static and Seismic Loads*, Ph.D. Thesis, McGill University, Canada, 1993.

Bhattacharjee, S.S. and Chebl, C., *Instrumentation de Stade Olympique de Montreal*, Report to SNC-Lavalin Inc., Montreal, Canada, 1997.

Bhattacharjee, S.S. and Leger, P. "Seismic cracking and energy dissipation in concrete gravity dams", *Journal of Earthquake Engineering and Structural Dynamics*, 22(11): 991–1007, 1993.

Bhattacharjee, S.S. and Leger, P., "Application of NLFM models to predict cracking on concrete gravity dams", *ASCE Journal of Structural Engineering*, 120(4): 1994.

Borrvall, T., Bhalsod, D., Hallquist, J.O., and Wainscott, B., *"Current status of sub-cycling and multi-scale simulations in LS-DYNA"*, *Proceedings of the 13th International LS-DYNA Users Conference*, 1–14, 2014.

Broek, D., *The Practical Use of Fracture Mechanics*, Springer Netherlands, 2012.

BSSC (Building Seismic Safety Council), *National Earthquake Hazards Reduction Program (NEHRP) Recommended Seismic Provisions for New Buildings and Other Structures*, 2020, FEMA P-1050-1/2015 Edition, https://www.nibs.org/page/bssc_pubs.

Budynas, R.G., *Advanced Strength and Applied Stress Analysis*, 2nd edition, McGraw-Hill, 1999.

Chopra, A.K., *Dynamics of Structures*, 5th edition, Pearson Education Inc., 2017.

Clough, R.W. and Penzien, J., *Dynamics of Structures*, McGraw-Hill, 1975.

Cook, R.D., Malkus, D.S., and Plesha, M.E., *Concepts and Applications of Finite Element Analysis*, 3rd edition, John Wile & Sons, 1989.

Courant, R., Friedrichs, K., and Lewy, H., "Über die partiellen Differenzengleichungen der mathematischen Physik", *Mathematische Annalen* (in German), 100(1): 32–74, 1928.

Cowper, R.G., "The shear coefficient in Timoshenko's beam theory", *Journal of Applied Mechanics*, 33: 335–340, 1966.

Crandall, S.H., *Engineering Analysis, A Survey of Numerical Procedures*, McGraw-Hill Book Co, 1956.

CSIAmerica.com, *Computers and Structures Inc., SAP2000 v22*, https://www.csiamerica.com/products/sap2000.

Dassault Systems, *ABAQUS* Student edition, 2020a, https://edu.3ds.com/en/get-software.

Dassault Systems, SIMULIA User's Guides, 2020b. https://help.3ds.com/HelpProductsDS.aspx.

Desai, C.S., Zaman, M.M., Lightner, J.G., and Siriwardane, H.J., "Thin-layer element for interfaces and joints," *International Journal for Numerical and Analytical Methods in Geomechanics*, 8(1): 19–43, 1984.

Dow Automotive Systems, *"Adhesives & Sealants"*, "Structural-adhesives", 2021, https://www.adhesivesmag.com/articles/96128.

Fung, Y.C., *Fundamentals of Solid Mechanics*, Prentice Hall, 1965.

GRABCAD.COM, 2020, https://grabcad.com/dashboard.

Hilber, H.M., Hughes, T.J.R., and Taylor, R.L., "Improved numerical dissipation for time integration algorithms in structural dynamics", *Earthquake Engineering and Structural Dynamics*, 5: 283–292, 1977.

Hughes, T.J.R., *The Finite Element Method*, Prentice-Hall Inc., 1987.

Irwin, G.R., "Analysis of stresses and strains near the end of a crack traversing a plate", *Journal of Applied Mechanics*, 24: 361–364, 1957.

ISO, International Standards Organization, *Standard 10303, STEP (Standard for the Exchange of Product model data)*, 2020, https://www.iso.org/standards.html.

Irons, B., and Ahmad, S., *Techniques of Finite Elements*, Ellis Horwood Limited, Halsted Press, John Wiley and Sons, Chichester, England, 1980.

Johnson, G.R. and Cook, W.H., *"A constitutive model and data for metals subjected to large strains and high strain rates"*, *Proceedings of the 7th International Symposium on Ballistics*, 541–547, 1983.

Kennedy, J.B. and Madugula, M.K.S., *Elastic Analysis of Structures: Classical and Matrix Methods*, Harper & Row Publishers, 1990.

Lemaitre, J. and Chaboche, J.L., *Mechanics of Solid Materials*, Cambridge University Press, 1994.

Lewis, W.J., "Computational form-finding methods for fabric structures", *Proceeding of the ICE. Engineering Computational Mechanics*, 161: 139–149, 2008.

Lewis, W.J., "Modelling of fabric structures and associated design issues", *ASCE Journal of Architectural Engineering*, 19(2): 2013.

Logan, D.L., A First Course in the Finite Element Method, Cengage Learning, 2012.

LSTC.COM, *LS-DYNA Theory Manual, ANSYS/LST*, 2021. https://www.lstc.com/download/manuals.

Mahmood, H.F. and Paluzsny, A., "Design of thin walled columns for crash energy management-their strength and mode of collapse", *SAE Transactions*, 90(4): 4039–4050, 1981. https://www.jstor.org/stable/44725016.

Mindlin, R.D., "Influence of rotatory inertia and shear on flexural motions of isotropic elastic plates", *ASME Journal of Applied Mechanics*, 18(31): 38, 1951.

Miner, M.A., "Cumulative damage in fatigue", *Journal of Applied Mechanics*, 12: 149–164, 1945.

Modak, S. and Sotelino, E.D., "The generalized method of structural dynamics applications", *Advances in Engineering Software*, 33: 7–10, 2002.

nCode, Software and solutions for fatigue and durability analysis, *HBM Prenscia Inc*, https://www.ncode.com/.

Newmark, N.M., "A method of computation for structural dynamics", *ASCE Journal of Engineering Mechanics*, 85: 67–94, 1959.

NHTSA, National Highway Traffic Safety Administration, US Department of Transportation, *Crash Simulation Vehicle Models*, 2010 Toyota Yaris, 2020, https://www.nhtsa.gov/crash-simulation-vehicle-models.

Paris, P.C. and Erdogan, F., "A critical analysis of crack propagation laws", *Journal of Basic Engineering*, 85(4): 528–533, 1963.

Popov, E.P., *Mechanics of Materials*, 2nd edition, Prentice-Hall Inc., 1978.

Press, W.H., Teukolsky, S.A., Vetterling, W.T., and Flannery, B.P., *Numerical Recipes – The Art of Scientific Computing*, 3rd edition, Cambridge University Press, 2007.

Reissner, E., "The effect of transverse shear deformation on the bending of elastic plates", *ASME Journal of Applied Mechanics*, 12: 68–77, 1945.

Rice, J.R., "A path independent integral and the approximate analysis of strain concentrations by notches and cracks", *Journal of Applied Mechanics, ASME*, 35: 379–386, 1968.

Siemens.COM, Products and Services, FEMAP Pre- and Post-processor, 2021, https://www.plm.automation.siemens.com/global/en/products/simcenter/femap.html

Tada, H., Paris, P.C., and Irwin, G.R., *The Stress Analysis of Cracks Handbook*, 3rd edition, American Society of Mechanical Engineers, 2000.

Tedesco, J.W., McDougal, W.G., and Ross, C.A., *Structural Dynamics, Theory and Applications*, Addison Wesley Longman Inc., 1999.

Timoshenko, S.P. and Goodier, J.N., *Theory of Elasticity*, 3rd edition, McGraw-Hill Book Co., 1982.

Timoshenko, S.P. and Gere, J.M., *Theory of Elastic Stability*, McGraw-Hill Book Co, 1963.

Timoshenko, S.P. and Woinowsky-Keiger, S., *Theory of Plates and Shells*, 2nd edition, McGraw-Hill International Editions, 1970.

Tinawi, R., Leger, P., Ghrib, F., Bhattacharjee, S.S., and Leclerc, M., "Structural safety of existing concrete dams: influence of construction joins. Final Report", Canadian Electricity Association; Montreal, PQ (Canada); CEA-9032 G 905, 1994.

Uflyand, Y.S., "Wave propagation by transverse vibrations of beams and plates", *Journal of Applied Mathematics and Mechanics* (in Russian), 12: 287–300, 1948.

Ugural, A.C. and Fenster, S.K., *Advanced Mechanics of Materials and Applied Elasticity*, 5th edition, Prentice Hall, 2012.

World Auto Steel (WorldAutoSteel.org), *Future Steel Vehicle Results and Reports*, 2015. https://www.worldautosteel.org/downloads/futuresteelvehicle-results-and-reports/.

Wikipedia.org, *Timoshenko-Ehrenfest Beam Theory*, 2020. https://en.wikipedia.org/wiki/ Timoshenko-Ehrenfest_beam_theory.

Zehnder, A.T., *Fracture Mechanics*, Springer Netherlands, 2012.

Zienkiewicz, O.C. and Taylor, R.L., *The Finite Element Method, Volume 1, Basic Formulation and Linear Problems*, 4th edition, McGraw-Hill Book Company, 1989.

Zienkiewicz, O.C. and Taylor, R.L., *The Finite Element Method, Volume 2, Solid and Fluid Mechanics, Dynamics and Non-linearity*, 4th edition, McGraw-Hill Book Company, 1991.

Index

Printed in the United States
by Baker & Taylor Publisher Services